Hazardous Materials:

Strategies And Tactics

Hazardous Materials:

Strategies And Tactics

David M. Lesak

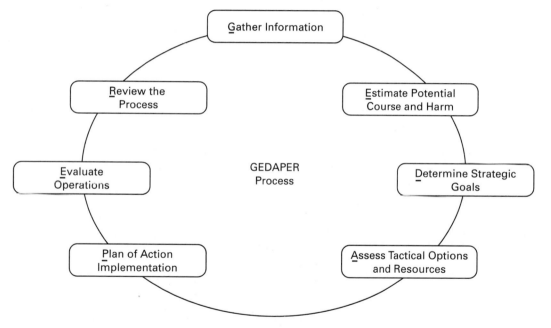

Brady / Prentice Hall
Upper Saddle River, NJ 07458

Library in Congress Cataloging-in-Publication Data

Lesak, David M.
    Hazardous materials   :   strategies and tactics / David M. Lesak.
      p.   cm.
    ISBN 0-8359-5209-6
    1. Hazardous substances--Accidents--United States--Management.
  I. Title.
  T55.3.H3L47   1999
  628.9'2--DC21   98-18780
                                   CIP

Publisher: *Susan Katz*
Acquisitions Editor: *Sherene Miller*
Editorial Assistant: *Carol Sobel*
Marketing Manager: *Tiffany Price*
Marketing Coordinator: *Cindy Frederick*
Director of Production and Manufacturing: *Bruce Johnson*
Managing Production Editor: *Patrick Walsh*
Senior Production Manager: *Ilene Sanford*
Creative Director: *Marianne Frasco*
Cover Design: *Bruce Kenselaar*
Cover Photograph: *David Lesak*
Composition: *BookMasters, Inc.*
Presswork/Binding: *R.R. Donnelley & Sons, Harrisonburg, VA*

© 1999 by Prentice-Hall, Inc.
A Simon & Schuster Company
Upper Saddle River, New Jersey 07458

Printed in the United States of America

10 9 8 7 6 5 4 3 2 1

ISBN 0-8359-5209-6

Prentice-Hall International (UK) Limited, *London*
Prentice-Hall of Australia Pty, Limited, *Sydney*
Prentice Hall Canada Inc., *Toronto*
Prentice-Hall Hispanoamericana, S.A. *Mexico*
Prentice-Hall of India Private Limited, *New Delhi*
Prentice-Hall Japan, Inc., *Tokyo*
Simon & Schuster Asia Pte. Ltd., *Singapore*
Editor Prentice-Hall do Brasil, Ltda, *Rio de Janeiro*

Contents

Preface

This text provides an in-depth examination of operational decision making through the use of the GEDAPER Process. This process was developed over many years, with much thought, trial and error, work, and application. This copyright protected process was developed by David M. Lesak, Chief of the Lehigh County Pennsylvania Hazmat Team. Chief Lesak gave permission for the process to be used as the model operational decision making process in all present generation field and resident program hazardous materials courses at the Federal Emergency Management Agency's (FEMA) National Fire Academy, at the National Emergency Training Center in Emmitsburg, Maryland.

This process has been successfully used for many years by both public and private sector emergency responders as part of an overall training approach. Specifically, the process provides a systematic and scientific approach to hazmat response and decision making. Rather than following the traditional shotgun effect topic by topic training, the GEDAPER process seeks to put the pieces together in a operationally based approach. Further, this text applies operational perspectives, approaches and examples to help hazmat responders meet the regulatory and legal standard of care needed in today's society.

There is no attempt to indicate that the GEDAPER process is the only suitable process available for emergency responders. Rather it is used to provide an example of the fundamental ingredients required for the highly systematic and structured operational environment needed in today's hazmat emergency response. It is hoped that this approach will assist other emergency responders by putting all the hazmat pieces together in their daily operations.

There is an essential point that needs to be addressed. This process was not created, refined or streamlined in a vacuum. Many others have played a significant role. Some of the most influential and thought provoking follow.

Acknowledgments and Dedications

There are many people who have helped me gain the knowledge and experience needed to develop the **GEDAPER** process. Several have been my instructors, a few have been mentors, still others have been colleagues with whom I have worked. Over the many long hours that we worked, contemplated and sometimes fought, you scrutinized, questioned and required me to "present and defend" crucial concepts and approaches. Your thoughtful questioning and friendship have been invaluable. To all of you, my heartfelt thanks. Hopefully, this meager attempt helps make up for the previous lack of recognition that was beyond all of our control.

Darrel Begnaud—Teach Inc., LA.

Dan Civello—Jefferson Parish Fire Department, LA.

Phil McArdle—Fire Department, New York City, NY.

Mike Shannon—Oklahoma City Fire Department, OK.

Pat Walsh—Washington, D.C. Fire Department (retired), DC.

Noel "Chris" Waters—Charleston County, SC.

Jim Covington—National Fire Academy, MD.

Burt Phelps—Ann Arundle Fire Department (retired), MD.

Ed McDonnald—Fire Department, New York City (retired), NY.

Gene Chandler—Pouder Fire Authority, CO.

Dan Fabian—Miami Fire Department (retired), FL.

Romey Brooks—National Fire Academy, MD.

Gerald Boyd—U.S. Department of Energy, MD.

Ed Shea—Fire Department, New York City (retired), NY.

Ken Stewart—Stewart Associates, PA.

John Pearson—Summit Associates, PA.

To the memory of **George Lewis**—Fire Marshal, Delaware County, PA.

A special thanks to **Jan Kuzma** for sharing his knowledge of chemistry, and the friendship and insights he provided during our many years of instruction together.

A special thanks to Chief **Mike Shannon** and Fire Fighter **Phil McArdlle** for sharing their knowledge and vision as we have worked and taught together. Both of these friends provided different operation perspectives from my own as well as providing a reality check to identify what is needed, what the realities are in their jurisdictions and what **works on *ALL of our streets.***

To M. S. "**Sandy**" **Bugucki,** MD, Ph.D., Assistant Professor, Section of Emergency Medicine, Department of Surgery, Yale University School of Medicine, for sharing in her knowledge and perspective on Hazmat EMS issues and her shared belief that knowing how and why are as critical as knowing what.

To **Alice Lesak** and **Myra Sell** for their editorial review and support.

To the members of the Boise, Idaho; Cleveland, Ohio; and Rochester, Minnesota Fire Department Hazmat Teams for their thoughts and recommendations.

Finally, to my wife **Lora** and my children **Jana** and **Scott,** for their loving support and help over all these years, and, "Yes, the book is finished."

I would like to thank the following reviewers:

Charles Butler—Sau-Tech Arkansas Fire Academy, AR.
Tim Peebles—Hall County Fire Services, GA.
John Rasmussen—Greenville County Emergency Medical Services, SC.
Paul Maniscalco—New York City Fire Department, NY.

1 An Introduction to Hazardous Materials and the Chemical Revolution

CHAPTER OBJECTIVES

Upon completion of this chapter, you will be able to:

1. Define the term *alchemy*
2. Describe the development and role of synthetic substances in the chemical revolution
3. Identify the three primary federal agencies involved in the classification and regulation of chemicals
4. Define the four primary classifications of hazardous chemicals used by the three federal agencies
5. Identify the three primary titles found in the Code of Federal Regulations that involve hazardous chemicals

CHEMICALS IN OUR SOCIETY

The human race has studied and used chemicals in various forms for thousands of years. There is documented evidence that the ancient Greeks named, studied, and found uses for many different chemicals available in their time. As the human race progressed in knowledge and technology, so did our knowledge and use of chemicals.

By the time we reached the Dark Ages, our curiosity and desire to manipulate the world around us led to the early ancestry of modern chemistry. The nobility of the time looked for new ways of acquiring wealth. To that end, a group of specialized employees, the alchemists, arose. The alchemists had one primary goal: the creation of gold from base materials, primarily base metals (lead, iron, copper, etc.).

The work of the early alchemists was most definitely hit or miss. They knew little about what they were doing. There are some accounts of extremely violent and even lethal experiments taking place. It is safe to say that as these tinkerers tried different chemical concoctions, there was a very high potential for turnover in their positions. However, as the alchemists became more and more methodical and scientific in their approach, they made tremendous strides in their understanding of the world around them. The alchemists and their trial-and-error investigations evolved into chemistry.

In 1869, Siberian scientist D. I. Mendeleyev (1834–1907) established the principle of periodicity. As his work continued, Mendeleyev developed the first Periodic Table of Elements that was the forerunner of today's periodic chart. So accurate was his principle that Mendeleyev predicted the existence of several elements not proven to exist until much later. Mendeleyev's contribution to modern chemistry cannot be understated.

During the same period, the industrial revolution was gathering tremendous momentum. New industries evolved and older industries matured. As a result of this development, new ideas and processes flourished. While attempting to address shortcomings and problems, many new processes were developed. What were once considered waste products now became the raw materials for new processes and products. Unfortunately, the potential hazards and risks were not known and thus not appreciated.

The Role of Synthetics

Over the centuries, chemicals have played an ever-increasing role in our lives. By the beginning of the twentieth century, it was clear that this trend would continue and accelerate. The bench mark of a true chemical revolution started in 1938 when society started to develop its "chemical dependency."

This year marked the beginning of a massive, yet relatively unnoticed, revolution in the nature of the material goods found in society. Prior to 1938, the goods found in our homes, schools, work places, churches, and elsewhere had one thing in common: All of them were produced from some naturally occurring material derived from animals, plants, or minerals.

In 1935, an amazing material that would forever change many modern material goods was developed. By 1938, this material, nylon, became commercially available and heralded a totally new, complex, and ever-changing era. There are many reasons for this:

1. Chemistry and technology had reached a point that allowed the efficient and cost-effective production of ever-increasing quantities of raw chemicals and their synthetic creations.
2. These new synthetics could be used to produce items (goods) that were impossible to produce from natural materials.
3. Although there were some earlier synthetic materials, most were difficult and expensive to use practically.

Nylon was very different. There was an almost unbelievable number of different uses that could be found for it: stockings, rope, cloth, gaskets, seals, valve assemblies, and so on. Initially, the impact of this material was relatively small. However, its development started a race to produce more and different synthetics.

Soon afterwards, World War II swung into full gear. Because of the global nature of the fighting and location of the natural raw materials, it was difficult or impossible for the United States and countries from the rest of the world to obtain these materials. However, many of the synthetics could be made from components found in fossil fuel, particularly oil, which was plentiful in the U.S. As a result of these factors, there was an even greater demand for research and development of new and more versatile synthetics.

As the newly developed synthetics went into production, the manufacturers found that there were relatively large quantities of waste products generated in the manufacturing process. Not understanding the possible dangers associated with these wastes, many of the manufacturers simply found some nonproductive area or hole in the ground into which these wastes could be dumped. This was really the start of our country's hazardous waste disposal problems.

The disposal of these wastes did produce some very interesting and unanticipated side effects. The manufacturers found in the areas where these waste materials had been dumped, certain kinds of weeds, insects, and other living things died off. In some other dumping areas, they noticed accelerated growth. This led to the question of whether there would be any use for materials that can kill plants and animals. Obviously, the answer was yes. Today these types of synthetic substances are known as agricultural chemicals, and they are responsible for a whole new era in agriculture.

As scientific and technological knowledge continued to grow, new possibilities and horizons came into focus. Research and development continued to identify new synthetics and possible applications. By the late sixties and into the early seventies, there was a veritable explosion in the development, application, and use of synthetic materials and their new technologies. When the eighties rolled around, we found that an average of over two tons of synthetic materials were produced annually for every family of four in the United States. Stop and look around any home or office. It is difficult to find anything that is made of only naturally occurring materials.

What kind of materials are these "synthetics"? Predominantly, today's synthetics are found in the form of polymers (poly: many; mer: piece). Polymers are nothing more complex than extremely large molecules sometimes referred to as macromolecules (macro: large). These macromolecules are made through a chemical reaction process known as polymerization. The polymerization reaction involves the chemical combination of many of the same very small molecules known as monomers (mono: one; mer: piece). Monomers contain a specific type of bond within their molecules known as pi bonds. These pi bonds are quite reactive and have the ability to react with the pi bonds found in a surrounding monomer. As the polymerization reaction takes place, individual monomers join together, creating an ever larger molecule. An analogy is the process of making an ever larger object by attaching multiple small pieces found in Lego building blocks.

It is also important to understand that polymers can be either natural or synthetic. Some examples of natural polymers include:

1. wool
2. feathers
3. cotton
4. linen
5. wood

Some examples of synthetic polymers include plastics, such as:

1. PVC (polyvinyl chloride)
2. vinyl
3. ABS (acrylonitrile/butadiene/styrene)
4. acrylics
5. phenolics
6. nylons
7. aremids
8. polyethylene
9. polyester

Today society has reached a point of development that would have totally astounded Mendeleyev and his peers. As of 1995, there were approximately thirteen million different chemicals registered with the Chemical Abstract Service (CAS) of Columbus, Ohio.

In 1988 alone, CAS added over 600,000 substances to its files. That means that in 1988, an average of 50,000 substances were added per month. Over the eight year period, from 1981 to 1988, an average of over 500,000 substances have been added annually to the CAS listing. This is not to say that all of these chemicals are hazardous or found in commercial production, but rather that they are being developed and may some day find their way into production. Nor should this be interpreted to mean that all of these substances are brand new chemical compounds. Rather, many of the listings are mixtures of two or more different compounds. Gasoline, for example, is composed of many different individual chemicals, such as octane, hexane, benzene, toluene, xylene, alcohols, ketones, and ethers.

To more fully understand the phenomenal growth in the number of synthetic substances, let us examine some figures. In 1965 (the first year that such information is available), CAS listed 211,934 substances on file. As of 1992, CAS had listed over 11 million. In essence, the number of chemicals listed with CAS has increased over 51 fold in only a 27 year period.

CLASSIFICATION SYSTEMS

As a result of the overwhelming number of different chemicals and mixtures of chemicals, there was a need to develop different systems for classifying or grouping them. Most of the systems have evolved since about 1970 as our knowledge, experience, and need to regulate these substances has grown.

Most of the commonly utilized classification systems have been developed by regulator governmental agencies, such as the United States Department of Transportation (DOT), the Environmental Protection Agency (EPA), and the Department of

Labor's Occupational Safety and Health Administration (OSHA). Each of these regulatory agencies has a different set of problems with which they are faced with and mandated to control. As a result, there are some major differences in how chemicals are classified, depending upon who is doing the classification. Such a situation can easily lead to much confusion and misinterpretation. So, let's take just a moment to examine the different terms that may be encountered and some specific information about each.

Hazardous Materials

The term *hazardous materials* was first defined by DOT in 1975. Hazardous materials are defined and regulated in the Code of Federal Regulations (CFR), Title 49, Parts 100 to 199 (49 CFR 100–185). Each hazard class is identified in Section 171.8 and referenced to its specific Section in 173. Further, Section 171.8 states that a hazardous material is "a substance or material, including a hazardous substance, which has been determined by the Secretary of Transportation to be capable of posing an unreasonable risk to health, safety, and property when transported in commerce and which has been so designated."

This definition is quite vague and broad. To more clearly define and identify what types of substances are included, DOT expands the definition within the same Section. They further define a hazardous substance as follows:

"A material, including its mixtures and solutions, that–
1. is listed in the appendix to Section 172.101 of this subchapter;
2. is in a quantity, in one package, which equals or exceeds the reportable quantity (RQ) listed in the appendix to Section 172.101 in this subchapter;
3. when in a mixture or solution-
    i. for radionuclides, when the quantity (in curries or terabecquerels) meets the specified levels found in paragraph 6 of the appendix to 172.101.
    ii. for other than radionuclides, is in a concentration by weight which equals or exceeds the concentration corresponding to the RQ of the material as listed in a corresponding table."

A specific listing of all DOT hazardous materials is found in 49 CFR 172.101 and its appendices. Remember, in this definition, the term hazardous materials refers only to materials in transportation.

Hazardous Wastes

The term *hazardous waste* is defined and regulated by the EPA in 40 CFR, sections 261 and 265. In essence, a hazardous waste is any waste material which is "ignitable, corrosive, reactive, or toxic" and "which may pose a substantial or potential hazard to human health and safety and to the environment when improperly managed." Basically this means that a hazardous waste has the ability to cause damage to living organisms and/or the environment.

Hazardous Substances

The term *hazardous substances* is defined by EPA, DOT, and OSHA, and in Federal legislation including the Comprehensive Emergency Response and Compensation Liability Act (CERCLA) of 1980. Because the definition of hazardous substance found in OSHA's 29 CFR Part 1910.120 is so global in nature and the regulation applies to emergency response, it will be the definition that applies to this term throughout this text. The 1910.120 definition states that a hazardous substance is any substance,

> exposure to which results or may result in adverse affects on health or safety of employees:
> 1. any substance defined under section 101(14) of CERCLA;
> 2. any biological agent and other disease-causing agent which after release into the environment and upon exposure, ingestion, inhalation, or assimilation into any person, whether directly from the environment or indirectly by ingestion through food chains, will or may reasonably be anticipated to cause death, disease, behavior abnormalities, cancer, genetic mutations, physiological malfunctions (including malfunctions in reproduction) or physical deformation in such a person or their offspring;
> 3. any substance listed by the U.S. Department of Transportation as a hazardous material under 49 CFR 172.101 and appendices; and
> 4. hazardous wastes . . . or combination of wastes as defined in 40 CFR 261.3 or . . . 49 CFR 171.8.

Extremely Hazardous Substances

The term *extremely hazardous substances* (EHS) was established by the Superfund Amendment and Reauthorization Act (SARA) in 1986 and, as the name implies, refers to substances that have a high degree of toxicity. Additionally, EHSs present significant inhalation hazards. EHSs were originally classified as Acutely Toxic Substances by EPA and defined as substances that are so hazardous that they have the potential to make routine response mechanisms ineffective at protecting the public.

Following the passage of SARA, most of the Acutely Toxic Substances wound up on the EHS list. The number of listed EHSs has varied from about 360 substances to slightly over 400. There are provisions in SARA for individuals or governmental agencies to petition EPA for the inclusion of additional substances. As a result of such petitions and the generation of further scientific information, it is anticipated that the number of EHSs will rise.

To emphasize the high degree of toxicity of EHS, consider the SARA mandates. SARA requires facilities with specified quantities of these chemicals (known as threshold planning quantity or TPQ) to identify themselves to the appropriate state emergency response commission (SERC). The TPQs of the various EHSs are found in an EPA regulation and a document known as the "List of Lists." TPQs range from as little as one pound to as high as 10,000 pounds. Facilities that have a TPQ of an EHS must further identify a facility coordinator who will assist the Local Emergency Planning Committee (LEPC) develop a facility-specific preplan.

The overall classification system is based upon regulatory requirements and the specific substance involved. In many instances the classification is based on the substance's intended use or location within their life times.

For example, gasoline being transported to a gas station in a cargo tank is classified as a hazardous material because it is involved in transportation and listed by DOT as a hazardous material. Once that gasoline is transferred into the storage tanks at the station, it is now considered a hazardous substance because its potential impact on people is the primary concern. If the storage tank now leaks, the gasoline would be considered a hazardous waste. However, gasoline is not considered to be an extremely hazardous substance because it does not meet that criteria.

Chlorine is similar, but regardless of any other classification that may apply, it is also an EHS. Chlorine being transported to a user facility in a rail tank car is a hazardous material because it is involved in transportation and is listed by DOT. Chlorine is also on EPA's list of extremely hazardous substances. When the chlorine reaches its destination, it is no longer a hazardous material but now simply classified as an EHS.

In either case, although the classification of the substances may change, their chemical behavior and hazards remain the same.

Obviously, there have been substantial changes in the world over the past decades, and the chemical revolution has played a major part. As the knowledge of chemicals and their hazards has grown, the government and the private sector have identified appropriate approaches to dealing with these problems. The result has been the development of the hazardous materials emergency response standard of care.

SUMMARY

This chapter discussed the following topics:

1. The early historic development of chemistry
2. The development and role of synthetic substances in the chemical revolution including reasons for the shift away from naturally occurring substances
3. The roles of the Department of Transportation, the Occupational Safety and Health Administration, and the Environmental Protection Agency in the classification and regulation of chemicals
4. The four primary classifications of hazardous chemicals including hazardous materials, hazardous substances, hazardous wastes, and extremely hazardous substances
5. The role of the primary Code of Federal Regulations involving hazardous chemicals including 29 CFR, 40 CFR, and 49 CFR

2 The Hazardous Materials Emergency Response Standard of Care

CHAPTER OBJECTIVES

Upon completion of this chapter, the student will be able to:

1. Define the term *standard of care* regarding hazardous emergency response
2. Explain the role of historical incidents and the growth of knowledge as they apply to the formation and revision of the standard of care
3. Identify four laws, three regulations, three standards, and five guidance documents that impact the standard of care
4. Explain the relationship between laws and regulations
5. Describe the changing relationship between the three levels of government on the basis of the four laws discussed
6. Identify the key requirements found in the following laws that affect hazardous materials response:
   - Clean Water Act and its Amendments
   - Comprehensive Environmental Response Compensation and Liability Act
   - Superfund Amendment and Reauthorization Act
   - Hazardous Materials Transportation, Uniform Safety Act
7. Describe six primary emergency response implications of SARA
8. Explain the facility reporting and planning requirements found in SARA
9. Describe the employer's emergency response plan
10. Explain the five primary level of competency and training required in 29 CFR 1910.120
11. Explain the relationship between regulations and standards

HAZARDOUS MATERIALS EMERGENCY RESPONSE

Prior to the early 1970s, the primary types of hazardous substance emergencies encountered by emergency responders involved flammable or combustible liquids and gases, or corrosives. Responders periodically heard of "weird" incidents involving chemical explosions or fires, but most seemed remote and someone else's problem.

Then in the early to mid 1970s a new, and in some instances, almost catastrophic type of incident started to occur on a routine basis: rail tank car BLEVEs (Boiling Liquid Expanding Vapor Explosions). Most firefighters from those days can remember the first time they read the accounts and saw films of actual BLEVEs. For the first time, hazardous substance incidents became identified as something out of the ordinary that required different strategic and tactical approaches.

Over the following decade, a significant number of incidents occurred that resulted in serious injury to and death of responders. In-depth investigation and, in some cases, scientific investigation led to greater understanding of the problem. These lessons learned from such incidents helped to identify appropriate and inappropriate response actions. The public became further sensitized to hazardous chemicals as a result of media coverage of fiascoes such as what happened in Love Canal, New York.

The increasing awareness and understanding of the nature and magnitude of the problem stirred political and social pressure—and action. As a result, all levels of government identified and confirmed the importance of planning, training, and developing integrated response systems.

Ideas from such diverse fields as chemistry, firefighting, public health, environmental management, industrial hygiene, toxicology, integrated emergency management, and many others, came together to identify the need for a systems approach to the problem. The system required the integration of chemical emergency response planning, training, competency development, and response option identification that was safe and effective for the management of chemical emergencies. Without such a system, the safety and welfare of the public and responders would be in jeopardy. As such, a loose, haphazard set of planning, training, and response criteria formed.

THE STANDARD OF CARE

In 1984, two incidents sent shock waves through industry, government, and the public. The first incident took place in Bhopal, India. A release of methyl isocyanate (MIC) led to approximately 2,000 deaths and 200,000 to 300,000 long-term, negative health impacts. (By the mid 1990s the death toll had reached approximately 6,000 as a result of these long-term effects.) Shortly after Bhopal, a release of MIC also occurred in Institute, West Virginia. Fortunately this second release resulted in no fatalities. However, it provided evidence that a disastrous incident could happen in this country.

Because of these two incidents and the knowledge gained over the previous decade, social and political forces joined to identify the need for formalized hazardous materials planning and response requirements. Congress and other federal agencies embarked upon a series of studies designed to identify the level of preparedness and response competencies that existed for chemical emergency response. The results of these studies indicated that there was generally a woeful lack of both at the state and local levels.

As a result, legislative and regulatory solutions began to evolve. Further, private organizations such as the National Fire Protection Association (NFPA) started the development of consensus standards and governmental agencies developed guidance documents as well. The combination of laws, regulations, standards and guidance form the framework to identify minimum standards required in the planning for, and the response to, chemical emergencies. These minimum standards form the basis of the chemical emergency response standard of care (Table 2-1). *Standard of care* is the level of competency anticipated or mandated in the performance of one's duties. This definition provides only part of the picture, however. To define standard of care is straightforward, but to fully understand what it implies is something more elusive.

Table 2-1   Standard of Care Matrix

	Responder health and safety	Right-to-know and planning	National response system	Training requirements	Response procedures	Transportation
Federal legislation						
Clean Water Act			X		X	
CERCLA (Superfund)			X		X	
SARA	X	X		X	X	
HMTUSA				X		X
Federal regulations						
40 CFR Part 300			X		X	
29 CFR Prt. 1910.120	X			X	X	
40 CFR Part 311	X			X	X	
49 CFR Part 100-199				X		X
Consensus standards						
NFPA 471	X	X			X	
NFPA 472	X			X		
NFPA 473	X			X		
Guidance documents						
NRT-1, NRT-1A, NRT-2		X				
CPG 1-8, CPG 1-8A		X				

Original table courtesy of John Pearson.

KEY ABBREVIATIONS:

Clean Water Act—Water Pollution Control Act of 1970 and Amendments.

CERCLA—Comprehensive Environmental Response, Compensation and Liability Act of 1980.

SARA—Superfund Amendment and Reauthorization Act of 1986.

HMTUSA—Hazardous Materials Transportation Uniform Safety Act of 1990.

40 CFR Part 300—National Oil and Hazardous Substance Contingency Plan.

29 CFR Part 1910.120/40 CFR Part 311—Hazardous Waste Operations and Emergency Response (HAZWOPER).

NFPA 471—National Fire Protection Association—Recommended Practices for Responding to Hazardous Materials Incidents.

NFPA 472—Standard for Professional Competence of Responders to Hazardous Materials Emergencies.

NFPA 473—Competencies for EMS Personnel Responding to Hazardous Materials Incidents.

NRT-1, 1-A, 2—Hazardous Materials Emergency Planning Guide, Plan Review, Exercise Program Guidance.

CPG 1-8, 1-8A—FEMA Civil Preparedness Guide and Plan Review.

First, *competency* must be defined. In most cases, there are many different factors involved. One factor is the "accepted practices" found within a profession, trade, etc. Such accepted practices are reflected in the form of professional, industrial or governmental standards, and licensing requirements. Another factor involves legislative requirements set forth in the form of laws or ordinances that reflect what society or a segment of society feels is appropriate. Yet another factor includes court interpretations and opinions regarding the standards and legislative intent.

Second, it is vital to understand that the standard of care for any given situation or profession is not static but dynamic; in other words, it is capable of continuous change. The change in standard of care is usually due to the change in what is deemed "competent." As our knowledge and understanding change and expand, practices deemed competent change too. Further, judicial interpretations, reinterpretations and definitions will change acceptable practices.

An example of knowledge-based change in the standard of care involves the use of aspirin for the reduction of fever. As recently as the late 1970s, the standard of care stated that children suffering from a fever should routinely receive aspirin to lower the fever. However, the medical profession eventually found a high correlation between the use of aspirin and a potentially fatal neurological complication known as Reye's Syndrome. This new information resulted in a drastic modification in the standard of care. Now, children no longer receive aspirin but rather a non-aspirin pain reliever, such as acetaminophen or ibuprofen, to reduce the fever.

An example of moral/ethical-based changes in the standard of care involves certain punishments. For example, years ago a thief might have been hanged for stealing a horse. Today society would reject this punishment due to the change in what is acceptable.

Finally, a court mandated change in the standard of care involves an individual's right to remain silent. Before the famous Miranda decision, police did not inform a person of his or her rights at the time of arrest.

However, after the Miranda decision, police were required not only to read a suspect his or her rights but to make sure that the individual fully understood those rights.

The past two decades have seen a tremendous increase in our knowledge about hazardous materials and their associated problems. A result has been the development of laws, regulations, and standards aimed at meeting the challenges presented by hazardous material emergencies. In essence, the laws, regulations, and standards have come together to form the framework for the standard of care by which planning and response operations can and will be judged.

Thus, it is important to identify and discuss the key laws, regulations, and standards that compose the chemical emergency planning and response standard of care. This text will only examine some of the components of the standard of care. There are many other laws, regulations, and standards that also apply but are too numerous to address in this text.

The preceding matrix helps provide rapid identification of the primary components of the standard of care. Further, it helps to identify the general aspects of the standard of care addressed by each component.

The Legislative Basis of the Standard of Care

- Clean Water Act of 1970 and Amendments (CWA)
- Comprehensive Environmental Response, Compensation and Liability Act of 1980 (CERCLA)
- Superfund Amendment and Reauthorization Act of 1986 (SARA)
- Hazardous Materials Transportation Uniform Safety Act of 1990 (HMTUSA)

Clean Water Act of 1970 and Amendments.   The Clean Water Act (CWA) of 1970 was a landmark act because it mandated the establishment of a federal system for the response and management of chemical releases, primarily oils, in navigable waters and their tributaries. As part of this legislation, Congress mandated the development of the National Oil and Hazardous Substance Contingency Plan (NCP). The NCP was to "provide for efficient, coordinated and effective action" by the federal government to minimize the damages resulting from the release of oil and hazardous substances. As a result of this act, the NCP became part of the Code of Federal Regulations and was originally found in 40 Code of Federal Regulations (CFR) Part 1510.

Comprehensive Environmental Response Compensation and Liability Act of 1980 (CERCLA).   Most people are familiar with CERCLA because it started the original Superfund targeting the clean-up of hazardous waste dump sites around the country. CERCLA also mandated changes to the NCP and expanded the role and responsibility of the National Response Team (NRT). Of equal importance, CERCLA extended the financial liability of the spiller to cover the costs of the clean-up and recovery. The act broadened the scope and scale of federal involvement in hazardous materials incidents. However, the primary focus was still hazardous waste dumps and oil spills in navigable waters.

The changes required by CERCLA led to the present approaches used by the Nation Response System (NRS) as found in 40 CFR Part 300. Specific information on the present National Response System and Part 300 are found in a later section of this unit.

Superfund Amendment and Reauthorization Act of 1986 (SARA).   As the name implies, SARA amended and reauthorized CERCLA. Again, a major focus was on hazardous waste site clean-up, but this act added quite a few requirements that never existed before. Regarding the standard of care, SARA establishes a national baseline. There are four sections, known as Titles, found in SARA.

Title I—Provisions Relating Primarily to Response and Liability.   Title I primarily addresses amendments to hazardous waste site clean-up response by responsible parties and outlines their financial liability. One caution to remember: although the name of this Title includes the term *response,* it does not address incident response. Rather, the term *response* refers to superfund hazardous waste site operations. Section 126 contains the only reference to true emergency response. Specifically, Section 126 mandates OSHA and EPA to develop identical regulations covering training competencies for Hazardous Waste Operations and Emergency Response (HAZWOPER). OSHA's 29 CFR Part 1910.120 and EPA's 40 CFR Part 311 contain these regulations.

Congress required the development of identical regulations so all hazardous waste and emergency-response operations are covered by some regulation. This was necessary because of limits on the scope and authority of OSHA. The Occupational Health and Safety Act addresses private sector employees while excluding governmental employees from its protection. This approach allows states to establish their own OSHA authority. When a state does so, it covers governmental employees at the state and local level. At present, only 23 states have state OSHA plans in enforcement. This means that governmental employees in the other 27 states have no coverage.

This is where EPA comes in to the picture. EPA has no restriction for enforcement on governmental entities. As such, it was the obvious choice to become the second enforcement authority for these regulations. Through this requirement, Congress assures coverage of *all* emergency-response personnel in all states.

There is only one difference between the EPA and OSHA regulations. In 311, EPA specifically identifies that "compensated and non-compensated workers are considered to be employees of state or local government." As such, it does not matter whether personnel are career, paid on-call, or volunteer; the regulations apply to all of them. Further, the regulation specifies that the governmental jurisdiction for which the service is rendered is the employer of these personnel and thus responsible to meet the regulatory mandates.

Title II—Miscellaneous Provisions. Title II deals with miscellaneous provisions of the Superfund program and has little impact upon emergency response.

Title III—Emergency Planning and Community Right-To-Know. Title III, sometimes referred to as the Emergency Planning and Community Right-To-Know Act (EPCRA), establishes specific requirements that effect industry and state and local governments. These requirements involve emergency-response planning, inventory and release reporting, and community access to information.

Section 301 requires the establishment of an SERC, the identification of local planning districts, and the establishment of LEPCs. The LEPC is the focal point for many Title III requirements. As such, there is much emphasis placed on how the Governor of each state is responsible for oversight of LEPC identification and member appointment. Each LEPC is to consist of a minimum of 15 members representing each of the following groups:

1. State-elected officials
2. Local-elected officials
3. Law enforcement
4. Civil defense/emergency management
5. Firefighting
6. First-aid/emergency medical
7. Health care departments
8. Local environmental
9. Hospitals
10. Transportation

11. Broadcast media
12. Print media
13. Community groups
14. Owners and operators of subject facilities

Generally, the only time an LEPC should have less than fifteen members is when specific agencies do not exist. For example, not all local jurisdictions have local environmental or health agencies.

Section 302 identifies substances and facilities covered by this section and notification requirements. This section requires EPA to develop a list of EHSs and identify threshold planning quantities (TPQ). Facilities with a TPQ require the development of a site specific emergency plan. Additionally, each facility with a TPQ of EHS must identify itself to the SERC which must then notify the EPA administrator.

Section 303 addresses comprehensive emergency-planning requirements. Each LEPC must develop an emergency plan and review it at least annually. The plan shall include the following nine components as a minimum:

1. A listing of facilities subject to the site planning, a list of EHSs present, and transportation routes to and from each facility. Further, facilities contributing or subject to additional risk require identification. Such facilities could include hospitals, nursing homes, schools, other EHS facilities, and others that require special concern and consideration.
2. Response procedures to be followed by facility, local emergency, and medical personnel to manage an emergency at the facility.
3. Designation of community and facility emergency coordinators.
4. Emergency release and incident notification procedures for coordinators, emergency responders and the public.
5. Methods for release detection and determination of the area or population at risk.
6. Description of emergency equipment found at the facilities in the community.
7. Evacuation plans and alternative traffic routing.
8. Training programs and schedules for emergency responders and medical personnel.
9. Methods and schedules for exercising the plan.

Section 303 also mandates the National Response Team to develop a guidance document for the preparation and implementation of the plans. The name of the document is the Hazardous Materials Emergency Planning Guide (NRT-1).

Section 304 addresses emergency notifications if there is a release of an EHS or other specified substance. Notification is provided to the community emergency coordinator and to the SERC as well as federal notification requirements found in CERCLA.

Section 305 addresses training and emergency system review. Part of this section is the $5,000,000 for annual grants to support state and local training.

Section 311 requires facilities to provide a Material Safety Data Sheet (MSDS) or a list of all substances requiring MSDSs by OSHA to the LEPC, local fire department, and the SERC.

Section 312 continues with the requirements of 311 but includes the use of Tier I or Tier II reporting forms. It identifies the specific types of information required on the Tier I and Tier II forms. Finally, 312 specifies that Tier II information must be provided to all levels of government, response agencies, medical personnel, and the public upon request.

Section 313 covers non-emergency toxic chemical releases from facilities. This section covers the annual release of such chemicals to the environment. For example, a waste water treatment facility that chlorinates the water before its release must report the total amount of chlorine so released.

Sections 321 through 330 cover such diverse topics as the provision of information regarding trade secrets; the provision of information for health care professionals; and civil actions, enforcement issues, fines, and exemptions.

Section 324 is a critical section because it addresses the availability of all plans, data sheets, forms and follow-up notices. All such information "shall be made available to the general public" upon request.

Hopefully, when considered in this context, these laws identify the pattern that has developed over the past 25 years. Initially, legislation only existed on the Federal level, even though local government and its response agencies are the first line of defense against most hazards including hazardous materials incidents. However, as the true magnitude of the problem became more apparent and major incidents occurred, federal agencies and Congress realized the need to involve local jurisdictions.

By the early 1980s, it became apparent that the federal government could not manage the problem alone. Rather, hazardous materials incidents are a local problem that can generate national impacts. As a result, SARA mandates state and local governments and their agencies to become involved with the development of a response and planning system—a system that local and state governments are capable of developing.

Other Right-to-Know Legislation and Planning Legislation.   Many state legislatures have passed laws mandating access to information about hazardous materials as well as various planning approaches and requirements. After the enactment of SARA, many states passed their own legislation that closely follows the mandates set forth in SARA. It is important to understand the requirements set forth in such legislation in one's own state.

Hazardous Materials Transportation Uniform Safety Act of 1990 (HMTUSA). Congress passed HMTUSA (now simply called the Hazardous Materials Transportation Act or HMTA) in 1990. Its primary purpose is to update the original Hazardous Materials Transportation Act of 1975. This Act requires some major changes in the regulations governing the transportation of hazardous, and in some instances, non-hazardous materials. Specifically, the Act intends to provide more uniformity between the United States and International hazardous materials transportation requirements.

Additionally, it intends to provide greater safety during the transportation and transportation related handling of hazardous materials.

One of the most fundamental changes involves the training and competency of individuals in the transportation industry. First, all commercial highway vehicle operators (with only a few exceptions) must get commercial driver's licenses (CDL) that include written and actual driving tests. Second, drivers who transport hazardous materials must have additional training and testing to receive a CDL that will allow them to transport these ladings. Further, not just drivers but all employees who handle the hazardous materials ancillary to its transportation must receive training in hazardous materials, their safe handling, and procedures to follow if there is an emergency.

In another area, DOT made significant modifications to its classification, placarding, and labeling systems. DOT introduced a new classification system that involves Classes and Divisions. Several classifications changed and all the Other Regulated Materials (ORMs) were dropped except ORM D. Several old placards and labels (e.g., Class A Explosives) became obsolete while several new ones appeared (e.g., Dangerous When Wet placard). The number of substances always requiring placards, regardless of mode or quantity, went from 5 to 7.

The container specification system changed as well. No longer may containers simply meet design specifications, but they must also successfully meet performance tests. Highway cargo tanks must undergo not just visual but pressure and leak tests on a routine basis. In general, all container specifications are more stringent than before.

Finally, the Act provides a revenue source for local-level hazardous-materials training and planning. Originally, SARA provided an annual pot of 5 million dollars for hazardous-materials training and planning. This SARA funding must be periodically reauthorized. To provide an additional source of funding, HMTUSA provides approximately 13 million dollars annually for training and planning. At least 75% of these funds must pass through from state to local government.

One result of this funding is the development of the Hazardous Materials National Curriculum. Through NRT's Training Committee (now DOT's), a group representing response, planning, and training organizations from the local, state, and federal levels was formed. This group provides input from around the country. A smaller group, known as the authors' group, held a series of meetings and work sessions to develop the curriculum.

The response curriculum follows the competencies identified in 29 CFR Part 1910.120 and the NFPA Standards 471, 472, and 473. The planning section involves a more difficult task because there were no pre-existing national competencies. State governments use the final curriculum as a device to measure whether a training program qualifies for Federal Funding.

The Regulatory Basis for the Standard of Care

- 40 CFR Part 300—The National Oil and Hazardous Substance Contingency Plan (NCP)

- 29 CFR Part 1910.120/40 CFR Part 311—Hazardous Waste Operations and Emergency Response (HAZWOPER)
- 49 CFR Parts 100 to 199—Hazardous Materials Transportation Regulations

There are 50 separate sections, commonly called Titles, of the CFR. Specific laws require the writing (promulgation) of all of these regulations. For example, Section 126 of SARA requires the promulgation of OSHA's 29 CFR 1910.120. There are specific rules governing the promulgation of regulations. First the agency or department assigned the responsibility must publish a document known as a "Proposal of Rule Making" and must request public comment regarding the particular rule. The proposed rule making is published in the Federal Register, a daily publication designed to provide information about federal governmental activities.

After receiving and reviewing the comments, the agency or department writes "Interim Final Rules." These rules are then published in the Federal Register and comments are requested. However, the regulations, as found in the interim final form, go into effect at this point.

After receiving and reviewing the comments, the final regulations are written, an effective date is identified, and they are printed in the Federal Register. The regulations appear at the next publishing of the appropriate CFR.

Although regulations are not laws, they carry the weight of law. In other words, they are the enforcement mechanism used to assure compliance with a law's intent. Generally, regulations are quite technical, rather lengthy, and address areas in the law that require additional clarification to implement.

EPA: 40 CFR Part 300—National Oil and Hazardous Substance Contingency Plan (NCP).   The NCP is found in 40 CFR Part 300, and it establishes the role that the federal government and its agencies will play in major incidents involving hazardous materials, substances, and wastes. These regulations identify the roles, responsibilities, and components of the NRS.

The development and maintenance of these regulations are the responsibility of the NRT, which is a group of 14 federal departments and agencies responsible for designing and overseeing the federal response to major chemical emergencies. The NRT is chaired by EPA, vice chaired by the U.S. Coast Guard (USCG), and located in Washington, D.C.

The NRT does not respond to chemical emergencies. Rather, it delegates that role to the 13 Regional Response Teams (RRT), roughly following the standard federal regions. The RRTs are the level where actual response may occur. Federal On-Scene Coordinators (OSC) are pre-designated representatives of EPA or USCG and are located in each region. If the incident is large enough, an OSC may respond to the scene or provide other assistance. The OSC can provide assistance from the Technical Assistance Teams (TAT) for the region, the National Strike Force (NSF), the Public Information Assistance Team (PIAT), or other federal resources. Generally, the OSC will not take control of the incident but rather assist state and local authorities.

The NRT is also responsible for providing the mechanism to report releases of chemicals into the environment. Contacting the National Response Center (NRC) accomplishes such reporting. The NRC is a 24-hour hotline staffed by USCG personnel and designated the "One Number" federal government reporting center (only one number need be called to access the entire system). When reported to the NRC, the spiller is able to meet the initial release reporting requirements as well as access the governmental side of the response system. Additionally, emergency responders can call the NRC to gather information and to access governmental expertise and response capabilities.

OSHA: 29 CFR 1910.120, and EPA: 40 CFR 311—Hazardous Waste Operations and Emergency Response (HAZWOPER).   OSHA 1910.120 and EPA 311 are identical regulations mandated by Section 126 of SARA. Congress mandated both OSHA and EPA to promulgate these regulations so that all response personnel have coverage whether or not the state had its own OSHA plan.

Many of the provisions of these regulations apply specifically to hazardous waste site operations. However, paragraph (q) addresses emergency response. Emergency response is defined as any "response effort by employees from outside the immediate release area or by other designated responders (i.e., mutual aid groups, local fire departments, etc.) to an occurrence which results in, or is likely to result in an uncontrolled release of a hazardous substance." Responses to releases "where there is no potential safety or health hazards (i.e., fire, explosion or chemical exposure) are not considered to be emergency responses."

All employers who will have their employees involved in emergency-response activities must develop a written Emergency Response Plan (ERP) prior to any response activities. The ERP must be available for inspection and copying by employees, their representatives, and OSHA.

At the minimum, the ERP must address the following components:

1. Pre-emergency planning and coordination with outside parties
2. Personnel roles, lines of authority, training, and communication
3. Emergency recognition and prevention
4. Safe distances and places of refuge
5. Site security and control
6. Evacuation routes and procedures
7. Decontamination
8. Emergency medical treatment and first aid
9. Emergency alerting and response procedures
10. Critique of response and follow-up
11. Personal protective equipment (PPE) and emergency equipment

Section (q)(3) addresses specific procedural requirements for response. The most fundamental requirement is the use of an Incident Command System (ICS). The individual in charge of the ICS, the Incident Commander (IC), is responsible to identify to the extent possible all hazardous substances and conditions present on the site. Considering the hazards (substances and conditions), the IC shall implement appropriate emergency operations. This includes assuring the use of appropriate personal protective equipment (PPE). The minimum allowable PPE consists of full structural turn-outs (as defined in 29 CFR 1910.156(e)).

Personnel susceptible to inhalation route exposures or other types of potential inhalation hazards must wear positive pressure self-contained breathing apparatus (SCBA) while engaged in response activities. The only time the IC can allow the use of a lesser degree of respiratory protection is after direct air monitoring determines that its use presents no hazards to personnel.

The IC must limit the number of personnel in areas of actual or potential hazardous conditions (the hot zone) to those actively participating in control activities. However, this mandate includes the use of the buddy system (groups of two or more maintaining direct line of sight).

During personnel entries into the hot zone, back-up personnel must be designated. The back-up personnel require appropriate equipment (including PPE) to assist entry personnel or provide rescue should it be needed. Further, emergency medical (EMS) personnel must be standing by and able to provide first aid as needed. These EMS personnel require training to at least the level of advanced first aid. Further, EMS must have appropriate equipment available for treatment as well as the capability of transporting personnel as needed. This means if there are victims involved in the incident, a minimum of two EMS units is needed: one for responders and at least one for victims.

A Safety Officer is required. The Safety Officer must be *knowledgeable* in the operations to be performed. This means that the Safety Officer's training should be to at least the Hazardous Materials Technician level if the incident requires the involvement of a hazmat team and technician level activities. The only option is to have a Safety Officer who is thoroughly knowledgeable with the employer's ERP and standard operating guidelines (SOGs). This must be accompanied by operational check sheets that document whether the SOGs are followed and act as a benchmark for compliance with procedural options. The Safety Officer is responsible to identify and evaluate hazards found on the scene and to provide direction regarding the safety of operations to be performed.

If the Safety Officer determines that the situation or tasks will expose personnel to an Immediately Dangerous to Life and Health condition (IDLH) and/or an Imminent Danger condition, the Safety Officer has the authority to take immediate actions, including altering, suspending, or terminating the activities. Upon taking such actions, the Safety Officer must immediately notify the IC.

IDLH as defined in 1910.120 is "An atmospheric concentration of any toxic, corrosive or asphyxiant substance that poses an immediate threat to life or would cause irreversible or delayed adverse health effects or would interfere with an individual's ability to escape from a dangerous atmosphere." Imminent Danger as defined in 29 CFR is a situation in which "Conditions or practices . . . which could reasonably be expected to cause death or serious physical harm. . . ." When terminating emergency operations, the IC must assure that victims, responders, and equipment are appropriately decontaminated.

Should earth-moving, digging, lifting, or other hoisting equipment be needed and not designated as part of the response organization, their operators are considered "skilled support personnel." These skilled support personnel need not meet the training mandated in this section. However, before performing any work, these personnel are to receive an initial briefing. The briefing must include instruction on the use of PPE, the specific chemical hazards involved, and what duties they are to perform. To assure their health and safety, these personnel must receive all the safety and health precautions provided for other responders on the scene.

Specialist employees are another class of employees. Specialist employees are "Employees who, in the course of their regular duties, work with and are trained in the hazards of specific hazardous substances." These employees provide technical advice or assistance to the IC at the scene. This class of employees must receive training or demonstrate competency in their field on an annual basis.

Paragraph (q)(6) addresses the subject of training and competency of emergency responders. Specifically, this section states that employees must receive training in the duties they are expected to perform. Such training must be completed prior to their involvement in any response activities. It is important to note that if the SARA Title III plan or the employers ERP indicated that a given group or agency (such as Public Works, Health Department, etc.) are to play a role in an emergency response, they must receive appropriate levels of training.

However, before identifying and explaining these levels, it is advantageous to consider the overall realm of training. The regulations and standards that presently exist regarding training use the term *competency* to identify what an individual should achieve through a training program. As such, it is critical to define the term. Competency is the demonstration of a desired behavior. The behavior arises from a combination of knowledge and skills gained through educational processes.

To successfully train an individual to be competent at their job or any given aspect of their job, one must first identify what behaviors they are required to perform in that job. For emergency-response activities, operational tactical objectives and methods are the baseline behaviors that are desired. Consequently, the development of an effective training program requires the identification of specific tactics and methods needed to perform one's job function.

1910.120 provides a broad, general outline of these behaviors. Some have severely criticized OSHA because their requirements are so broad and general. Such criticism is baseless because no federal or even state government can identify the specifics because every employer, facility, and jurisdiction is different. New York is not the same as Los Angeles, which is not the same as Phoenix, which is not the same as Allentown, which is not the same as Sioux City. Each jurisdiction has unique hazards, needs, resource availability, and response systems. A cold storage facility is not the same as a chemical production facility or a semiconductor manufacturer. For example, a response system that has no available team requires the use of different tactical options than one with a team. A county in a larger metro with a single fire department has a different response system than a county that has forty individual departments. Further, a community that has little industry but major transportation corridors has a different problem than a community that has large petrochemical industries.

So how does one identify the specific training needs for a given facility or juris-diction? The answer is the employer's ERP and its associated SOGs. These documents (assuming they are based on strategies and tactics) provide the specific information needed to identify competencies. These competencies form the basis for an *effective* training program.

Unfortunately, very few training programs have been developed this way. When the regulations first appeared, almost everyone scrambled to find some training pro-gram to meet the mandates. Little if any thought was given to the nature of the compe-tencies needed by a given jurisdiction. Further, few understood hazmat strategies and tactics. As a result, we have many responders who passed an initial written test and pass their annual recertification test but do not demonstrate a modification in behavior (and thus, competency) when they respond to an incident.

Quite often initial, refresher, or recertification training uses a shotgun approach. In this approach the program talks about topics such as recognition and identifica-tion, ICS, decontamination, spill control, etc. However, there is no linkage between the topics or how they tie into operational situations and settings found out on the in-cident scene.

Further, many response agencies have a rudimentary ERP but no SOGs. This is truly unfortunate, because SOGs provide the linkage between any given topic in a train-ing program and the behavior that is desired on the scene. Personnel understand SOGs that form the basis of all operational strategies and tactics. They identify what the or-ganization wants personnel to do and how it wants them to do it.

Why do so many plans fail during an actual emergency? Most commonly it is be-cause there is no linkage between the plan and the behaviors required of the people who will implement the plan. That linkage is the SOGs. When the SOGs and ERP integrate the functional assignments found within the community's LEPC (SARA Title III) Plan, the plan will work because its functional designations are translated into understand-able and measurable behavioral expectations. This is the precise reason why OSHA specifies the employer as the certifying authority of competency and individual level of training. Specifically, since the SOGs and ERP are employer specific, only the em-ployer can identify what constitutes competency for its operation.

The response community has reached the point where simply training to meet the requirements is not effective and does not meet their unique needs. It is time for train-ing programs to graduate from simple compliance tools to true competency building tools. With the assistance of SOGs as well as training and evaluation approaches that require students to perform cognitive and psycho-motor (hands-on) skills needed on the street, the training will result in true competency.

One of the more successful approaches for developing and evaluating cognitive skills is through the use of scenario-based training and evaluation. In this type of ap-proach, the students are faced with a scenario in which they must answer a series of questions or perform a series of tasks based on the SOGs.

The following are just a few examples of questions that could be asked:

1. What is the first thing you need to do?
2. How would you do that?

3. What else must be done?
4. Identify the strategies and tactics that you would use.

The following are just a few of the actions that may be required.

1. From a slide, identify the DOT hazard class and division of the placard displayed.
2. Use the *Emergency Response Guide* to identify the appropriate Guide Number and pertinent emergency information found within.
3. Set-up an appropriate perimeter.
4. Identify and perform appropriate control options (strategies and tactics).

Psycho-motor skills are also relatively easy to check. For example, the ERP and SOGs indicate that operations level personnel shall be able to perform basic spill control tactics such as diking. Some of the specific methods range from the use of available soil, sand, or charged hose lines. Once the students have received classroom training on the strategy and its associated tactics, they proceed to an appropriate location and are tasked with stopping a simulated spill. Their evaluation centers on how safely and effectively they perform the task.

Training programs based on these types of concepts and approaches are truly effective at developing and evaluating competency of our response personnel. They also provide the benefit of translating training activities into identified and desired behaviors when personnel respond to an actual incident. Additionally, they provide a secondary benefit by enhancing operational safety because personnel know what to do in a given situation and how to go about doing it. Finally, once basic behaviors are mastered, subsequent training and evaluation can focus on new and more advanced skills and knowledge.

With this discussion in mind, it is time to examine the training requirements found in 1910.120. This regulation identifies five primary levels of training for emergency-response personnel. It further identifies two ancillary levels that generally fall in the realm of contractors or facility personnel. This discussion will confine itself to the five primary levels identified as the following:

1. First responder awareness level (awareness)
2. First responder operations level (operations)
3. Hazardous materials technician level (technician)
4. Hazardous materials specialist level (specialist)
5. On-scene incident commander (incident commander)

First Responder Awareness Level.   Awareness level personnel are "Individuals who are likely to witness or discover a hazardous substance release." They are trained to initiate an emergency response by notifying the proper authorities and will then take no further actions. This level employee includes personnel such as industrial workers, inspectors, code enforcement personnel, and law enforcement. Awareness personnel must demonstrate the following competencies:

1. An understanding of what hazardous substances are and their associated risks when involved in an incident
2. An understanding of potential outcomes of emergencies involving hazardous substances
3. The ability to recognize the presence of hazardous substances in an emergency
4. The ability to identify the hazardous substances, if possible
5. An understanding of their role within the ERP including site security and control
6. An understanding of the use of the U.S. Department of Transportation *Emergency Response Guide*
7. The ability to recognize the need for additional resources and to make appropriate notifications

First Responder Operations Level. Operations level personnel are "Individuals who respond to releases or potential releases of hazardous substances as part of the initial response." Further, they respond, "for the purpose of protecting nearby persons, property or the environment from the effects of the release." They respond in a defensive fashion without trying to stop the release (which is considered an offensive action). Their function is to contain the release from a safe distance, keep it from spreading, and prevent direct contact or exposures to themselves.

Herein lies a major area of contention. Based on a strict interpretation and various enforcement actions from around the country, the following examples identify actions deemed to be offensive in nature and thus beyond the scope of operations level personnel:

1. The application of "dumdum" or a golf tee to a hole in the gas tank of a vehicle
2. The closing of a valve during a propane fire

However, in 1993, OSHA released a document known as *OSHA Instruction CPL 2-2.59*. This document specifies that operations level personnel who have received appropriate, documented training and wear appropriate PPE may now shut off a propane or natural gas valve during a fire. However, the document indicates that should the incident involve gasoline instead, even appropriately trained and PPE-protected firefighters will commit a technical violation if they turn off the valve.

Typically, all fire personnel involved in actual emergency response must attain the operations level of training and competency. Due to the language, "part of the initial response," and, "for the purpose of protecting nearby persons, property and the environment," many (including various state agencies) feel that EMS personnel also require training to the operations level.

Further, there is a growing sentiment that all responders cannot be pigeonholed into the same operations level box. Rather, there is a need for separate tracks for the various disciplines involved in the emergency response. The National Hazmat Curriculum Guidelines developed by the National Response Team's Training Committee reflects these thoughts.

Along with the competencies identified at the Awareness level, operational personnel are to receive at least eight hours of training or have sufficient experience to objectively demonstrate the following competencies, and the employer shall so certify:

1. Knowledge of basic hazard and risk assessment techniques
2. Ability to select and use proper PPE provided
3. Understand basic hazardous materials terminology
4. Be able to perform basic spill control techniques within the scope of their training, available equipment and PPE on their unit
5. Know how to implement basic decontamination procedures
6. Understand relevant standard operating and termination procedures

Hazardous Materials Technician.   Technician level personnel are "Individuals who respond to releases or potential releases for the purpose of stopping the release." Technicians often perform offensive actions because they will approach the breach in the container to perform leak control activities. Technicians are to receive a minimum of training at the operations level (24 hours–1993). Additionally, technicians must have the following competencies and the employer shall so certify:

1. Know how to implement the employer's ERP
2. Use field survey equipment to classify, identify, and verify known and unknown substances
3. Function within an assigned role in the ICS
4. Properly use chemical protective clothing provided
5. Understand hazard and risk assessment techniques
6. Perform advanced, offensive control techniques including leak control
7. Understand decontamination and termination procedures
8. Implement decontamination
9. Understand basic chemical and toxicologic terminology and behavior

Technician level training is another area where there is a considerable amount of controversy. The first consideration is the amount of time required for an individual to become a technician. It is generally accepted that, on the basis of 1910.120, forty hours of training is the minimum required. However, is that enough training?

Again, the unique situations that exist in the real world are the primary tool needed to answer this question. Consider a fixed facility that has only one specific hazard class of substances found on the site (e.g., corrosives). They have an ERP that specifies the strategies and tactics that are available for their on-site responders. In essence, the technician for this facility needs to learn about a very narrow range of information and psycho-motor skills. As such, it may be possible that, based on the employer's ERP, in as little as twenty-four (24) hours an individual can obtain all the competencies needed to handle an incident safely and effectively *at this facility.*

However, suppose the facility has several hundred different hazardous substances with which to contend. Without question, the absolute minimum time required to acquire appropriate competencies is forty hours. This assumes that a high degree of inhouse technical assistance is available during a response, and that an ongoing competency development program is in place.

Now, suppose the technician will be part of a municipal hazmat team. Again, the obvious answer is a minimum of forty hours of training because the regulations identify that as the minimum.

However, is that enough? This question addresses the crux of this entire matter and its answer depends on the one's perspective. If we consider the technician as totally independent and working on his or her own, the answer is no. Yet, the reality of emergency response is that a technician is a member of a team.

Teams consist of many individuals with differing backgrounds, strengths, and weakness. Do these individuals have to know common practices and procedures? Yes. Then is it necessary or even possible, for all members of a team to be identical? No.

The reality of the situation is that some individuals on a team have strengths in psychomotor activities while others have strengths in cognitive skills. Some individuals simply have little or no understanding of chemistry and have no desire for the subject. Other individuals have no understanding of mechanics and have no desire to learn it. Does this mean such individuals must be excluded from a team because they do not meet the established template? If the answer is yes, in reality, most of the teams around this country will lose integral members and may become nonfunctional.

Further, this situation would result in a tremendous waste of personnel. Moreover, the strict recommendations found in various standards can and will have this detrimental effect if they are allowed to establish unattainable expectations and competency levels.

There is a basic flaw in the present approach because, on the basis of present and upcoming consensus standards, a technician must be a master of all trades and then develop various specialties. Further, these standards indicate that a technician is a technician is a technician. This approach is the same as saying a firefighter is a firefighter is a firefighter. Obviously, this is not true. This realization led to the idea of Firefighter 1, Firefighter 2, and in some locations, Firefighter 3. Why not use the same approach by establishing Technician 1 and Technician 2?

A Technician 1 is the baseline technician. The newly trained technician would start at this level, and just as the newly trained firefighter, would be on probation. During the probationary period, the new technician must receive additional training and develop additional skills. Technician 1 focuses on psycho-motor skills. Once the probationary period is passed, there is no demand for the Technician 1 to ascend to Technician 2.

A Technician 2 is an advanced technician. To attain this level, the technician must undergo yet more training that focuses on cognitive knowledge. At this level, the technician must have a moderate level of understanding about chemistry and containers. This individual would perform functions such as information gathering from technical sources (MSDS, computer programs, reference books, etc.). Further, the Technician 2 could fill all the ICS positions within the employer's system, except

Hazmat Group Supervisor or Branch Director. As such, the Technician 2 supervises various activities and functions performed by other members of the team or by operations level personnel.

Hazardous Materials Specialist.   Specialist level personnel are "Individuals who respond with and provide support to hazardous materials technicians." Although the specialist role parallels that of the technician, the specialist must have a more focused and in-depth level of learning and knowledge. Additionally, specialists act as the site liaison with federal, state, local, and other governmental authorities. Specialists must have a minimum of training at the technician level (24 hours–1993). Additionally, the specialist must have the following competencies and the employer shall so certify:

1. Know how to implement the employer's emergency response plan
2. Use advanced survey instruments and equipment to classify, identify, and verify known and unknown substances
3. Know the state emergency response plan
4. Properly select and use chemical protective clothing
5. Understand in-depth hazard and risk techniques
6. Perform specialized control, containment, and confinement operations
7. Determine and implement decontamination procedures
8. Develop site safety and control plans
9. Understand chemical, radiologic, and toxicologic terminology and behaviors

The specialist possesses a substantially more in-depth knowledge of cognitive skills than the Technician 2. These individuals fill the position of the Hazmat Group Supervisor or Hazmat Branch Director. They develop and implement the specific strategies and tactics for the group or branch. Further, they analyze and assess data from a myriad of sources to determine the most effective and efficient strategies and tactics for the given incident situation. Additionally, they act as the contact person for all governmental and outside agencies involved in the response.

To perform such tasks, the specialist needs extensive knowledge in chemistry and a strong knowledge of meteorology, geology, physics, and other sciences. These individuals must fully understand the process and implication of damage assessment for not only the container but the product and the environment. They must have the expertise to identify when gathered information is in conflict with the incident situation and when data is just plain wrong.

A case has been made in various quarters that there is no difference between a technician and a specialist. Further, that industry should provide the expertise to perform the skills just identified. Both assumptions require additional consideration to fully assess their merit.

First, there is a very definite difference between the knowledge level of Technician 1, Technician 2, and Specialist. The Technician 1 for all practical purposes is a mechanic. They know how to work in chemical and other special protective clothing. They perform extensive hands-on activities including spill control, leak control, air monitoring, and even product sampling. Most teams have many members who like this type of work and form the backbone of team operations.

Due to their greater depth of knowledge, the Technician 2 performs tasks such as information gathering, recommends appropriate PPE and decontamination procedures, and similar higher level cognitive skills.

Specialists perform even higher-level cognitive skills that involve analysis, interpretation, synthesis, and extrapolation. They approve or alter the recommended PPE and decontamination procedures. While technicians use instrumentation to perform air monitoring to collect raw data, specialists analyze and extrapolate the meaning and implications of these data to fully identify both the qualitative and quantitative implications of the readings.

Still to some, this position regarding competency holds no meaning. However, when viewed from the perspective of competencies, educational process, and levels of learning, there is a tremendous distinction. For example, consider air monitoring. A technician must know his or her available instrumentation and be able to identify its role in the monitoring program, turn it on, calibrate it, and retrieve accurate data while wearing various PPE and operating under a multitude of conditions. In other words technicians need to know how to use the instruments safely and effectively. Knowing how to use a piece of equipment is one of the baseline levels of learning and can be taught in a relatively short period of time. Educationally, it is a relatively simple, straightforward task.

So where does the specialist fit into the picture? The specialist takes this raw data, synthesizes it, and more importantly extrapolates from it. For example, the monitoring team uses a combustible gas indicator (CGI or % LEL meter), distilled water moistened pH paper, radiologic monitors, a photo ionization detector (PID), a flame ionization detector (FID), and an oxygen meter at an industrial outbuilding with unknown contents. They report negative "hits" on all of the instrumentation except the oxygen meter that has a reading of 18%. From this data, the specialist would extrapolate that there is some contaminant present and there is a high probability that there is a release. Most likely the release involves an asphyxiant such as an inert gas. The only other option is some type of oxidation process similar to the rusting of iron that has consumed the oxygen from the air.

Further, the oxidation process would only be a possibility if the readings came from a tightly sealed enclosure of some type. A key question for the monitoring team is, "Did you hear a hissing sound when the door or hatch was opened?" The question is essential because oxygen depletion due to oxidation will convert gaseous oxygen into some type of solid oxide. This transformation causes a decrease in total air volume, reduces pressure, and can lead to a partial vacuum and a hiss when the closure is opened.

The extrapolation of such information is a fundamental competency required of the specialist. Educationally, this entails one of the highest levels of learning, significant knowledge, and experience as well as interpretive skill. In education, this is the type of competency developed in 500 level courses (graduate and post graduate levels) while the technician competencies are the 100 level courses (introductory course).

Some may respond, "How can all members of a team reach these competencies?" They probably cannot. The bottom line is that not all people are capable of, or even desire to meet, the competencies of the higher levels. That is all right. Everyone can not be and do everything.

Through the use of this kind of effective personnel development and management, coupled with the previously described approaches to training and competency evaluation, more workable and meaningful training and competency development are achievable. Further, the translation of training and education into desired behaviors on the street is within the grasp of all organizations.

On-Scene Incident Commander.   On-scene Incident Commanders are individuals "Who will assume control of the incident scene beyond the first responder awareness level." They must receive a minimum level of training at the operations level (24 hours–1993). Additionally, the IC must have the following competency and the employer shall so certify:

1. Be able to implement the employer's ICS
2. Be able to implement the employer's ERP
3. Know and understand the hazards and risks to employees working in chemical protective clothing
4. Know how to implement the employer's ERP
5. Know of the state emergency response plan and the National Response Team
6. Know and understand the importance of decontamination procedures

Competency Maintenance.   As stated after each training level, the employer is responsible to certify that personnel indeed have the competency specified in each level. Further, once the initial training has been completed, personnel must maintain their competencies and the employer must recertify their employees annually. One option for competency maintenance is through annual or ongoing refresher training. The training must be long enough and cover topics needed to assure personnel will perform their assigned competencies safely and effectively during a response. Another option is the demonstration of competencies. In this case, personnel must demonstrate the performance of basic tasks and/or successfully complete a written examination. Actual response operations can also help meet this requirement. In any case, the employer must recertify personnel annually and maintain records of the methods used to demonstrate those competencies.

There is also a discussion of trainer qualifications as well. Specifically, all trainers must have successfully completed a train-the-trainer program for the course or subject material that they are to teach. An alternative to train-the-trainer programs is training and/or academic credentials that indicate the individual has knowledge of the instructional materials involved. Additionally, the trainer must have the instructional experience necessary to demonstrate competent instructional skills and good command of the subject matter to be presented.

1910.120 addresses additional requirements for hazmat teams. All hazmat team personnel and other specified personnel must have baseline physical examinations and be provided with a medical surveillance program as required in paragraph (f). Further, any responder who exhibits signs of exposure as a result of an emergency response must be provided with medical consultation as specified in (f)(3)(ii).

Hazmat teams must also meet the requirements of (g)(3) through (g)(5). These sections identify basic criteria for the selection of PPE, minimum types and levels of PPE available, and criteria for totally encapsulating chemical protective suits. Finally, it mandates the development of a Personal Protective Equipment Program. The PPE Program must address the following:

1. Selection of appropriate PPE based upon site hazards
2. Use and limitations of the types of PPE available for use by team members
3. Work mission duration criteria
4. Maintenance and storage of PPE
5. Decontamination and disposal of PPE
6. Training and proper fitting of PPE
7. Donning and doffing procedures
8. Inspection procedures prior to, during, and after use
9. Evaluation of program effectiveness
10. Limitations during temperature extremes, heat stress, and other appropriate medical considerations

Finally, paragraph (q) indicates that when cleanup activities occur after the emergency response is complete, personnel performing the clean-up must meet one of two sections of OSHA regulations. If the response and cleanup are not on facility property, the requirements of paragraphs (b) through (o) of 1910.120 take precedence. If these activities occur on facility property and facility personnel will perform the cleanup 29 CFR 1910.38(a), 1910.134, 1910.1200, and all other appropriate health and safety training requirements take precedence.

DOT: 49 CFR Part 100 to 185.   49 CFR addresses the DOT regulations involved in the transportation and classification of hazardous materials. Specifically, this part of the CFR identifies all hazardous materials, their hazard classification, and transportation requirements. Further, it identifies all shipping requirements for each of the five modes of transportation (highway, rail, air, water, and pipeline). Some of the shipping requirements include:

1. The types of cargo containers allowed for the transportation of each hazardous material and the container's specifications
2. Placarding and labeling
3. Shipping papers, product information, and transportation emergency actions
4. Driver and product handler competencies

Although 49 CFR is the primary enforcement tool used in the transportation of hazardous materials it has a tremendous amount of information that can be invaluable to emergency-response personnel. Unfortunately, not many responders, other than those directly involved in enforcement, know much about the information in these regulations. Included in the Appendices is an index of the various sections of this regulation to help take away some of the mystery in finding one's way around the regulations. Also, individual sections will be discussed later in this text.

The Standard Basis for the Standard of Care

National Fire Protection Association Standards
- NFPA 471—Recommended Practices for Responding to Hazardous Materials Incidents
- NFPA 472—Standard for Professional Competence of Responders to Hazardous Materials Incidents
- NFPA 473—Competencies for EMS Personnel Responding to Hazardous Materials Emergencies

The primary types of standards to be discussed here are consensus standards, which are developed by and for people involved in a given type of activity. They identify appropriate procedures and levels of competency. In determining procedures and competencies, there is a degree of compromise involved. When approved, the final document represents a baseline for conducting the activity.

Although consensus standards are not mandatory, unless adopted by the jurisdiction having authority, they are still quite important. Because their development is through a group process involving people performing the activity, they become legally important. Specifically, consensus standards help to identify how the "reasonable person," defined as anyone involved in performing the same activity, would perform the activity. If this basic standard is not met, there are potential legal implications.

The role of consensus is only important if the standard is realistic and is developed by a true cross-section of persons involved in the field. Both of these litmus tests must be considered before the applicability of the standard is truly fixed.

National Fire Protection Association (NFPA) 471—Recommended Practices for Responding to Hazardous Materials Incidents.   NFPA 471 is a standard generated by the NFPA Technical Committee on Hazardous Materials. Specifically, this standard outlines minimum operational considerations and guidelines for incidents involving hazardous materials. This standard addresses considerations such as:

1. Incident response planning
2. Response levels
3. Control options
4. Personal protective clothing

5. Chemical protective clothing
6. Decontamination
7. Safety and communications

NFPA 472—Standard for Professional Competence of Responders to Hazardous Materials Incidents. NFPA 472 is also generated by the NFPA Technical Committee on Hazardous Materials. Specifically, this standard establishes minimum knowledge and competence levels that response personnel need to safely and effectively respond to hazardous materials incidents. In essence, this standard identifies basic knowledge and competencies needed for hazardous materials response.

The standard classifies response personnel by four basic levels, which are identical to those in OSHA 1910.120: first responder awareness, first responder operational, hazardous materials technician, and on-scene incident commander. The committee chose to drop the hazardous materials specialist level and instead identifies specialty areas. Any training program based upon the goals and objectives set forth in this document will, in most cases, comply with the OSHA mandates.

As with most NFPA training standards, this particular standard is competence based. This means that the personnel receiving the training must demonstrate the performance of specified tasks or functions as well as demonstrate that they have attained a given level of knowledge regarding specific terminology, concepts, procedures, and activities.

NFPA 473—Standard for Professional Competence of Emergency Medical Personnel to Hazardous Materials Incidents. NFPA 473 follows the same principles as NFPA 472 in that it specifies various levels of knowledge, tasks, and functions performed by EMS personnel involved in hazardous materials response. NFPA 473 was developed because there was little uniformity in training and competency requirements.

Guidance Documents
- National Response Team (NRT-1)—*Hazardous Materials Emergency Planning Guide*
- National Response Team (NRT-1A)—*Criteria for Review of Hazardous Materials Emergency Response Plans*
- National Response Team (NRT-2)—*Developing a Hazardous Materials Exercise Program*
- Federal Emergency Management Agency (FEMA)—*Civil Preparedness Guide: State and Local Emergency Operations Plans (CPG 1-8)*
- FEMA—*Civil Preparedness Guide: Review of State and Local Emergency Operations Plans (CPG 1-8A)*

In addition to laws, regulations, and standards, there are many guidance documents available that provide direction or outline specific procedures regarding various aspects of this entire problem. These documents are mentioned for additional reference.

National Response Team (NRT) Planning Documents—NRT-1, NRT-1a, and NRT-2.   As mandated in SARA Title III, the National Response Team (NRT) published two separate guidance documents, NRT-1 and NRT-1a. NRT-1 provides "Unified Federal guidance for hazardous materials emergency planning and presents a Federal consensus upon which future guidance, technical assistance and training will be based." NRT-1a provides a systematic method for the analysis of hazardous materials emergency plans. Finally, NRT-2 provides guidance on the development of a comprehensive plan and system exercise program.

Civil Preparedness Guide (CPG)—CPG 1-8 and CPG 1-8a.   These two guidance documents are publications of FEMA. CPG 1-8 is the Guide for Development of State and Local Emergency Operations Plans and provides information on FEMA's concept of emergency operations planning by using the Integrated Emergency Management System (IEMS), and describes the processes involved. CPG 1-8A provides a systematic method for the analysis of all-hazard emergency operations plans.

Legal Implications of the Standard of Care

These and related laws, regulations, standards, and guidance combine to identify a rather well defined standard of care for hazardous materials emergency response. When a defined standard of care exists, there are always potential legal implications because the standard of care identifies minimum expectations for the provision of the service. If the service is, or is thought to be, substandard or less than anticipated, legal impacts are possible.

The initial legal implication is liability. Webster defines liability as, "the state of being liable." Being liable means that some individual, group, or agency is legally bound or responsible to perform or provide some function, duty, or service.

In the case of hazardous materials, legislation (particularly SARA Title III) establishes the responsibility for the development of response plans, capabilities, systems, and public information. These responsibilities are delegated to the states and the local planning districts as well as private facilities.

NRT-1 further specifies the approaches, procedures, and mechanisms needed to establish an effective plan. It is important to note that although NRT-1 is not a regulation or standard, it still may carry substantial legal weight and consideration. A similar situation exists for CPG 1-8.

OSHA 1910.120 and EPA 311 mandate each employer, whether governmental or private, to develop specific response plans or use the plans required by Title III. They also establish employer responsibility for the management, safety, and training of emergency-response employees.

NFPA 471, 472, and 473 all build upon 1910.120 and 311. They clarify and identify specific competencies and procedural options involving personnel training and emergency-response activities.

All of these documents establish not only a standard of care, but also duties and responsibilities. As the result of assigning responsibility, there is now liability. The primary concern regarding liability is negligence.

Negligence is the failure to perform one's duty or responsibility without reasonable regard for foreseeable harm to another or performing the duty outside the guidelines established by the standard of care. Gross negligence is the willful or, in some cases, almost willful failure to perform one's duty or responsibility, or willfully performing outside the standard of care. To prove negligence in a civil proceeding, a party must prove the following:

1. A duty or responsibility is owed.
2. A failure to perform that duty occurred within the realm of a standard of care.
3. Damage occurred.
4. The failure to perform the duty or responsibility resulted in the damage.

In most negligence lawsuits there are only the following four avenues of defense available to the defendant:

1. Try to prove that no duty or responsibility exists. Obviously, this would be a difficult task if the suit revolved around one or more of the responsibilities identified in these documents.
2. Try to prove that no standard of care exists. Again, this could be very difficult to prove considering these documents.
3. Try to prove compliance with the standard of care. This defense will require good substantiation.
4. Try to prove no injury occurred, that any injury did not result from a violation of the standard of care, or that injury was not foreseeable.

Liability can affect the responsible party or parties in a number of different ways depending upon the type of enforcement action undertaken. There are three types of enforcement actions: regulatory, civil, and criminal.

First, regulatory enforcement occurs through some enforcement agency. The agent responsible for enforcement of that regulation could conduct an investigation and, through an administrative authority, issue a citation or levy a fine. This type of action is similar to a simple building or fire code violation. Such an action may occur before any loss or injury takes place.

Second, civil action occurs through the civil court system as a tort action. Regulatory bodies can bring such action, but more commonly employees, injured parties, or potentially injured parties bring them. This type of action normally requires that some type of loss or injury has already occurred or that the situation is such that a loss or injury is highly probable.

Third, and least likely, criminal action occurs through the criminal judicial system. Regulatory or prosecutorial agencies bring such actions. Most commonly this type of action occurs only after there has been a major loss, a large number of injuries, or death. Further, such actions most commonly require some type of willful misconduct, in other words, gross negligence or malfeasance.

MEETING THE STANDARD OF CARE

Obviously, various aspects of the hazardous materials standard of care are beyond the sole control of response agencies and their personnel. As such, a coordinated and cooperative effort is required. The primary place for this to occur is the LEPC through the planning process.

Yet, only individual agencies can meet other sections of the standard of care. Each agency must develop emergency response plans that integrate with the LEPC plans, train their personnel in accordance with the regulations, and work to develop personnel competencies in line with the standards.

What it all comes down to is how effective is the actual response. If someone is injured or worse, or if there is excessive damage, business interruption, or a multitude of other possibilities, there may be major impacts. When operational problems occur, the door is open for the examination and investigation of the entire system and areas covered by the standard of care. Ask any emergency responder who has been through such a situation and he or she will indicate that even when cleared of any misdeeds, it is still one of the worst experiences of his or her life.

One of the new regulatory philosophies is quite simple and straightforward. If there are no problems, there is obvious compliance. If there are problems (i.e., someone is hurt, there is excessive damage, etc.) there is obvious non-compliance.

This entire situation means that the strategies and tactics used to control hazardous materials incidents must be determined through a systematic and hopefully scientific approach. Further, the rationale used to choose the strategies and tactics can be easily documented. In short, the use of a systematic process helps to document that operational management was in accordance with the standard of care.

The remaining units of this text provide a model process for operational decision making in the command sequence that uses just such a systematic and scientific approach. It is based on standard firefighting decision-making that is easily understood. Further, the process is the model decision-making process used in a series of national level training programs.

SUMMARY

This chapter discussed the following topics:

1. The concept and implications of the standard of care regarding hazardous emergency response
2. The role of historical incidents and the growth in knowledge as it applies to the formation and revision of the standard of care
3. Laws, regulations, standards, and guidance documents that impact the standard of care
4. The relationship between federal laws and regulations

5. The changing relationship between the three levels of government based on the basis of the laws discussed

6. Identify the key requirements found in the following laws that affect hazardous materials response:
   - Clean Water Act and its amendments
   - Comprehensive Environmental Response Compensation and Liability Act
   - Superfund Amendment and Reauthorization Act
   - Hazardous Materials Transportation, Uniform Safety Act

7. The primary emergency response implications of SARA

8. The facility reporting and planning requirements found in SARA

9. The employer's emergency response plan

10. The primary levels of competency and training required in 1910.120

11. The relationship between regulations and standards

3 Operational Decision Making: The Command Sequence and the GEDAPER Process

CHAPTER OBJECTIVES

Upon completion of this chapter, the student will be able to:

1. Explain the relationship between operational decision making and the ICS
2. Explain the incident management triangle
3. Describe the traditional command sequence
4. Define and describe the four steps found in the traditional command sequences
5. Relate the implications of size-up, determining strategies, identifying tactics, and the review process
6. Identify the seven steps found in the GEDAPER process
7. Explain each of the seven steps in the GEDAPER process
8. Compare and contrast the GEDAPER process with the traditional fire service command sequence

OPERATIONAL DECISION MAKING AND MANAGEMENT

Making decisions about anything can be a difficult and complex process. In essence, decision making is a problem-solving process. The more complex the problem, the more difficult the problem-solving process. Obviously, response personnel are often called upon to make operational decisions about many types of diverse and complex operations. Further, they periodically confront situations that are neither familiar nor common to them. As a result, the use of a decision-making (problem-solving) system can be of tremendous benefit. This unit provides an overview of operational decision making, its relationship to the command sequence, and how they form the basis of the GEDAPER process. (The steps of the process form the framework for the following units.)

Albert Einstein once noted that to effectively solve a problem, the problem-solver must fully and thoroughly understand the true nature of the problem. This means that the problem-solver must identify and understand the fundamental features of the problem, how they work, what they mean, what they can do, and how they interact with each other and their surroundings. Although this sounds like a deeply profound scientific insight that applies to quantum physics and the like, it truly applies to any type of problem solving. This includes solving problems at the scene of an emergency.

Many people feel that the use of the Incident Command System (ICS) alone will guarantee effective operational decision making and thus effective control of the incident. This brings up a vital distinction. The ICS is just that—a command system. A command system identifies management functions and responsibilities needed to command and control resources (e.g., apparatus, personnel, and equipment). The ICS will provide the ability to control multiple resources for any type of incident. ICS, by itself, is only part of an overall incident decision-making and management process.

In essence, incident decision making and management involve the incident management triangle (Figure 3-1). The triangle includes the Incident Commander, the ICS, and Operational Decision Making. An appropriate analogy is an automobile. The auto is the ICS, the driver is the Incident Commander, and the fuel is operational decision making. Take any one of the three parts of the system away and the system will not function correctly, if at all. By using the ICS and a systematic, scientific operational decision-making process, the IC and the subcommands have a sound means of managing incidents of all types.

Command Sequence

The command sequence is a process the fire service has used for operational decision making. It involves a series of steps based upon structural or wildland firefighting. The process involves size-up, determining appropriate incident specific strategic goals, assessing tactical objectives, and tactical methods needed to meet the strategic goals, and then evaluate the effectiveness of the operation (Figure 3-2). The process is spiral in that evaluation leads back to continual size-up. This process will and does work quite effectively on routine incidents.

However, in today's world, the decision making occurs in situations of unbelievable diversity and complexity ranging from mass casualty incidents, to urban search and rescue, to hazardous materials incidents. Responders often face incidents that are so different in nature that there seems to be little or no commonality on which to base decisions. Further, responders must sort through tremendous volumes of often highly technical information.

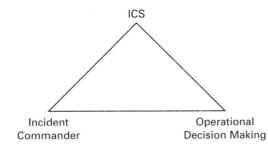

ICS

Incident
Commander

Operational
Decision Making

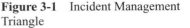
Figure 3-1   Incident Management Triangle

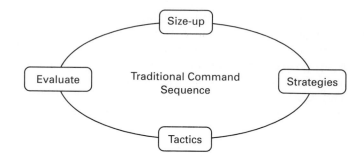

Figure 3-2   Traditional Command Sequence

This information ranges from engineering prints for high-rise buildings to chemical data sheets for hazardous materials. Such situations often threaten to overwhelm the decision maker's ability to use information to determine strategic goals and tactical objectives. How is the information sorted and managed? What information is "need to know" and what information is "nice to know"? What are the appropriate strategic goal systems for different types of incidents? What tactical objective and method options are available to meet the strategies?

The answer to this quandary is the use of a totally integrated and sequential system for decision making. To fully comprehend the requirements of such a system, it is necessary to analyze and understand the steps involved in any type of problem solving (because operational decision making is problem solving). However, first it is necessary to define *analysis*. Next it is necessary to analyze and understand the traditional command sequence. Finally, it is necessary to identify each step in the process.

Analysis involves breaking an object, chemical, process, or other entity into its primary components. Such a breakdown allows one to identify the important features of the components. Next, the interrelationships between the components must be identified. When all the components, their features, and interactions are considered as a whole, a better understanding of the strengths, weakness, interactions, and implications is gained.

The first step in the traditional command sequence is the size-up (Figure 3-3). Size-up involves the gathering and analyzing of data to determine the nature, extent, and probable course and harm of the incident. It further establishes the foundation for determining appropriate strategies and tactics for the incident. As such, size-up is a

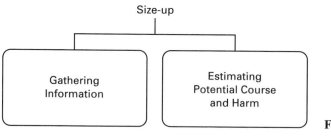

Figure 3-3   Size-up

multiple-step process. The steps include the gathering of information, and the estimating of potential (and most probable) course and harm of the incident. The information gathered helps to identify the incident situation, while the estimate of course and harm helps to identify potential implication on incident priorities.

Identifying the incident situation involves determining the specific type of incident that is occurring. Is this incident a fire, a hazmat incident, an EMS incident, a combination of several types of incidents, or some other type of incident? Next, it is very helpful to determine what has happened. Is it a structural fire? Was there a collapse? Was it a vehicle accident? Are hazardous materials involved? This is an essential step because each type of incident requires its own unique set of strategies and tactics. If the wrong strategy-tactic system is used for the incident, the results can be disastrous.

Once the incident situation is identified, consideration must be given to its potential implications on the incident priorities. There are three incident priorities: Life safety (of the public and responders), Incident stabilization, and Protection of property and the environment. The acronym LIP is a useful device to help remember them (Figure 3-4).

It is important to understand that these incident priorities are the same for all emergency-response services, e.g., fire, police, and EMS. These priorities are etched in stone and require consideration in all emergency operations. However, the importance of each priority may vary depending upon the exact incident situation and the probable course and harm. For example, the degree of life safety concern for an apartment fire at 3 A.M. is very different from that of a dumpster fire behind a supermarket.

This is not to say that life safety is not a concern in the dumpster fire. Rather life safety of occupants is generally not a paramount concern in this type of fire. However, the life safety of response personnel is.

Finally, to fully address incident priorities, the estimating (hypothesizing) of potential course and harm (impact) of the incident is critical. Has the worst already happened? What is expected to happen? How long will it take for that to happen? How will the incident progress until actions can be implemented? If an accurate estimate is not forthcoming, operations are strictly reactive as opposed to proactive. In essence, responders are simply waiting for the incident to "do something" before they take any

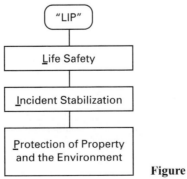

Figure 3-4   Incident Priorities

actions. This means that the incident is managing the responders instead of the other way around.

Once this multi-step size-up operation takes place, the identification of appropriate strategic goals can occur. This implies that a goal system is in place. Personnel next identify tactical objective, methods, and resources needed to meet the goals (Figure 3-5). When identified and implemented, command personnel must evaluate the effectiveness of the strategies and tactics chosen.

In this text, we will use a decision-making process that is simply a refinement of the existing command sequence that consists of size-up, strategies, tactics, and evaluation process. It incorporates a systematic and scientific decision-making process that is usable regardless of the type or magnitude of the incident. Once understood, personnel have a unified system available for making decisions about all types of incident. The process helps decision making by outlining the basic thought process followed by identifying specific strategic and tactical options available to meet the needs of hazardous materials response.

Close examination of operational decision making indicates that there is a series of sequential steps that must occur. These steps break the decision-making process into a concise scientific approach. Again, the process is nothing more than a refinement of

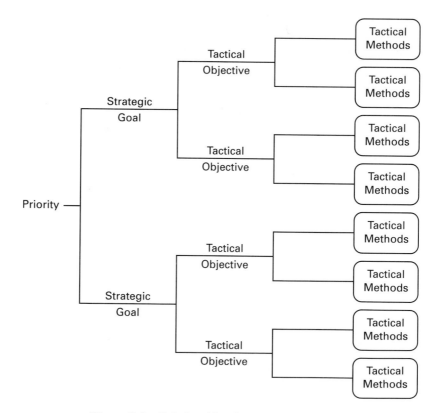

Figure 3-5   Relationship of Strategies and Tactics

the standard command sequence. These sequential steps apply to all foreseeable incident situations. Such systems are profoundly beneficial to many professions faced with the need for rapid and decisive decision making. The military and the medical profession (in their SOAPIER process) both use systems based upon the same sequence of steps examined within this text.

THE GEDAPER PROCESS

The acronym GEDAPER is used to help identify the seven steps in this incident decision making process (Figure 3-6). GEDAPER (D. M. Lesak, 1988) stands for the following:

1. *G*ather information
2. *E*stimate potential course and harm
3. *D*etermine appropriate strategic goals
4. *A*ssess tactical options and resources
5. *P*lan of action implementation
6. *E*valuate the effectiveness of the plan
7. *R*eview the process

At the most fundamental level, operational decision making involves the analysis, comparison, assessment, and evaluation of incident information. Further, it

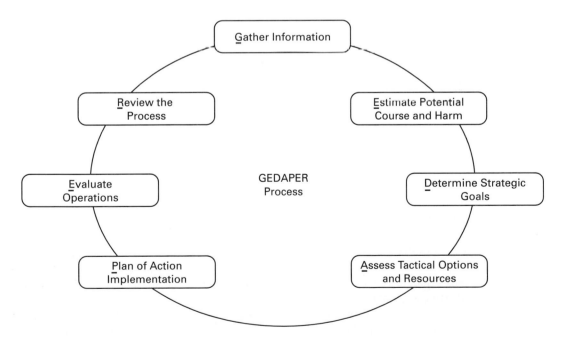

Figure 3-6   The GEDAPER Process

involves consideration of resource needs and their availability plus the identification of both strategic and tactical options and alternatives.

The whole process starts with gathering information, or data. There are three specific types of information that must be gathered: physical, cognitive, and technical. Physical data is collected by direct observation of the incident. Cognitive data is collected from the individual's knowledge, training, and experience. Technical data is collected from outside information sources such as pre-incident surveys, plans, or other types of references. Specifically, the data is collected about the primary "players" of the incident. These "players" are incident components, the fundamental things about which information is gathered. For hazmat emergencies, there are three incident components: product, container, and environment.

Once obtained, the basic information is used to estimate the potential course and harm of the incident. The estimation step is performed by forming a series of predictions about expected incident occurrences, their potential, and most probable impacts. These estimates allow responders to anticipate actions and resources that may be necessary to stabilize and manage the incident.

The estimating of course and harm allows command personnel to determine appropriate strategic goals needed for the particular incident. Once the strategic goals are identified, an assessment of tactical options and resources is needed. In other words, one must specify tactical options and the resources that are available or needed to meet the incident goals.

When the tactical assessment is complete, the incident's plan of action is identified. Next, the plan of action (POA) requires documentation and communication to appropriate personnel. Plan of action implementation occurs when the plan is communicated and implemented by responders. After implementation, the POA must be evaluated to assure that it is both effective and efficient. If the plan is neither, i.e., strategic goals are not accomplished or the incident is of an extended duration, a review of the incident must be undertaken to assure the response effort is on track.

To understand the steps of the GEDAPER process more fully, we will examine each in more detail.

Gathering Information

The information-gathering process is one of the most critical steps in operational decision making for one simple reason: all subsequent steps in the process find their foundation in the gathered information. Without solid information, responders will have no assurance that the strategic goals, tactical objectives and tactical methods chosen will be effective or even appropriate for the incident. If there is insufficient or incorrect information, or if the information is misinterpreted, it is highly possible to choose inappropriate strategic goals and tactical objectives. Further, the strategies and tactics may be ineffective and potentially unsafe.

The three types of information are based on the way they are received: physical, cognitive, and technical. The process of gathering specific incident information often starts before the receipt of the call, with pre-incident planning. The gathering of pre-incident information starts by gaining familiarity with one's first due response district. However, this basic familiarity requires supplemental information gained through target hazard preplanning. Such data is stored in one's mind, on paper or in a computer until needed. The data stored in one's mind is cognitive while that stored on paper or in the computer is technical.

When an incident occurs, information gathering swings into full gear. Initially, it involves the gathering of physical data about the incident scene. This physical data is any information about the incident's occurrences detected by direct observation through one's bodily senses. The most commonly used and safest senses are those of sight and hearing. Although the senses of touch, smell, and taste may come into play, they generally result in potentially hazardous exposure to the product. Thus, they must be prevented. Further, from a legal perspective, such exposures generally are noncompliant with OSHA regulations.

At the time of dispatch, physical data, such as the location of the incident and the general type of incident, will start to confront responders. Such data will now trigger the recall of mentally stored cognitive data. If the incident is of any magnitude or complexity, responders may face a tremendous influx of data. This influx can be detrimental to an individual's ability to sort, assess, and comprehend all the data.

The potential influx of data identifies a fundamental principle of effective problem solving, command, and management. That principle is the span of control (Figure 3-7). In ICS, span of control is the number of operations, sections, units, or people that one individual can effectively supervise and manage. ICS generally identifies the span of control as five. In simple or straightforward operations, the span of control can approach seven or eight, and sometimes even higher. In complex and difficult operations, the span of control can fall to two or three.

At the most basic level, span of control is not just the number of functions or units one can manage but rather it is the maximum number of cognitive mental functions, ideas, or concepts that an individual can effectively address at any given time. As such, span of control is a significant consideration while gathering information.

To assist with the management of incoming data, it is extremely helpful to break an incident into its primary component parts. By doing this, an analysis of the incident begins. (Remember, analysis is defined as breaking something into its constituent parts. This is followed by the identification of their interrelationships.) Further, the

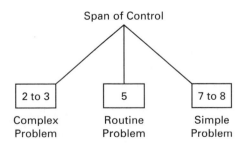

Figure 3-7 Span of Control

identification of components allows responders to start "pigeonholing" information so data management can be more effective.

Many incidents handled by the fire service have three components. Each component requires that information be gathered. For example a fire requires information about the fire itself (e.g., its location, size, smoke conditions), the environment in which the fire is occurring (e.g., time, weather, exposures), and the fuel (e.g., class, structure, wildland). An EMS incident involves information about the victim, the environment in which the victim is located, and where the injury occurred, including the nature, extent, and location of the injury itself. A hazardous materials incident also has three components: the product or products involved, the environment in which the incident occurs, and the container or containers (Figure 3-8).

This is not to imply that these components are completely separate and independent. To the contrary, all the components interact with each other and have fundamental interrelationships. However, for the purpose of gathering information, it is very useful to separate them so each component may be examined separately. By identifying incident components, the rapid sorting and shifting of incoming information to the appropriate "box" is possible.

Further, the gathering of incident information occurs not only at the outset of the incident. Rather, it occurs with an initial burst followed by dribs and drabs as the incident unfolds. As subcommands, units, entry teams, or reconnaissance teams further investigate the incident or accomplish specific tactics and strategies, there may be hectic bursts of updated or new data. As a result, the gathering of information is a dynamic process that must continue throughout the incident. New and updated information must be factored into the incident estimate as well as the strategic and tactical choices that have been made.

As information is gathered, responders must assess it. During assessment, personnel identify the hazards, vulnerability, and risk of the incident. Hazard assessment occurs during the gathering of information phase of the process, while vulnerability and risk assessment occur during the estimating phase.

Hazard assessment is the process of identifying what hazardous substance(s) is present, how much there is, and its physical and chemical properties. Hazard assessment also identifies the chemical, physical, and health hazards present on the scene. For fixed facilities, LEPC site, or facility specific plans, pre-incident surveys and similar types of pre-plans provide much of this information. Obviously, this is part of the information gathering step.

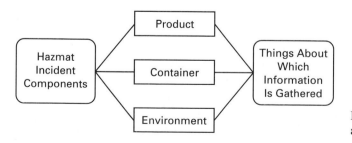

Figure 3-8   Incident Components of an Emergency

Estimating the Potential Course and Harm of the Incident

Vulnerability assessment is determining how large an area may be affected as well as who and what are located in that potentially affected area. A key factor in the vulnerability assessment is identifying how the released product will travel, specifically in what direction and how far it will go. The LEPC's facility-specific plans, the isolation table in the DOT Emergency Response Guidebook, computer plume models, and other resources can be used to provide such data.

Finally, risk assessment is identifying the probability that a specific outcome will occur. Generally, emergency responders have been trained to consider the worst case scenario when estimating the potential course and harm of a hazardous materials incident. However, most have not been trained to also factor in the probability that the "worst case" event would occur. As such, risk assessment (probability) is a key factor in identifying the incident estimate (Table 3-1).

Table 3-1   The Potential Courses and Harm of an Incident

Hazard assessment	Vulnerability assessment	Risk assessment
Identify hazardous materials present, their location, quantity, and what they can do	Who and what will be affected if a release occurs	The probability of various scenarios occurring

For example, consider an incident that involves a 55-gallon drum of super unleaded gasoline. The drum has fallen from the rear of a truck, landed on the street, and has a small hole in the side. There is a small trickle of product coming from the hole. Reference sources indicate that this drum could explode in the heat of a fire. The explosion could result in a large fireball with shrapnel flying up to 300 feet. From the hazard and vulnerability assessment it would indicate that this is potentially a rather substantial incident and responders should isolate and evacuate the area. The risk assessment, however, should indicate that the probability (risk) of an explosion is relatively small in this particular incident situation. Further, if the release is diked in conjunction with the application of foam or an adsorbent, control of the incident will occur and an explosion will be prevented.

Responders must make a series of predictions regarding the anticipated course and harm of various aspects of the incident and identify a generally probability of their occurrences. If these predictions do not occur, responders will be in the unenviable position of playing catch-up with unanticipated occurrences. Such oversights can lead to disastrous results, especially if insufficient resources are requested or the chosen tactical operations do not consider arrival and set-up time, for example.

Some of the many predictions include the following anticipations:

1. Magnitude of the incident
2. Spread of the fire or hazardous material

3. Life hazards
4. Mode of operation (e.g., offensive, defensive)
5. Effects upon exposures (including the environment)
6. Safety factors and considerations

When considered together, all the predictions provide an overall estimate of the potential course and harm of the incident.

Remember, the two distinct steps of gathering information and estimating course and harm are lumped together in the traditional command sequence and called size-up. Hopefully this discussion has indicated that size-up can be a complex and intensive process, depending upon the individual incident.

Determining Appropriate Strategic Goals

Once the estimate of the incident's probable course and harm is complete, responders must now determine the appropriate strategic goals needed to successfully intervene. The specific strategic goals depend upon the specific type(s) of incident situations involved. Strategic goals are broad, general outcomes needed to control and stabilize a specific type of incident. Just as each type of incident requires its own specialized type of equipment, materials, and knowledge, each type of incident has its own possible outcomes.

For example, structural firefighting systems have between five and eight strategic goals, depending upon the exact system used. Such goals normally include Rescue, Exposures, Confinement, Extinguishment, Overhaul, Ventilation, and Salvage (known as Lehman's RECEO VS); or Rescue, Exposures, Ventilation, Attack, and Salvage (known as the REVAS system) (Table 3-2).

Hazardous materials systems generally include between seven and ten goals, again depending on the system used. Such goals include isolation, notification, identification, protection, spill control, leak control, fire control, and recovery/termination (Table 3-3).

Table 3-2 Firefighting Goal Systems

RECEO VS	REVAS
Rescue	Rescue
Exposures	Exposures
Confinement	Ventilation
Extinguishment	Attack
Overhaul	Salvage
Ventilation	
Salvage	

Table 3-3 Hazardous Materials Goal System

INIPSLFRT
Isolation
Notification
Identification
Protection
Spill control
Leak control
Fire control
Recovery/Termination

Consider the following example. You respond to a structural fire at a large mill frame factory. Upon arrival, you find a heavy body of fire involving more than 50% of the structure. There are exposures on sides B, C, and D (i.e., sides 2, 3, and 4). Under such circumstances, the most probable initial estimate must be that the volume of fire in the original fire building is so great that:

1. The apparatus to provide the fire flow is unavailable until additional alarms arrive.
2. The needed fire flow is probably unavailable.
3. The fire building is, for all intents and purposes, a write-off (loss).
4. A defensive mode of operation is the only appropriate approach.
5. If the exposures are not addressed immediately, they will also be consumed.
6. More resources (alarms) are needed to address the exposures.

When considered together, the series of predictions provide the basis for the estimate of the incident's course and harm in this initial phase. In turn, the estimate allows for the determination of appropriate strategic goals. Returning to the mill frame example, it should be obvious that extinguishment of the original fire building is not an appropriate strategic goal but confinement and exposure protection are much more appropriate. Rescue may not be a primary goal in the original fire building (due to the advanced degree of involvement) but it may be the priority goal for the exposures.

Assess Tactical Options and Resource Requirements

Once appropriate strategic goals are determined for the incident, the specific tactical objective and method options needed to meet those goals now become more apparent. Additionally, consideration of the resources needed to accomplish any of the specific tactical options is necessary. If there are insufficient or no resources available, responders must either determine the method to acquire those resources or choose another option.

For example, during a wildland fire in a mature stand of mixed pine and fir, the determination is that confinement is an appropriate strategic goal. One of the tactical options available to meet this goal is to bulldoze a fire line along a ridge. The prediction indicates that the fire will reach the ridge in about 2 1/2 hours under present conditions. The closest available dozer will take 2 hours to reach the ridge, and it will require 45 minutes for the dozer to cut the break. As such, an alternative tactical option must be identified because the dozer (an essential resource for the tactics) will not arrive in time.

Plan of Action Implementation

With the identification of appropriate and attainable tactical options, a Plan of Action (POA) starts to develop. Command must now identify, assign, and communicate the specific tasks to tactical units. Most commonly, the POA is first a mental, then a verbal plan. However, written POAs, using printed check lists, can provide a timely and concise way to document actions taken and greatly assist in the evaluation or review process. In complex or long duration situations, a written plan is vitally important. This is especially true if the operation requires the transfer of command. It also makes the processes of evaluation and review easier.

At this particular point, the ICS starts to flex its muscle and swing into full gear. ICS is one of the most effective and efficient methods available to implement the chosen plan of action. Further, ICS provides a system that allows the IC and sub-commands to direct, track, and control all the resources available and needed to initiate and achieve the goals and objectives specified in the POA.

Evaluate

When implemented, command personnel must continually evaluate the effectiveness of the POA. As personnel perform their assigned tasks, command personnel will expect to see or hear reports of a positive impact on the situation. If there is a positive impact and the POA is producing the desired effect, the plan was appropriate.

However, if there is no positive impact or the desired effect does not occur, there is a flaw in the POA. Such flaws are often traced to changes in the status of the incident itself, especially those of longer duration. In other cases, such flaws in the POA are the result of some problem in one or more of the first four steps. In either case, the flaws may be the result of insufficient, incorrect, or outdated information, which can lead to an inaccurate estimation, the use of an inappropriate strategic goal, or the inaccurate assessment of tactical options and resource needs.

Review

If there is a major flaw in the POA or as the incident evolves (which is most common in longer duration operations), responders must now go back and review the POA. The review process involves an examination of every step of the process starting with the updating and confirming information previously gathered. Is some vital piece of information missing, or has the length of the operation changed the over-all incident situation? Is the basis for the estimate of incident course and harm inaccurate or dated, or has previously unidentified information been obtained? Were the incident's course and harm over- or under-estimated? Are the strategic goals inappropriate or was a tactical option chosen to meet those goals incorrect? Were anticipated resource needs less than those actually required?

In any event, when the POA is not working due to a major flaw, responders face a very difficult task indeed. Such situations often compromise the safety of operating personnel as well as the outcome of the incident. Operations, command staff, and planning personnel often need to discuss the situation to identify the specific problem. The problem may simply be a result of insufficient feedback from operating personnel regarding changes or variations of conditions, which can be quickly addressed with only a brief consultation, or the problem may be major and require a full regrouping.

Once the review process is complete, the implementation of a modified version of the POA is possible. Again, once implemented, the new POA requires evaluation and review to assure that the modifications meet their intended goals.

By this time one may think that hours have elapsed and surely the "building has burned down." This is not true. The first five or six steps are part of a routine use by thousands of firefighters each day. In practice, these steps often take from 30 seconds to a minute. The performance of these steps is generally so rapid and automatic that most responders are totally unaware of the process.

However, in incidents that are not routine or that are outside one's prior experience, any operation decision making process must provide systematic methods that will work in a wide variety of different operational situations. Emergency operation management is not an art form, but rather a science. GEDAPER provides the scientific tool needed to make concise and systematic decisions about any critical incident situation as part of the over-all incident management process.

SUMMARY

This chapter discussed the following topics:

1. The relationship between operational decision making and the Incident Command System
2. The incident management triangle and the traditional command sequence
3. The relationship and implications between size-up, determining strategies, identifying tactics, and the review process
4. An introduction to the seven-step GEDAPER process and its relationship to the traditional fire service command sequence (see Figure 3-6)
5. An overview of each of the seven steps in the GEDAPER process

4 Gathering Information

CHAPTER OBJECTIVES

Upon completion of this chapter, the student will be able to:

1. Identify the need for information in any problem-solving process
2. Explain the three types of data used in the information gathering process
3. Define the three hazardous materials incident components about which information must be gathered
4. Identify four primary product data that must be gathered
5. Identify and describe the two primary categories into which all containers are classified
6. Explain the four pressure ranges into which all containers fall and the implications these ranges have on response operations
7. Identify the five modes of transportation
8. Identify and describe the primary types of containers used in each mode of transportation
9. Identify appropriate sections of 49 CFR Parts 100 to 185 that apply to each mode of transportation and its corresponding containers
10. Define and differentiate between bulk, nonbulk, and intermodal container
11. Identify and describe the various types of fixed containers
12. Identify agencies, organizations, and groups that provide specifications for fixed containers
13. Identify nine environmental considerations about which information must be gathered
14. Describe the role of damage assessment in the gathering of information and in the estimating of potential course and harm
15. Explain the relationship between damage, matter, energy, and forces
16. Identify and explain the three primary types of stressors and their potential impacts on the incident components

17. Identify the four primary types of metals used in container construction
18. Identify the five primary properties and treatment methods for metals
19. Describe the relationship between these properties and treatment methods and their impact on the potential behavior of metals
20. Identify and describe the ten potential types of container damage
21. Identify additional damage assessment considerations

GATHERING INFORMATION—SECTION I
THE BASICS

In any incident situation, regardless of its type, the gathering of information is critical to the decision-making process and the overall management of the incident. When one understands an integrated decision-making process such as GEDAPER, the full importance of information gathering becomes very apparent. Granted, the IC can make a decision without accurate, concise information. However, it should be obvious that decisions made under such dubious conditions will be questionable at best and may be downright deadly.

At this point the question arises of why such decisions would be questionable. The answer is straightforward. Information, in conjunction with experience, knowledge, and in some cases the input of advisory personnel and technical references, forms the basis for the estimate of potential course and harm anticipated for the incident. The information and the estimate of the incident course will then be essential for the determination of strategic goals, assessment of tactical options and resources, and thus aid in the development of the POA. Finally, information is essential during the evaluation and review process in determining the effectiveness of actions undertaken. This unit will discuss the specific types of information needed to start the entire process.

As identified in Unit 3, there are three types of data gathered, examined, and considered during any type of incident: physical data, technical data, and cognitive data (Figure 4-1).

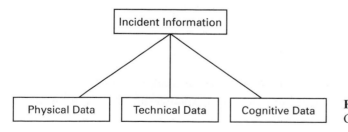

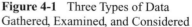

Figure 4-1 Three Types of Data Gathered, Examined, and Considered

PHYSICAL DATA

Physical data is specific information gathered by the senses about the incident situation, primarily sight and hearing. Unfortunately, the use of the other senses can easily result in chemical exposure. As such, their use generally is not recommended. Physical data includes dispatch information, visible vapors and smoke, audible sounds and witness reports, flame color and intensity, for example.

TECHNICAL DATA

Technical data is specific information gathered normally from reference sources such as pre-plans, computers, technical advisors, and so on. Granted, this information enters by way of the senses, but it is retrieved from an outside source.

COGNITIVE DATA

Cognitive data is specific information that comes from the experience, training, and education of the individual. Cognitive data and the ability of an individual to interpret and extrapolate based on physical and technical data is a major factor when identifying one's capability to make sound judgments and decisions on an incident scene. As implied by the definition, cognitive data relates directly to the depth of knowledge and experience an individual possesses as well as his or her ability to put that information to good use during an incident. The use of cognitive data as well as the depth of cognitive knowledge and abilities are the primary factors that separate technicians from specialists.

For example, the technician knows the different kinds of plugs available to minimize or prevent the further release of product from a container's breach. However, it is the specialist who analyzes the overall data and determines, as a result of his or her in-depth knowledge, when it is appropriate to insert a plug.

The importance of cognitive data cannot be over emphasized. Cognitive data about the product, chemistry, physics, containers, container behavior, stressors, stressor implications, environment, meteorology, topography, geology, and so on are essential to the safe and effective handling of major hazardous materials incidents. Does this imply that every technician needs to have a degree in chemistry or engineering? No. However, they do need to have a good rounded understanding of the basic concepts. Does this mean that the specialist needs to have an in depth knowledge of chemistry, containers and their interactions? Absolutely.

Consider the following two examples:

1. Two products involved in a box trailer roll-over are both oxidizers. One is sodium peroxide and the other is aqueous potassium permanganate. What is the potential of this incident? How should responders proceed?

The sodium peroxide is a strong oxidizer that is highly water reactive. Aqueous potassium permanganate is a water solution. If these two products come in contact, a violent and possibly explosive reaction could occur. Should personnel make an entry to perform a reconnaissance? If so, what type of protective equipment should they wear?

2. The incident involves a 20,000-gallon capacity DOT 111A100W5 tank car found standing on a siding. It is carrying fuming sulfuric acid (oleum). There is liquid pouring from the bottom of the tank car. The liquid is emitting a very large white vapor cloud. What is the potential problem? How should you proceed?

   First, the tank car is mandated to have a protective rubber liner, and it is forbidden to have a bottom outlet. Liquid pouring from the bottom of the car indicates a potentially major container problem that could result in the complete failure of the car. There is a high probability that the liner, if present, is compromised and the lading has eaten through the tank shell.

   Second, based on DOT regulations, this specification car can only legally transport fuming sulfuric acid with less than 30% sulfur trioxide. If the sulfur trioxide concentration is greater than 30%, this tank car is not the proper package for the lading. In such a case, should personnel enter to perform additional reconnaissance? If so, what type of protective clothing should they wear? What could happen to personnel or the surrounding environment should the car fail? What steps should be taken to minimize the potentially massive vapor cloud?

These are the types of questions that need answers early in the incident, often well before industrial representatives are available. These are the types of questions and answers a specialist must be able to provide. These are the types of decisions often required to handle real incidents safely and effectively. This is the type of cognitive data that decision makers must possess to do the job correctly. It illustrates the vital role that the gathering of information plays and the in-depth knowledge needed by some response personnel on the incident scene.

Although the examples stressed cognitive information, all three types of data are very closely interconnected. The case can be made that all the data comes from the physical realm because technical and cognitive data originally come to the individual through one of the senses. However, it is important to understand that during an incident the three individual types of data and their sources are most important to assure their thorough and effective use.

PRE-INCIDENT PLANNING

The process of gathering information about an incident should normally begin before the actual occurrence of the incident. Specifically, the gathering of pre-incident information starts with response district familiarization and pre-incident planning. Further, planning activities of the LEPC are invaluable sources of further information. Every responder should have a basic knowledge of his or her own response district and also be familiar with the pre-incident surveys or plans prepared for specific target hazards, including transportation corridors.

SARA Title III reporting and planning mandates, state and local right-to-know legislation, and private sector programs such as the Chemical Manufacturer's CAER and Chemical Transportation Association's TRANSCAER Programs work to identify hazardous materials locations within the community. Further, their purpose is to develop effective emergency response plans needed by the local response agencies. For these programs to be effective, all response agencies and personnel have a responsibility to be knowledgeable about the following:

1. The locations of such hazards
2. Applicable departmental SOGs
3. Pre-plan information and procedures

Unfortunately, anyone who reads the trade journals has read accounts similar to the following: First-alarm companies are dispatched to Smith's Greenhouse. Upon arrival, first-due companies find a working structural fire and go into standard structural-fire ground operations. Initial-responder interviews often contain comments such as, "It was only after we had three lines in operation and saw purple, green, and blue flames that it became evident that this was not a routine structural fire." Such situations and statements indicate a lack of even basic knowledge about one's response district, appropriate procedures, or hazmat response in general.

Do not underestimate the negative impact such inappropriate actions can have, nor the potential nightmare it can cause for hazmat and command. Ineffective identification of the incident situation indicates a failure to obtain cognitive data and results in the compromise of the overall information gathering process. This further results in inappropriate strategies and tactics that often produce highly counterproductive results. Most often they threaten the life safety of firefighters and possibly the public, render available resources unusable due to contamination, expand the magnitude of the incident by spreading contaminants over larger areas, and threaten to escalate environment impacts. Basically, such operations run counter to the three fundamental incident priorities: life safety, incident stabilization and protection of property and the environment (LIP).

DISPATCH AND INITIAL RESPONSE PHASE

With the dispatch of units to any given incident, the information gathering process goes into full swing. During this phase, the focus is primarily on physical data and secondarily on cognitive data.

The dispatch process should provide some basic factual, physical data such as the incident's location, nature, and in some cases the magnitude (e.g., received multiple calls, reported from across the street, second alarm). Such specific physical data may also lead to the gathering of cognitive data, such as "123 Main St., XYZ Chemical

Company, an odor investigation." From these physical data, personnel should know there is the possibility of chemical involvement because of the address (district familiarity) and the name of the company (chemical companies do not bake bread). Additionally, personnel would also know from pre-planning that there are toxic chemicals stored in and used at that location.

During the response, personnel gather additional physical data. Does the dispatch situation information agree with that found on the scene? Are smoke or vapor clouds visible? How large is the affected area (including exposures, both interior and exterior)? Are people injured? Has product been released? Is the release ongoing? Where is the product going? And so on.

OPERATIONAL PHASE

If a chemical incident is in progress, very specific additional information must now be gathered. Normally, there will be a substantial amount of confusion and data bombarding personnel. Span of control problems, resulting from the large amount of incoming information, are quite common at this point.

To help alleviate the information related span of control problem, it is helpful to break the incident into its component parts. Remember, in incidents involving chemicals, the three incident components are product, container, and environment. To help manage the influx of information, it is very helpful to have simple checklists to help first responders gather, sort, and log their information. When the hazmat team arrives, generally one or more members perform this function. These team members are often called research, science, reference, or resource. Through the delegation of this function, span of control is maintained.

At this point the question arises, what types of information must be gathered about the individual incident components? To answer that question more fully, it will be helpful to examine each component individually (Figure 4-2).*

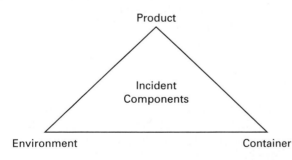

Figure 4-2 Gathering Information into Individual Components

*NOTE: Sample check sheets or lists for each ICS position within the hazmat group or branch are located in the supplemental materials at the end of this text.

GATHERING INFORMATION—SECTION II
PRODUCT INFORMATION

There are specific bits of physical, technical, and cognitive product data that must be gathered. Primarily this information revolves around the physical and chemical properties and behaviors of the products involved. This section will briefly touch upon some of the more critical properties. However, this is not intended to provide in-depth coverage of chemistry. (For a better understanding of chemistry, it is recommended that the reader attend a course such as the National Fire Academy's Chemistry of Hazardous Materials.)

The product information helps hazmat and command personnel determine the specific primary and secondary hazards associated with the products involved. In other words, this information helps identify what the products can do, how they will do it, how bad they will do it, and how personnel can best protect themselves and the public from them. This is not as simple a task as it may seem because all products have multiple hazards and behave differently under different conditions.

For example, a product classified as a corrosive will exhibit the primary hazards associated with the corrosive materials (i.e., the ability to destroy living tissues and metals). However, corrosiveness may not be its only hazard. Corrosive materials may have secondary hazards such as toxicity, fuming, mild to explosive water reactivity, moderate to strong oxidizing ability, initiation of spontaneous combustion to hypergolic ignition, and explosivity when mixed with fuels. As a result, it is vitally important to gather information needed to determine all the hazards of a given product.

One of the most crucial bits of data needed is the specific name of the products involved. If the products' names or United Nations (U.N.) identification numbers are unavailable, it is impossible to use any technical data sources (including DOT's North American Emergency Response Guide) to determine its specific properties, characteristics, and behaviors. In others words, it is impossible to determine hazards such as flammability, toxicity, density, flash point, and solubility. The lack of such basic product information makes all operational decisions questionable.

Depending upon the specific circumstances of the incident, gathering this vital data can be very easy, very difficult, or almost impossible. Regardless of the ease or difficulty in obtaining this data, it is absolutely imperative that the products be accurately identified or at least classified. The identity should include the specific chemical name of the product, the U.N. identification number, the CAS number, or any other identification tool. All too often responders receive nebulous information like, "No problem; this stuff is something like lime." On its merit, the statement "something like lime" provides responders with no specific information about the product, only information about a product *not* involved.

Once the products are identified, it is vital to gather data about their physical and chemical properties. These are behaviors and characteristics unique to a given product, whether the product is an element, compound, or mixture.

PHYSICAL AND CHEMICAL PROPERTIES

Physical properties are characteristics and behaviors observed without the product's chemical interaction with other materials. Many physical properties are detectable through the senses, i.e., sight, hearing, taste, touch, and smell. Such physical properties include quantity, color, odor, taste, physical state (i.e., solid, liquid, gas, or plasma), and physical form (e.g., particulate, slurry, sludge, gel, paste, liquefied compressed gas, cryogenic liquid, or molten solid). Some other physical properties not as readily identified include boiling point, condensation point, freezing point, melting point, as well as vapor pressure, flash point, solubility, density, specific gravity, and vapor density.

Chemical properties are characteristics and behaviors observed when the product interacts with other chemicals or an energy source and results in one or more new substances. Such chemical properties include flammability, ignition temperature, flammable range, polymerization, toxicity, corrosivity, pH, oxidizing capability, water reactivity, pyrophoric nature, decomposition conditions, and radioactivity.

Some of the properties of a given product may be readily identifiable during an incident because they are readily identifiable through the senses of sight or sound. Other properties require the use of cognitive data about a given substance. Still others require the use of technical data sources to glean additional information. In any case, it is vital to obtain specific physical and chemical data about the products involved.

Consider the following situation: Personnel arrive on an incident scene and find a pool of burning liquid. There are multiple victims down in the immediate area. The container has a red DOT flammable liquid label. From the information provided, personnel could infer that the product is flammable and possibly may produce toxic by products and vapors. Efforts to extinguish the fire with firefighting foam result in a violent reaction. This generates a massive, choking, white gas cloud. It turns out the product was methyltrichlorosilane, a toxic and reactive substance that is highly water reactive, resulting in the generation of hydrogen chloride gas and phosgene. Obviously, the DOT classification alone does not provide enough information for the identification of appropriate strategies and tactics.

The following are the minimum physical and chemical properties that should be identified during an emergency:

Physical Data
1. Quantity released and available for release
2. Physical state and form
3. Solubility in water or other substances
4. State changes (e.g., boiling, condensation, freezing, melting points, and sublimation)
5. Density (most helpful values are specific gravity and vapor density)
6. Vapor production (e.g., vapor pressure, vapor density, flash point, and sublimation)
7. Miscellaneous (e.g., viscosity, electrical conductivity, heat conductivity, deliquescence, color,* and existence of release)

*NOTE: Odor and taste can be important but are not recommended properties for which to check.

Chemical Data

1. Flammability—ignition temperature, pyrophoresis (spontaneous ignition on contact with air), and flammable range (upper and lower explosive limits)
2. Toxicity—IDLH, STEL, TLV, routes of exposure, degree of toxicity, target tissues, environmental toxicity (biologic oxygen demand [BOD], which is the toxicity to plants and animals), etiologic (biologic hazard), carcinogenicity (ability to cause cancer), and teratogenicity (ability to cause birth defects)
3. Reactivity—pyrophoresis, polymerization, spontaneous decomposition, spontaneous detonation, incompatibilities, instabilities, slow oxidation, hypergolic ignition, and radioactivity
4. Energy sensitivity—thermal, pressure, photo, electrostatic, shock or impact lability, auto-pressurization, and spontaneous polymerization

It is important to remember that all of these properties are based upon the standard conditions of 21% oxygen and normal atmospheric pressures and temperatures. Any incident situation that can change any of these "normal conditions" may dramatically influence any or all of a product's properties. Incidents involving oxidizers such as oxygen, chlorine, fluorine, oxysalts, peroxides, and non-metal oxides are prime examples. Atmospheres containing 24% or more oxygen are considered oxygen enriched. Such a situation is a potential imminent danger condition because properties such as ignition temperature, flash point, and flammable range may undergo radical modification.

The Use of Product Information

A question that often arises is why must personnel know these properties and behaviors. The answer is not "because they are in the regulations and standards." Rather, the answer is that response personnel must understand and be able to extrapolate what each property and behavior means in the context of a hostile, emergency response environment.

In other words, the regurgitation of a textbook definition means nothing if personnel can not interpret and identify what "this stuff" can do, how it can do it, and how personnel can protect themselves and others from its effects. This is the essence of hostile or emergency scene chemistry that responders need to know or have available to them.

For example, the definition of flash point is, "The minimum temperature to which a substance must be heated for it to release enough vapor for a vapor flash to occur if an ignition source is present." What does this mean to a responder? First, at the flash point, ignition of the vapor will not result in sustained combustion, but it will result in a rapid vapor flash. When the flash is over, the flame goes out. However, the liquid will continue to produce vapor until enough is present to again produce a flash. This process can occur multiple times.

Generally, liquids with flash points above 120°F are relatively minimal hazards (considering fire) in spill situations. However, one must consider when and where the spill occurred. In many parts of the country it is not uncommon to find road surface temperatures in the range of 130° to 150°F or higher during part of the year. If a product with a flash point of 120°F spills onto a surface that is 150°F, the surface can heat the product above its flash point. Obviously, this could be a major problem during a response.

It is highly recommended that at least some members of a hazmat team be well versed in the basic product properties and behaviors. Programs such as the National Fire Academy's "Chemistry of Hazardous Materials," the "Fire Chem I," and "Fire Chem II" are invaluable in gaining this knowledge. As a refresher the Fire Chem Series Study Guide is also very beneficial.

PRODUCT INFORMATION FOR TERRORISM INCIDENTS

Unfortunately, terrorism has become an increasingly important issue in this country over the past years. Incidents such as the bombings of the World Trade Center in New York and the Alfred P. Murrah Building in Oklahoma City, the Sarin attack in Tokyo, as well as many lower profile situations have introduced a whole new realm to the arena of hazmat response.

The acronym B NICE is used to identify the predominant terrorism options. B NICE stands for Biologic, Nuclear, Incendiary, Chemical, and Explosives. A system that focuses on the more nonconventional options is NBC (Nuclear, Biologic, and Chemical). The primary difference between routine hazmat and terrorism is that terrorism incidents, by design, will be hazmat incident and a crime scene that also includes mass casualties, mass fatalities, or possible mass rescue incident. Also, responders may be a target of the attack, as well.

With the possible exception of a few incendiaries, all of these options involve chemical hazards within the realm of existing hazardous materials response (with some variations). Further, when the nature of the agents is understood, one finds that standard hazmat procedures with some modifications will work quite well in handling the agents (if responders are not the terrorist's target).

As in any hazmat incident, one of the key issues associated with terrorism is recognition, specifically recognition that the incident is not a standard EMS run, gas explosion, or minor chemical release. Incidents in locations such as high-rise buildings, transportation centers, shopping malls, conventions, convention centers, or any place where large numbers of people congregate are potential targets. As such, all responders must approach incidents in such locations with some added caution.

Symptomology also plays a big role. The appearance of multiple victims down, seizing, or exhibiting SLUD (Salivation, Lacrimation, Urination, and Defecation) syndrome, devastating explosions (e.g., Oklahoma City) must be treated as potential terrorism incidents. Consideration must be given to the possibility of secondary devices (sometimes called sucker punches). Additionally, response agencies must develop plans and procedures to address such incidents by modifying their existing hazmat and mass casualty/fatality/rescue procedures, since responders may be the target.

Furthermore, just as with any hazmat incident, dispatchers play a vital role in starting the recognition process. Dispatchers must be directed to gather additional information for reported incidents in potential target occupancies and locations. The report of a person or persons down at such locations requires additional questioning regarding where, how many victims, any unusual occurrences, etc. Further, this information must be forwarded to responders so they are not walking into the situation blindly. It is the consensus of most responders familiar with this type of response that alert and efficient dispatchers are one of the best means of preventing injury to first arriving responders to terrorism incidents.

To better understand the nature of the threat, it is essential to understand the nature of the substances potentially involved in a terrorism incident. To accomplish this end, the options associated with B NICE will be considered.

CONVENTIONAL AGENTS—EXPLOSIVES AND INCENDIARIES

By far, explosives and incendiaries have been the favorite tools of the terrorists for centuries. In Europe, Asia, and this country, the bombing or burning of buildings and symbolic locations is a standard method of expressing the dislike of a situation or form of government.

Explosives fall into three primary categories: commercial, military, and homemade (improvised). For a long time, commercial and military explosives have been readily available by theft, coercion, or purchase. Until recently, homemade explosives were usually seen only in the small pipe bombs or letter bombs. However, the Oklahoma City bombing changed the entire complexion of this problem.

ANFO (ammonium nitrate and fuel oil) has been a common blasting agent on farms, in mining, and highway construction for many years. As a blasting agent, ANFO requires initiation by a number 8 (or larger) blasting cap or some type of initiating charge. The components are readily available almost anywhere, and the process is simple and relatively safe.

Incendiary devices range from the rudimentary Molotov cocktail to self-initiating devices to combination devices incorporating explosive with incendiary liquids and gels. Civil disturbances such as riots and wars (e.g., the Northern Ireland situation) have involved many of these devices.

Should a suspect object be found, responders must know how to access expertise, such as bomb squads and bomb technicians, in the handling of explosive and incendiary devices. Should military ordinance or explosive be found, responders must know how to access military explosives ordinance disposal (EOD) personnel. In most localities, such expertise has already been identified as part of the overall response system developed to handle hazardous materials. If it has not, planning must be initiated to address this need.

NONCONVENTIONAL AGENTS—NUCLEAR, BIOLOGIC, AND CHEMICAL AGENTS (NBC)

Starting primarily in World War I, nonconventional agents were identified for their potential use in wartime situations. The first type of agents identified and used in this context were chemical agents. Initially, these agents were industrial chemicals such as chlorine, phosgene, and cyanides. This group was expanded to include substances such as mustard agents and irritants.

Based on the World War I experience, German scientists developed additional highly effective wartime agents. Before, during, and after World War II, extensive research and development occurred focusing on nuclear weapons and their application. Further, research and development involving biologic agents also took place. Needless to say, all of these nonconventional agents present horrific potentials. To gain a better understanding of their potentials, a discussion of each type follows.

Nuclear Agents

Traditionally, nuclear weapons have been considered the primary nuclear terrorism threat. However, the likelihood of such a terrorist event is relatively low in comparison to the potential for other types of terrorism. There are several reasons for this lower potential.

First, the construction of a "homemade" device is rather complex and expensive. Second, the purchase of an existing device is extraordinarily expensive and complex. Third, both homemade and purchased devices are difficult to obtain by anyone other than a government-sponsored terrorist organization. Fourth, the use of such a device would result in some major response by the United States that few are willing to risk. Finally, the use of such a device would alienate the American public, thus drying up one of the biggest money sources used by many terrorist groups.

A more likely scenario is the spreading of a radioactive material to produce a detectable contamination. Even if the contaminant was only a low-level radioactive and presented a nominal hazard, the presence of radioactive contamination would create a public reaction of fear, mistrust, and most likely hysteria. The mistrust and hysteria would probably focus on the federal government (an outcome desired by the terrorists). Further, such a situation would be fertile ground for the conspiracy mongers, tabloid media, and others who would like to blame the federal government for society's ills, and thus aid the terrorists in achieving their goal. (Remember the conspiracy theories regarding the CIA and drugs, UFOs, the JFK assassination, and the Montana Freemen.) Hazmat responders must be fully aware of such potential situations and consider their implications, should an emergency response to an incident involve any radioactive materials.

Biologic Agents

Biologic agents involve living organisms in some fashion. The two primary types of biologic agents are biologic toxins and pathologic microorganisms.

Biologic toxins are chemical substances derived from living organisms. Many of these agents are produced by bacteria or fungi in the form of biologic wastes (the equivalent of urine and feces). Others are made from plant seeds or sap. Some of the more common types of toxins and the type of organism from which they are derived are found in the following table.

Toxin	Organism
Botulinum toxin	Clostridium botulinus
Staphylotoxin	Staphylococcus
Enterotoxins	Bacteria of the gut
Mycotoxins	Fungi (molds)
Ricin	Protein from the castor bean

The pathogenic microorganisms are bacteria, viruses, or similar living organisms that can produce disease in humans, other animals, or plants. Many of these diseases are lethal, while others are merely incapacitating. Some of the more common types of pathogenic microorganisms are found in Table 4-1.

Of these agents, only pneumonic plague and small pox are transmitted by droplet and thus are non-contact contagious. For all the others, universal precautions are effective in the prevention of transmission.

Incidents involving biologic agents are unique from the other types of terrorism discussed in this section. This is because the latency period between exposure to these microorganisms and the onset of symptoms is often one to four days. As a result, these incidents will not be a typical emergency response. Rather, they will be more of a public health response. For response agencies, one of the biggest problems may be providing staffing to cover all those who become ill.

Table 4-1 Common Pathogenic Microorganisms

Disease	Effect
Anthrax	Severe respiratory distress, edema, shock, and high fatality rate
Tularemia	Plague-like infection involving lymphatic system and septicemia
Cholera	Inflammation of the entire small bowel leading to severe diarrhea, dehydration, and death
Encephalitis	Inflammation of the brain; includes Venezuelan Equine Encephalitis (VEE)
Plague (bubonic & pneumonic)	Bubonic involves lymphatic system; pneumonic involves respiratory system
Small pox	Fever followed by eruptions and then pustules

Chemical Agents

Chemical agents can injure or incapacitate exposed people. There are five primary categories of chemical agents:

1. Nerve agents
2. Blister (skin) agents
3. Blood agents (systemic asphyxiants)
4. Choking agents (poison gases)
5. Irritating agents

Nerve Agents.   Three of the four present-day nerve agents were developed by a German scientist before World War II. All of these agents are organophosphates with an additional functional group(s). Organophosphates are also cholinesterase inhibitors and produce the same symptomology. All four are liquids with relatively low vapor pressures and high boiling points. V agent is a thick, oily liquid with an extremely low vapor pressure because it was designed as a persistent contact agent as opposed to inhalant. The following tables provide basic information on each of the four primary agents.

Name	Symbol designation
Tabun	GA
Sarin	GB
Soman	GD
V agent	VX

The second table (Table 4-2) indicates that all of these agents are extremely toxic through inhalation of the vapor and skin contact with the liquid. However, test data indicate that percutaneous exposure (vapor to skin) requires at least fifty times the concentration of vapor in air as opposed to inhalation exposure to produce effects. Furthermore, recent testing has provided indications that full structural turn-outs (e.g., pants, coat, hood, gloves, boots, and positive pressure SCBA) provide percutaneous exposure protection comparable to the military level four MOPP (biologic-chemical warfare) Gear used by the armed forces.

The general symptoms of exposure include miosis (abnormal contraction of the pupils), runny nose, nausea, vomiting, diarrhea, and shortness of breath for mild exposure and loss of consciousness, convulsions, apnea, and paralysis for higher exposure.

Table 4-2   Properties of the Four Primary Nerve Agents*

Agent	Formulation	LC_{50}	Vapor density	Boiling point	Vapor pressure
GA	Aminated organophosphate & cyanide	3 to 6 ppm	5.63	428°F	0.037 mm @ 68°F
GB	Fluorinated organophosphate	1.5 to 3.5 ppm	4.86	316°F	2.10 mm @ 68°F
GD	Fluorinated organophosphate	0.7 ppm	6.33	388°F	0.40mm @ 75°F
VX	Aminated & sulfonated organophosphate	0.9 to 0.5 ppm	9.2	>500°F	0.0007 mm @ 68°F

*NOTE: Much data provided for these and other chemical agents is provided in mg/m^3 as opposed to ppm. To convert from mg/m^3 to ppm use the following formula:

$$ppm = \frac{mg/m^3 \times 24.45}{the\ molecular\ weight\ of\ the\ substance}$$

As with all chemical agents, symptomology helps identify the type of agent involved in the same way instrumentation does.*

Blister or Skin Agents.   Blister agents (vesicants) are designed to produce severe irritation and blistering to the skin. Further, they damage the eyes and respiratory system. Once the agent is absorbed it may produce systemic effects that vary with the specific agent. Some examples of systemic effects include damage to bone marrow, gastrointestinal tract, hepatic, renal, and the central nervous system. There are three primary types of blister agents: mustard, lewisite, and phosgene oxime (Table 4-3).

On exposure, mustard is asymptomatic for hours while lewisite and phosgene oxime cause almost immediate pain. The mustards and lewisite are liquids with an oil-like viscosity, while phosgene oxime is a white solid below 95°F. Generally, treatment involves rapid decontamination and treatment of symptoms and lesions. Because these agents cause blistering, necrosis, or both, they all produce the equivalent of second or third degree chemical burns. As such, hypovolemia and hypotension are common clinical presentations. Table 4-4 provides specific data on each agent.

Because all blister agents are absorbed so rapidly, decontamination is often of little value. If clothing is wetted by an agent, the clothing must be removed by appropriately protected personnel. The underlying skin should be flushed with large amounts of water or hypochlorite solution followed by water.

Blood Agents.   The military uses the term *blood agent* for two cyanide compounds, hydrogen cyanide and cyanogen chloride. The term is archaic because the military first thought these agents were the only ones to enter the blood stream. This is obviously untrue, yet the name still persists. Both cyanide compounds are highly volatile liquids used extensively for industrial applications (over 250,000

Table 4-3   The Primary Blister Agents and Symptoms

Agent	Designation	Onset of symptoms
Mustard	H, HD, HS, and HT	Tissue damage within seconds to minutes, no clinical effect for hours
Nitrogen mustard	HN_1, HN_2, and HN_3	Similar to straight mustard
Lewisite	L	Pain on contact; visible tissue damage quickly
Phosgene oxime	CX	Pain on contact; more necrotic and just as fast acting as lewisite

*NOTE: Decontamination of victims involves the removal of contaminated clothing by *properly protected personnel.* This is followed by flushing with a hypochlorite solution and water. The hypochlorite solution is a 10:1 dilution of household bleach (1 cup per gallon of water). For personal protective equipment, a hypochlorite solution is also recommended.

Table 4-4   Specific Data for Each Blister Agent

Agent	Description	Systemic targets	Antidotal options
Mustards	Yellow to brown oily liquid	Bone marrow, CNS, and gastrointestinal tract	None; absorbed and produces damage in seconds to minutes; this rapid transformation means there is no residual agent in body fluids
Lewisite	Oily, colorless liquid	Capillaries (increased permeability), renal, and hepatic	None that works on lesions; BAL (dimercaprol) for intramuscular produces tissue necrosis
Phosgene oxime	Solid below 95°F that sublimes	No systemic implications identified	None; treat symptomatically; produces more tissue necrosis than mustard or lewisite

tons annually). Some specific information on these substances is found in the following table.

Agent	Designation	Boiling point	TLV-TWA	IDLH	Vapor density
Hydrogen cyanide	AC	72°F	5 ppm	50 ppm	0.9
Cyanogen chloride	CK	56°F	0.3 ppm	50 ppm	2.1

Cyanides are systemic asphyxiants that inhibit a cytochrome oxidase enzyme which enables the cells to use oxygen. As a result, the cells cannot produce energy and thus die quite rapidly. Inhalation of high concentrations of cyanide results in short term (3 to 5 breaths) deep, rapid respiration because cyanide stimulates the respiratory center. After approximately 30 seconds, the victim loses consciousness, and convulsions occur. Breathing stops in three to five minutes, and the heart stops in five to eight minutes. The symptoms from ingestion occur more slowly than by inhalation and include anxiety, agitation, weakness, nausea, possibly vomiting, and muscle tremors. When consciousness is lost, respiration decreases and convulsions, cardiac dysrhythmia, and apnea follow.

Unlike nerve agents, cyanide does not result in miosis but rather produces possibly papillary dilation from hypoxia (poor oxygenation of the tissues). Many victims will display cherry red skin, however approximately 30% show cyanosis. This occurs because cells cannot use oxygen so they do not remove it from oxygenated blood (arterial). As such, the venous blood has unusually high levels of oxygen and thus maintains a redder color. Cyanide victims may also have the smell of bitter almonds on their breath. Unfortunately, only about 50% of the population can smell this odor in the first place, which says something in general about the use of smell in any suspected cyanide incident.

There is a very effective cyanide antidote kit, but it must be administered very soon after exposure. It consists of a three-step process starting with the inhalation of amyl nitrate from pearls or ampoules. The nitrite causes the formation of methemoglobin for which the cyanide has a great affinity (unlike standard hemoglobin for which cyanide has no affinity). The methemoglobin acts as a "sponge" that absorbs and holds the cyanide and prevents its entry into the cells. This is followed with intravenous sodium nitrite. Finally, sodium thiosulfate ($Na_2S_2O_3$) is administered intravenously to detoxify additional cyanide.

Choking Agents.   Choking agents (sometimes called lung agents) produce severe respiratory distress accompanied by pulmonary edema and possible necrosis. The primary military agent is phosgene, also called carbonyl chloride, and has a designation of CG. Although chlorine is not used or classified by the military as an agent, it was effectively used in World War I and is readily available for terrorists.

Agent	Boiling point	TLV (TWA)	IDLH	Odor threshold	Vapor density
Chlorine	−30°F	0.5 to 1 ppm	25 ppm	3.5 ppm	2.4
Phosgene	47°F	0.1 ppm	2 ppm	0.5 ppm	3.4

Both of these highly volatile liquids have low boiling points. They produce lung damage by attacking the membrane lining the alveoli (tiny air-sacks in the lungs). As the tissue breaks down, fluid leaks from the capillaries and starts to fill the alveoli. This produces pulmonary edema that can be fatal. If sufficient, non-lethal injury can occur, but permanent lung damage will result.*

Irritating Agents.   Irritating agents are also known as riot-control agents, tear gas, and lacromators (Table 4-5). They produce irritation of the skin and mucous membranes to the point of incapacitation. Exposure to these agents does not normally result in serious injuries. However, these agents have been known to be lethal in high concentrations and to individuals with pre-existing conditions, the elderly, and the very young.

Unlike most other agents, irritants are extremely fine powders. For dispersal, they are mixed with a liquid that is then atomized as an aerosol. In some instances, it is also possible to disperse these agents in the simple powder form although this is difficult and less effective.

Table 4-5   Properties of Irritating Agents

Agent	Designation	TLV (TWA)	IDLH
Mace	CN	0.05 ppm	16 ppm
Tear gas	CS	0.05 ppm	0.3 ppm
Adamsite	DM (vomit gas)	0.04 ppm (Arsenic)	X
Pepper spray	none	10,000 gm/m^3 (LC$_{50}$)	X
Chloropicrin	none	0.1 ppm	4 ppm

*NOTE: Decontamination is often not required since these agents are gases. However, should a victim sustain liquid or heavy vapor contact, all contaminated clothing must be removed by personnel wearing appropriate PPE. The skin should be thoroughly flushed with a hypochlorite solution and water. It is purported that patients exposed to liquid chlorine should have bed linens changed at least every eight hours to minimize the impact of off-gased product.

Decontamination for irritants is different from the other chemical agents in that hypochlorite solutions must never be used because it exacerbates lesions. Again, contaminated clothing must be removed and the affected tissues flushed with water or saline. Skin can then be washed with a mild alkaline soap or bicarbonate solution.

DETECTION

There is a variety of detection options available. One of the simplest is the use of military M8 and M9 detector papers. Although there is more sophisticated military detection equipment available, they are either colorimetric tube, photoionization detectors, or flame ionization detectors (FID). Table 4-6 provides a basic listing of the agents and the types of colorimetric tubes that should detect them.*

FIDS will not detect hydrogen cyanide and cyanogen chloride. However, they should provide qualitative (yes/no) information on many of the others. PIDs will work, but data on the ionization potential and relative response are needed for the various agents.

ADDITIONAL REFERENCES

There are two references that are extremely helpful for a better understanding of the issue of chemical agents and NBC terrorism in general. The Commonwealth of Virginia's Department of Emergency Services has developed a short text written by John Medici and Steve Patrick titled, *Emergency Response to Incidents Involving Chemical and Biological Warfare Agents.* This text is available through the Virginia Department of Emergency Services' Technological Hazards Division, 310 Turner Road, Richmond, VA 23225-6491.

The second reference is written by Frederick R. Sidell, MD., and is titled *Management of Chemical Warfare Agent Casualties.* This text is available through HB Publishing, P.O. Box 902, Bel Air, MD 21014.

Table 4-6 Detectors for Agents

Agent	Tube type
Lewisite	Organic arsenic compounds
Sulfur mustard	Thioether
Nitrogen mustard	Organic basic nitrogen compounds
Sarin, soman, tabun	Phosphoric acid ester
Cyanogen chloride	Cyanogen chloride
Phosgene	Phosgene
Hydrogen cyanide	Hydrocyanic acid

*NOTE: This listing comes from Drager literature so check with other manufacturers if their systems are to be used.

Presently, the National Fire Academy, FEMA, the FBI, the Department of Defense, the International Association of Fire Chiefs, and the National Fire Protection Association are working jointly to prepare a series of informational programs and delivery systems to provide additional information on this topic. For further information on the programming and materials available, contact the National Fire Academy Hazmat Program at 16825 South Seton Avenue, Emmitsburg, MD 21727.

GATHERING INFORMATION—SECTION III
CONTAINER INFORMATION
Module I—General Container Information

Container information plays an equally important role to product information in operational decision making. There are specific bits of physical, technical, and cognitive data that emergency-response personnel should know about containers. One important difference between the container and the product is that there are fewer container types than there are products. However, there is a substantial amount of data needed regarding these containers.

It is essential for responders to be as familiar as possible with the more common containers that they may encounter. For example, it is extremely helpful to be able to recognize a DOT 407 specification cargo tank or a DOT 103 Class rail tank car. The exact identification of the container will provide specific information as to its strengths, weaknesses, problem areas, and so on.

Some of the most readily available physical data about the container involves its basic physical characteristics. Although such characteristics may seem self explanatory, the size, shape, and configuration are all key tools to identify specific types of containers. Such data help estimate product capacity, internal pressure, types of product, and so on.

DOT covers the specification for any container designed to transport hazardous materials. Specific information is found in 49 CFR Part 100 to 185. DOT generally uses the term "packaging" rather than "container." *Packaging* refers to a container that meets DOT requirements. A package or outside package is the packaging plus its contents.

Further, DOT also identifies *Packing Groups* based upon the danger of the hazardous material present. Packing Group I indicates great danger, Packing Group II indicates medium danger, and Packing Group III indicates minor danger.

Standards and specifications developed by various industrial groups and organizations are referenced by DOT. The Association of American Railroads (AAR), the American Society of Mechanical Engineers (ASME), and the American Society of Test Methods (ASTM) are the most common of these organizations.

Organizations such as the National Fire Protection Association (NFPA), the American Petroleum Institute (API), the Compressed Gas Association (CGA), and similar industrial associations provide specifications for many of the fixed containers. For example, API developed specifications for bulk petroleum storage tanks and periodically updates them. NFPA also identifies allowable containers for many types of products.

Just as it is essential to identify the specific product so that its properties and behaviors can be identified, it is also essential to identify the specific type of container involved so its properties and behaviors can also be identified. To prevent bogging down with detail in this section of the text, there is a more thorough discussion of portable tanks, highway cargo tanks, rail cargo tanks, and fixed tanks later in this chapter.

Unfortunately, there is a wide variety of container parameters. The following categories help identify some of the most common types of containers:

1. Materials of construction
2. Bulk or nonbulk packaging
3. Design pressure and maximum allowable working pressure
4. Portable containers
5. Transportation containers
6. Fixed containers

CONTAINER TYPES AND MATERIALS OF CONSTRUCTION

The materials used to construct a container are a fundamental piece of information and a major typing parameter. A huge number of different materials are available. The specific materials chosen depends upon variables such as the intended capacity, working pressure, type of lading, and so on. Some of the materials and construction types include paper bags, plastic-lined paper bags, cardboard boxes, glass jars, metal cans, fiber drums, plastic lined fiber drums, wooden crates, 55-gallon drums, plastic receptacles in 55-gallon drums, plastic jerricans, metal jerricans, plastic receptacle in plywood drums, and many more (Figure 4-3).

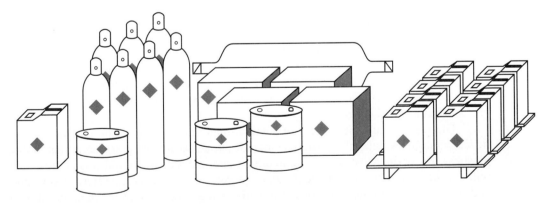

Figure 4-3   Various Portable Containers

Containers are listed as combination, composite, or single packaging. Composite packaging as defined by DOT consists of an inner and outer container that form, after assembly, an integral single package. A composite container is a permanently joined container within a container. As such, DOT classifies these composite units as a single container. Combination packaging as defined by DOT consists of one or more inner packaging secured in a non-bulk outer packaging, except for composite packaging. In other words, a combination container is one or more containers within another where one or more inner containers are removable.

An example of a combination package is one in which the product is placed within an inner package constructed of glass, earthenware, plastic, or metal. These inner containers can be in the form of cans, jars, ampoules, etc. The inner container or containers are then placed inside the outer container for later removal. In the case of plastic inner containers, the plastic may be a single, large container that fits completely within the outer container.

For example, a fiber drum may be fitted with a plastic inner container and used to handle a flammable liquid. The inner container holds the product while the outer container protects the inner container and its contents from damage during shipment and handling. Examples of combination containers include a paper bag with an attached plastic bag liner, a removable top 55-gallon drum with a plastic inner drum, and so on.

Single packaging is constructed of only one container or a composite container. The single container, or a system that forms a single container, holds and protects the product. Examples of single packaging include standard 55-gallon drums, cryogenic dewars, paper bags, and so on.

So what does this mean? By identifying the material of construction and whether the container is combination or single packaging, it is possible to predict potentials, such as:

1. The behavior of a container under various incident conditions
2. How its content (lading) may affect the container
3. Possible leak-control tactics that would be effective with the container
4. When intervention is inappropriate for the situation
5. When a container may be ready to fail
6. How the container may react to mechanical, thermal, and chemical stressors

Further, the material of construction is a major factor in the damage assessment of the container. Damage assessment is critical to accurately estimate the incident's potential course and harm.

The two primary categories of container materials of construction are nonmetallic and metallic. Nonmetallic materials include paper, cardboard, glass, ceramics, wood, plastics, and so on. Most of the nonmetallic materials will withstand less energy than will the metallic. They are subject to damage such as splits, tears, shattering, burning, and so on. As a result, many of the nonmetallic containers are used in conjunction with other materials to form combination or composite packaging.

There are four primary metal materials: mild steels (MS), high strength low alloy steels (HSLA), stainless steels (SS), and aluminum (AL). Other less common metallic materials include low-alloy, low-carbon steel and magnesium alloys.

Mild Steel is an alloy of iron and carbon. Sometimes called low carbon steel, it contains from 0.02% to 0.3% carbon. Medium carbon grades contain from 0.3% to 0.7%, and high carbon grades contain from 0.7% to 1.5% carbon. As part of the production process, MS undergoes a treatment process known as annealing.

Annealing is the process of heating a metal to a specified temperature, holding it at that temperature, and then slowly cooling it. It results in a customary bluish-black color for the metal. Annealing provides the MS with some important properties. During the production of steel, crystals form within its structure. Without annealing, these crystals have haphazard arrangements. Annealing allows the crystals to form uniform, matching patterns. These uniform patterns cause the steel to be ductile.

Ductility simply means that the MS can give and stretch. As a result, when exposed to a force (i.e., stressor), the steel will give and elongate a given amount without breaking. Elongation is the ability of a material to stretch. When the material stretches, it also thins. Upon impact, the container dents, and stretching occurs. If stretched enough, the MS will fail.

MS is not compatible with corrosives or exposure to very high pressures. Because MS will give, it tends to hold plugs well. As the plug is inserted into the breach, the MS gives. Then it also springs back to "bite down" and hold the plug.

High Strength Low Alloy steel, as the name implies, is an extremely strong grade of steel. The primary use of HSLA is when strength, not weight, is the concern. Its most common application is when the container needs to hold lading under high pressure. This is the reason HSLA is used so commonly in the construction of standard propane cargo tanks (MC 331). HSLA is not suitable for use with corrosives.

Some HSLA is quench tempered (QT). As with annealing, QT involves heating the metal to a specified temperature and holding it there for a specified time. However, unlike annealing, which involves gradual cooling, quench tempering involves rapid cooling usually by immersing the metal in a liquid bath. This liquid is most commonly an oil often called quenching oil. This rapid cooling not only causes the metal to become stronger but also more brittle. As a result, although very strong, QT HSLA is subject to cracking when exposed to impact or another mechanical stressor. These cracks often occur on the INSIDE of the container so they are difficult or impossible to identify by visual inspection.

Other HSLA is non-quench tempered (NQT). Since it is not quench tempered, NQT HSLA is somewhat less strong but is also less subject to cracking from impact. NQT HSLA generally requires a slightly thicker shell wall of the container.

Most high pressure highway cargo tanks, such as the MC 331 (see highway vehicles), use QT HSLA. Since 1966, pressure rail tank cars (see rail vehicles) use NQT HSLA. Before 1966, however, their construction was of QT HSLA. It is noteworthy that one of the reasons the rail companies switched to NQT was because of cracking problems associated with QT during derailments.

In either case, HSLA does not plug well because it is so stiff that it does not give and thus can not bite the plug. Further, since the use of HSLA is common in the construction of pressurized containers, such a breach indicates the potential for the complete compromise of the tank's structural integrity. As a result, violent failure could occur at almost any time.

Stainless steel is a specialized alloy of steel and chromium. Based on the alloy's components, there are three primary types: austenitic, ferritic, and martensitic. Austenitic stainless steel is the primary type used in the construction of bulk containers. It contains a minimum of 16% chromium and a minimum of 7% nickel. Special stress-corrosion resistant types contain 2% silicon. None of these austenitic SS forms are ferromagnetic. In this text, SS will refer to austenitic stainless steel unless otherwise noted.

SS is used for applications requiring a high degree of resistance to corrosion. As a result, many containers designed for the transportation of acids and bases use SS. An alternative to SS construction for the transportation of corrosives is the use of internal container liners for non-corrosion resistant materials of construction. SS is very similar to HSLA, however it is not quench tempered.

Aluminum is finding its way into more and more containers all the time. The reason is quite simple. Although AL is not the strongest metal, it is light in comparison to the others, and is still relatively strong. However, AL is not as strong as the steels, and thus requires a thicker plate in the construction of the container's shell. AL is not as hard as HSLA or SS so it will dent more easily and looses thickness due to scraping against hard surfaces. Aluminum does not plug well even though it gives because it tends not to spring back and bite the plug.

Hopefully it has become apparent that each material of construction has its own strengths and weaknesses. Further, as found in many other situations, the beneficial properties of one option are routinely offset by the negative properties of another.

DESIGN PRESSURE AND MAXIMUM ALLOWABLE WORKING PRESSURE

The design pressure (i.e., the maximum allowable working pressure of a container) is one of the first and possibly most important typing parameter to consider. The manufacturer designates the design pressure of a container. Further, the container must meet all appropriate DOT specification requirements for that container. Design pressure is the pressure at which the container is to function under normal conditions. The maximum allowable working pressure (MAWP) designation is set by DOT as part of its design pressure specifications and performance requirements for all cargo tanks (tank trailers) used to haul liquids. In various sections of 49 CFR, DOT identifies the MAWP as the design working pressure of the container.

Design pressure or the MAWP of a container is helpful in two respects. First, it helps to identify the normal ambient state of the contained substance (i.e., solid, liquid, or gas). Second, it helps to identify the potential violence of container failure.

Unfortunately, there is *no* uniform system for the designation of pressures. For the purposes of this text, there are four pressure categories. These categories are as follows based upon internal container design pressure or MAWP in pounds per square inch, gauge:

1. Atmospheric—0 to 5 psi
2. Low pressure—5 to 100 psi
3. High pressure—100 to 3,000 psi
4. Ultra high pressure—greater than 3,000 psi

At this point, it is important to understand some of the different pressure designations that may be encountered. First there is pounds per square inch, absolute (psia). This type of pressure reading starts from a complete vacuum, thus including the approximate 14.7 psi for normal atmospheric pressure. Second, there is pounds per square inch, gauge (psig). This type of pressure reading starts at zero at normal atmospheric pressure, making it approximately 14.7 psi less than psia. Finally, many international packages measure pressure in bars. One bar equals one atmosphere (approximately 14.7 psi). Further, bars start at zero at normal atmospheric pressure (the same as psig).

The ambient state information available for these general pressure groupings apply to all containers except pipelines, chemical reactor vessels, and a few other minor exceptions. For atmospheric and low-pressure containers, the normal ambient state of the products they contain is either solid or liquid. For high pressure and ultra high pressure containers, the normal ambient state of their products is gas.

This type of product ambient state information is very important when considering the spread of the product involved. For example, solids and liquids will only travel a finite distance in a given amount of time depending on the incident scene topography (i.e., the lay of the land). On the other hand, pound for pound, gases will generally travel over greater distances in the same period of time and may be affected only minimally by topography.

Further, all the gases will have some degree of expansion involved. Liquefied propane has an expansion ration of 273 to 1. This means for each cubic foot of liquid-filled container space, 273 cubic feet of gas will be produced.

The potential violence of failure is critical during any incident. Generally, the higher the container's design pressure or MAWP, the more potentially violent the container's failure will be. The reason for this fact is energy. Everything in the universe is either matter or energy. Matter occupies space, while energy is the ability to do work. There are two types of energy: kinetic and potential. Kinetic energy is energy of motion, and potential energy is the potential to do work.

All matter contains some amount of potential energy. This potential energy can take many forms. Four of the most important forms of potential energy in hazardous materials emergency response are nuclear, thermal, chemical, and mechanical (Figure 4-4).

Nuclear potential energy is the energy that holds atoms together. It is found within each particle of matter. In radioactive materials, the nucleus is unstable and has a tendency to break down over time. When such a material does break down, it converts potential

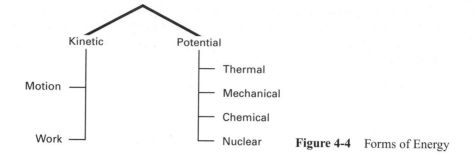

Figure 4-4   Forms of Energy

energy into kinetic energy by releasing alpha, beta, and/or gamma radiation. During nuclear fusion or fission, the nuclear reaction transforms some of the matter into a phenomenal amount of kinetic energy. The energy released is quantified by Einstein's formula $E = mc^2$, where E is the energy, m is the mass of the matter, and c is the speed of light.

Thermal potential energy is the potential heat found within an object or substance. Most commonly in emergency-response situations, thermal potential energy is in the form of chemical bonds within a material, or pressure within a container. When the thermal potential energy is released, it acts to heat or cool the product, container, or the surrounding environment.

Chemical potential energy is the energy found within chemical bonds holding atoms together in the molecules of a material. Every substance has its own unique amount of chemical energy depending upon the elements that make up the substance and the types of the bonds holding the molecules together. For example, sand is very difficult to get to react while nitroglycerine is not. Further, the total energy released from any such reaction is greater for the nitroglycerine. In other words, the nitroglycerine has more potential chemical energy.

Mechanical potential energy is energy in the form of position, location, or compression. For example, a container resting on top of a jumbled pile of other containers has more potential energy than one securely resting on the ground because it has the potential to release kinetic energy by falling.

In the same way, a container with its contents under pressure has additional mechanical potential energy than one with its contents at atmospheric pressure. Consider two compressed gas cylinders, one with an internal pressure of 2,200 psig and the other with an internal pressure of 0 psig. The cylinder containing 2,200 psig has a much greater amount of potential energy, due to the internal pressure, than does the cylinder containing 0 psig. If an impact shears the valves from both cylinders, the results will be completely different. The cylinder containing 2,200 psig will undergo a very rapid and violent energy conversion. The potential energy from the pressure transforms into kinetic energy of movement within the cylinder attaining a speed of 35 miles per hour in less than one second. The cylinder containing 0 psig undergoes no such energy transformation because it contains no potential energy in the form of pressure. When its valve shears, the cylinder will not move as a result of escaping gas.

Hopefully, the importance of these energy considerations is obvious. Specifically, they become very important in damage assessment and in estimating the potential course and harm of the incident. Damage assessment is discussed later in this chapter, and estimating is discussed in Chapter 5.

GATHERING INFORMATION—SECTION III
CONTAINER INFORMATION
Module II—Bulk and Nonbulk Packaging

Bulk packaging as defined by DOT, is any container other than a water craft or barge that meets the following criteria:

1. A maximum capacity greater than 119 gallons (450 liters) for liquids
2. A maximum net weight capacity greater than 882 pounds (400 kilograms) or 119 gallons for solids
3. Water capacity greater than 1,000 pounds (454 kilograms) for gases

In other words, large, portable tanks such as totes and IM 101s, highway and rail cargo tanks and many others are bulk containers.

Nonbulk packaging as defined by DOT is any container with a capacity of 119 gallons or less for liquids and solids, a weight of 882 or less for solids, and a water capacity of 1,000 pounds or less for gases (Figure 4-5).

PORTABLE CONTAINERS

Portable containers and tanks are either bulk or non-bulk packaging, ranging in size from relatively small to quite large. These containers are loaded onto or attached to a vehicle or ship for the purpose of holding their lading during transportation. Further, many of these containers act to hold their lading during storage and quite often even

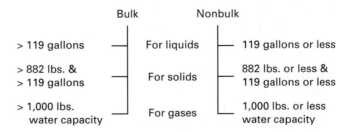

Figure 4-5   Criteria for Bulk and Nonbulk Packaging

during use. Many of these containers are used to transport hazardous materials, and thus must meet DOT specifications.

The primary exception is consumer commodities. As defined by DOT, these are materials packaged and distributed in a form intended or suitable for retail sales. This category includes not only household and small quantity chemicals, but also drugs and medicines.

Some examples of small, nonbulk, portable containers include aerosol spray cans, 55-gallon drums, wooden crates, plastic jugs, carboys, cardboard boxes, compressed gas cylinders, 4-liter jars, pint cans, 150-lb. propane cylinders, cryogenic cylinders, and cryogenic dewars. Also important is where these containers are found. Specifically, they are just as common in fixed locations as in transportation settings. Such fixed locations include industrial plants, shipping terminals, retail stores, or a neighbor's shed. In short, they can be found almost anywhere.

NONBULK PACKAGING TYPES AND MARKING CODES

DOT uses a system of packaging identification codes to provide information about nonbulk packaging. The code identifies the specific type of container, its materials of construction and its category. There are routinely three markings for each type of container. These markings consist of a number followed by one or two capital letters. The letters may then be followed by yet another number (i.e., category code). The following section explains each of these markings.

The first number in the marking indicates the type of container. The code for these numbers is as follows:

Type	Description
1	Drum
2	Wooden barrel
3	Jerrican
4	Box
5	Bag
6	Composite packaging
7	Pressure receptacle

The letter(s) in the marking indicate the material(s) of construction. One letter indicates a single material of construction container, while two letters indicate a composite container. For composite containers, the first letter identifies the material of construction for the inner receptacle, and the second letter identifies the material of construction for the outer container.

Type	Description
A	Steel
B	Aluminum
C	Natural wood
D	Plywood

F	Reconstituted wood
G	Fiberboard
H	Plastic
L	Textile
M	Paper, multiwalled
N	Metal, other than steel, or aluminum
P	Glass, porcelain, or stoneware

The final number, when present, indicates the packaging category. The code for these numbers is as follows:

Type	Description
1	Nonremovable head drum
2	Removable head drum

The following are examples of various container markings.

1A1 [1 (drum)—**A** (steel)—**1** (nonremovable head)]

This container is a steel drum with a nonremovable head. Most commonly this type of drum will be equipped with bungs for loading and discharging the product.

1N2[1 (drum)—**N** (metal other than steel or aluminum)—**2** (removable head)]

This container is a metal drum (other than steel or aluminum) with a removable head. This type of drum may have bungs or may require subpackaging, such as small, individual packages packed within the drum.

6HA1[6 (drum)—**HA** (plastic in steel)—**1** (nonremovable head)]

This container is a composite container. It consists of a plastic receptacle in a steel drum that has a nonremovable head.

4G[4 (box)—**G** (fiberboard)]

This container is a fiberboard box.

STANDARD NONBULK CONTAINERS

Steel Drums

There are two steel drum identification codes. All drums of a capacity greater than 11 gallons (40 liters) must have welded body seams. Those designed for solids or having a capacity less than 11 gallons are allowed welded or mechanically seamed body seams. Drums of a capacity greater than 16 gallons (60 liters) may have at least two expanded or separate rolling hoops (sometimes called chimes). The maximum bung sizes for fill-ing, discharging, or venting 1A1 drums can not exceed 3 inches (7.0 cm). If it does, the drum is considered a 1A2. The maximum capacity of these drums is 119 gallons (450 liters) or a maximum net mass of 882 pounds (400 kilograms). The following are the two types of steel drums.

Type	Description
1A1	Nonremovable head drum
1A2	Removable head drum

Aluminum Drums

There are two aluminum drum identification codes. Their body and heads must consist of aluminum at least 99% pure or aluminum alloy. All seams must be welded with chime seams with reinforcing rings. Rolling hoops, discharges, and maximum capacities are the same as steel drums. The following are the two types of aluminum drums.

Type	Description
1B1	Nonremovable head drum
1B2	Removable head drum

Metal Drums Other Than Steel or Aluminum

There are two identification codes for this type of drum. They must meet the same require-ments regarding chimes, rolling hoops, discharges, and maximum capacities as aluminum drums. The following are the two types of metal other than steel or aluminum drums.

Type	Description
1N1	Nonremovable head drum
1N2	Removable head drum

Plastic Drums or Jerricans

There are four identification codes for plastic drums and jerricans. All plastics must be resistant to ultraviolet radiation. Discharge openings follow the same requirements as the metal drums. The following are the four types of plastic drums and jerricans and their maximum capacities.

Type	Description	Maximum Capacity
1H1	Nonremovable head	119 gallons, 882 pounds
1H2	Removable head	119 gallons, 882 pounds
3H1	Nonremovable head	16 gallons, 265 pounds
3H2	Removable head	16 gallons, 265 pounds

Plywood Drums

There is only one identification for plywood drums: 1D. Although the body of the drum must be constructed of an appropriate plywood, the heads can be constructed of a material other than plywood having equivalent strength and durability. The maximum capacity of these drums is 66 gallons (250 liters) or a maximum net mass of 882 pounds (450 kilograms).

Fiber Drums

There is only one identification code for fiber drums: 1G. Fiber drums must be constructed of glued or laminated piles of heavy paper or fiberboard. The piles must be corrugated. Various materials such as waxed craft paper, metal foil, and plastic can be layered with the piles. The layering must not delaminate under normal conditions. The maximum capacity is 119 gallons (450 liters) or a maximum net mass of 882 pounds (400 kilograms).

Wooden Barrels

There are two identification codes for wooden barrels. Wooden barrels must have barrel hoops constructed of steel or iron. The 2C2 may have hoops made of hardwood. For the 2C1, the bung holes may not exceed one half the width of a stay. The maximum capacity of either type of wooden drum is 66 gallons (250 liters) or a maximum net mass of 882 pounds (400 kilograms). The following are the two types of wooden barrels.

Type	Description
2C1	Bung type barrel
2C2	Removable head barrel

Steel Jerrican

There are two identification codes for steel jerricans. The body seams and chimes must be welded in jerricans designed to carry more than 11 gallons (40 liters). For those designed for smaller capacities, mechanical or weld seaming of these joints is acceptable. Discharge openings may not exceed 3 inches for nonremovable heads. Their maximum capacity is 16 gallons (60 liters) or a maximum net mass of 265 pounds (120 kilograms). The following are the two types of steel jerricans.

Type	Description
3A1	Nonremovable head jerricans
3A2	Removable head jerricans

Boxes

There are eleven identification codes for boxes (Table 4-7).

Composite Packaging

There are twenty-one identification codes for composite packaging (Table 4-8).

Bags

There are nine identification codes for bags (Table 4-9). All nine have a maximum capacity of 110 pounds (50 kilograms).

Table 4-7   Identification Codes for Boxes

Type	Material of construction	Description	Capacity
4A1	Steel	Unlined and uncoated	882 pounds (400 kg)
4A2	Steel	Inner liner or coating	882 pounds
4B1	Aluminum	Unlined and uncoated	882 pounds
4B2	Aluminum	Inner liner or coating	882 pounds
4C1	Wood	Ordinary box	882 pounds
4C2	Wood	Sift-proof walls	882 pounds
4D	Wood	X	882 pounds
4F	Reconstituted wood	X	882 pounds
4G	Fiberboard	X	882 pounds
4H1	Plastic	Expanded plastic	132 pounds (60 kg)
4H2	Plastic	Solid plastic	882 pounds

Table 4-8   Identification Codes for Composite Packaging

Type	Receptacle	Outer packaging	Capacity—receptacle
6HA1	Plastic	Steel drum	66 gallons, 882 pounds
6HA2	Plastic	Steel crate or box	16 gallons, 165 pounds
6HB1	Plastic	Aluminum drum	66 gallons, 882 pounds
6HB2	Plastic	Aluminum crate or box	16 gallons, 165 pounds
6HC	Plastic	Wooden box	16 gallons, 165 pounds
6HD1	Plastic	Plywood drum	66 gallons, 882 pounds
6HD2	Plastic	Plywood box	16 gallons, 165 pounds
6HG1	Plastic	Fiber drum	66 gallons, 882 pounds
6HG2	Plastic	Fiberboard box	16 gallons, 165 pounds
6HH	Plastic	Plastic drum	66 gallons, 882 pounds
6PA1	Glass, porcelain, or stoneware	Steel drum	16 gallons, 165 pounds
6PA2	Glass, porcelain, or stoneware	Steel crate or box	16 gallons, 165 pounds
6PB1	Glass, porcelain, or stoneware	Aluminum drum	16 gallons, 165 pounds
6PB2	Glass, porcelain, or stoneware	Aluminum crate or box	16 gallons, 165 pounds
6PC	Glass, porcelain, or stoneware	Wooden box	16 gallons, 165 pounds
6PD1	Glass, porcelain, or stoneware	Plywood drum	16 gallons, 165 pounds

Table 4-8   Identification Codes for Composite Packaging (*continued*)

Type	Receptacle	Outer packaging	Capacity—receptacle
6PD2	Glass, porcelain, or stoneware	Wickerwork hamper	16 gallons, 165 pounds
6PG1	Glass, porcelain, or stoneware	Fiber drum	16 gallons, 165 pounds
6PG2	Glass, porcelain, or stoneware	Fiberboard box	16 gallons, 165 pounds
6PH1	Glass, porcelain, or stoneware	Expanded plastic	16 gallons, 165 pounds
6PH2	Glass, porcelain, or stoneware	Solid plastic	16 gallons, 165 pounds

Table 4-9   Identification Codes for Bags

Type	Material of construction	Description
5H1	Woven plastic	Unlined and uncoated
5H2	Woven plastic	Sift-proof
5H3	Woven plastic	Water-resistant
5H4	Plastic film	X
5L1	Textile	Unlined and uncoated
5L2	Textile	Sift-proof
5L3	Textile	Water-resistant
5M1	Paper	Multiwalled
5M2	Paper	Multiwalled and water resistant

Complete Markings Requirements

All nonbulk packaging (as well as intermediate bulk) must display markings that identify its exact specifications, design criteria, date of manufacture, and other pertinent data. These markings may be on a single line or multiple lines but must be preceded by a United Nations symbol as shown by the letters UN (Figure 4-6). These symbols indicate that the container meets the appropriate United Nations standards.

The following list identifies the required information in the order that it must appear on nonbulk packaging.

1. The letters "UN" or the UN circle symbol (as found on nonbulk packaging)
2. The container's coded number designation
3. The code letter designating the Packing Group of the container, as follows:
   • X—Packing Group I, II, and III test requirements
   • Y—Packing Group II and III test requirements
   • Z—only Packing Group III test requirements
4. The specific gravity of liquids or the total mass of solids in kilograms (kg) for which the container is designed (each kilogram (kg) equals approximately 2.2 pounds; 1 pound equals approximately 0.45 kg)
5. The test pressure is kilopascals for liquid containers and "S" for solids (each kilopascal (kPa) equals approximately 0.15 psi; 1 psi equals approximately 7 kPa)

6. The last two digits of the year when manufactured
7. The country of manufacture and authorization
8. Manufacturer's symbol
9. For metal or plastic drums and jerricans intended for reuse or reconditioning, the thickness to nearest millimeter
   - Multiple thicknesses respectively for the head and body
   - For example, a container marked 1.2—1.0 indicates a head thickness of 1.2 mm and a body thickness of 1.0 mm. A container marked 0.8—1.0—1.2 indicates a top thickness of 0.8 mm, a body thickness of 1.0 mm, and a bottom thickness of 1.2 mm.

The following are examples of the complete markings.

UN 4G/Y145/S/94
4 (box)—G (fiberboard) / Y (packing group II & III)—145 (kg) /
S (solid) / 94 (manufacture date)
USA/RA
USA (authorizing country)—RA (manufacturer's symbol)

This container is a fiberboard box that meets the Packing Groups II and III tests. Its maximum capacity is 145 kilograms of solids or inner packaging and was manufactured in 1994. The United States authorized its manufacturing by a company with the symbol RA.

UN 1A1/X1.4/150/95
1 (drum)—A (steel)—1 (nonremovable head)/ X (packing group I, II, & III)—
1.4 (specific gravity of liquid)/ 150 (test pressure in kPa)/ 95 (manufacture date)
USA/VL824
USA (authorizing country)/ VL824 (test by this organization)
1.0
1.0 (minimum thickness in mm)

This container is a steel drum with a nonremovable head. It meets Packing Groups I, II and III and holds liquids with a specific gravity of up to 1.4. It has a test

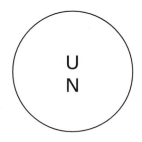

Figure 4-6   United Nations Symbol

pressure of 150 kPa (approximately 22 psi) and was manufactured in 1995. The United States authorized and was the location of its manufacturing. An organization with the code VL824 tested the container for appropriate compliance. Its minimum thickness is 1.0 mm.

BULK PORTABLE CONTAINERS

Another major group of portable containers is the bulk containers. These bulk containers include portable tanks, intermodal containers, and intermodal portable tanks. DOT specifically defines a portable tank as any bulk package (except a gas cylinder with a water capacity of 1,000 pounds or less) designed primarily for loading onto or temporary attachment to a transportation vehicle. Intermodal containers are freight containers designed and constructed for use in two or more modes of transport. Intermodal portable tanks are in a specification class of portable tanks designed for international intermodal transport.

In 1994, DOT instituted a new classification system for portable containers. Instead of simply having bulk containers, DOT introduced a subcategory identified as intermediate bulk.

Bulk packaging continues to include Specification 51, 60, IM 101, and 102 steel portable tank. Intermediate bulk packaging incorporated the prior Specification 56 and 57 containers. Further, intermediate bulk brought a whole host of different containers into the realm of hazardous materials transportation. Some of these containers were previously in hazmat service but only because DOT had granted them exemptions. This text covers both the old and new systems because containers manufactured under the old system are quite plentiful and will be seen for many years to come.

The smaller bulk portable tanks now found in the intermediate bulk category, range from rectangular to cylindrical to dual tapered to conical containers used to transport, store, and often dispense products. New designs including textile and plastic bags (bulk bags) are becoming common. A typical example is the metal tote (Spec 57). Other examples include tanks constructed of plastic. When constructed of plastic, the tanks most commonly have a protective skin made of various materials. The capacity of these various tanks can range to as much as 793 gallons. Most commonly, they are handled by fork lift. Many have a discharge spigot in the center of a bottom edge while others have funnel-like tops. When inverted, the funnel top container dispenses its powdered or granular contents.

Due to their capacity and ease of dispensing, totes are commonly used in production operations not requiring large quantities of the product for normal use. Commonly, they contain corrosive liquids and solids, poisonous liquids and solids, flammable liquids, water reactive solids, and so on.

The three primary tank types under the old system are Specification 56, Specification 57, and Specification 60 portable tanks (Figure 4-7). The term Spec commonly is used instead of specification.

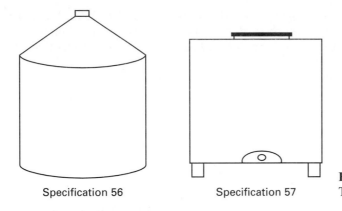

Specification 56   Specification 57

Figure 4-7   Spec 56 and 57 Portable Tanks

Specification 56

Spec 56 container requirements are found in CFR 49, 178.251, and 178.252 (pre-1994). Most commonly, Spec 56 containers transport powdered or granular solids. They must have an MAWP of less than 15 psi and a gross weight of 7,700 pounds or less. Materials of construction include AL, MS, HSLA, SS, and magnesium alloy (MG). This is the type of container similar to those used to transport calcium carbide. They have no mandate for reclosing pressure relief valves.

Both the Spec 56 and 57 portable tanks must have readily accessible certification plates permanently affixed to each tank. The plate must include the following information:

1. Tank manufacturer
2. Specification identification
3. Design pressure psig (for Spec 57 only)
4. Test pressure psig (for Spec 57 only)
5. Serial number
6. Original test date or leakage test date
7. Tare weight in pounds (the weight of the empty container)
8. Rated gross weight in pounds (when tare weight is subtracted, the weight of the lading is identified)
9. Volumetric capacity in U.S. gallons (or cubic feet)
10. Materials of construction

Specifications 57 and 60

Requirements for Spec 57 containers, commonly called totes (a trade name), are found in 49 CFR, 178.251, and 178.253 (pre-1994). Spec 57 containers commonly transport liquids of relatively low volatility. They must have a capacity of at least 110 gallons but no more than 660 gallons. One of the more common capacities is 475 gallons. Materials of construction include AL, MS, HSLA, SS, and MG. These containers require venting capabilities and protection for all fittings. They do not require reclosing pressure relief devices. As listed under Spec 56, Spec 57 containers required a certification plate. Commonly, the location of the plate is in the upper left hand corner of one of the sides. Spec 60 containers are similar to Spec 57s, except that they are only made of MS.

Spec 57 containers commonly have valving and piping that allow their use as portable storage tanks for production operations at users' facilities. Further, they generally have a design that allows stacking. All stacking attachments must securely mount to the tank and allow safe and stable stacking.

THE NEW INTERMEDIATE BULK CATEGORY

Intermediate bulk containers range in capacity from 119 gallons (5.3 cubic feet, 4.5 cubic meters, or 450 liters) to 793 gallons (35.3 cubic feet, 3 cubic meters, or 3,000 liters) and up to 882 pounds. A coding system identifies the exact container specification they meet. The code uses a two-digit number followed by one or two capital letters. Some of the specifications then follow the letters with a number (Tables 4-10 and 4-11). The numeric code identifies the following:

1. Whether the container can be used for solid or liquid service
2. Whether the solid cargo container empties by gravity or pressure
3. Whether the container is flexible or rigid

The letter codes identify the material of construction. For example, a container identified as 13H is for solid material service, is flexible, is made of plastic, and discharges its product by gravity. A container identified as 31N is for liquid material service, is rigid, and is made of a metal other than steel or aluminum.

Each intermediate bulk container designation based on its material of construction has various individual specifications. The next section identifies and describes these various specification options based on the material of construction, including composites.

Table 4-10 Numeric Codes for Intermediate Bulk Containers

Number	Liquid/solid	Gravity/pressure	Rigid/flexible
11	Solid	Gravity	Rigid
13	Solid	Gravity	Flexible
21	Solid	Pressure ($>$1.45 psi)	Rigid
31	Liquid	X	Rigid

Table 4-11 Letter Codes for Intermediate
Bulk Containers

Letter	Material of construction
A	Steel (all types and surface treatments)
B	Aluminum
C	Natural wood
D	Plywood
F	Reconstituted wood
G	Fiberboard
H	Plastic
L	Textile
M	Paper, multiwall
N	Metal (other than steel or aluminum)

Metal Containers

Metal containers are made for solid or liquid service. Under the old system, these containers were classified as Spec 56 or Spec 57. The metal containers have the following codes and use descriptions:

Type	Description
11A, 11B, 11N	Solids, gravity load, and discharge
21A, 21B, 21N	Solids at pressure > 1.45 psi
31A, 31B, 31N	liquids or solids

Rigid Plastic Containers

Rigid plastic containers are for solid or liquid service (Table 4-12). Those designed for liquid service must have some form of pressure-relief device that will prevent rupture should the internal pressure exceed its hydrostatic test pressure. These relief devices may include spring loaded, rupture disks or other means of construction.

Wooden Containers

Wooden containers are for solid service where loading and discharge is by gravity. The containers may be rigid or collapsible and must have a liner (not an inner package). The inner liner consists of a tube or bag inserted into the body of the container. The liner must have closures for all openings. Further, the liner shall not be integral to the container itself. Their codes and descriptions are as follows:

Type	Description
11C	Natural wood with inner lining
11D	Plywood with inner liner
11F	Reconstituted wood with inner liner

Table 4-12   Rigid Plastic Container Codes and Descriptions

Type	Description
11H1	Solids, fitted with structural equipment for stacking, with gravity load, and discharge
11H2	Solids, freestanding, with gravity load, and discharge
21H1	Solids, fitted with structural equipment for stacking, with pressure load, and discharge
21H2	Solids, free standing, with pressure load, and discharge
31H1	Liquid, fitted with structural equipment for stacking
31H2	Liquid, free standing

Top lifting devices are prohibited for any wooden container. They may be handled by a detachable pallet or an integral pallet base for lifting purposes. Wooden containers can be reinforced to allow stacking.

Fiberboard Containers

Fiberboard containers are for solid service to be loaded and discharged by gravity. The only specification in this category is 11G. They consist of a fiberboard body with or without top or bottom caps. If necessary, they can have an inner liner which must meet the same requirements as those specified for wooden containers.

Furthermore, fiberboard containers are prohibited from having top lifting devices. The fiberboard must meet water resistance and top and bottom puncture resistance requirements. They may have integral or removable pallet bases for lifting purposes. Fiberboard containers can be reinforced to allow stacking.

Flexible Containers

Flexible containers are for solid hazardous material service (Table 4-13). For all practical purposes, flexible containers are very large bags, sometimes called bulk bags. They are constructed of woven plastic, fabric, paper, or a combination of these. The woven

Table 4-13   Flexible Container Types and Descriptions

Type	Description
13H1	Woven plastic without coating or liner
13H2	Woven plastic with coating
13H3	Woven plastic with a liner
13H4	Woven plastic with coating and liner
13H5	Plastic film
13L1	Textile without coating or liner
13L2	Textile with coating
13L3	Textile with liner
13L4	Textile with coating and liner
13M1	Multiwalled paper
13M2	Multiwalled paper that is water resistant

plastic and fabric bags are similar to a large burlap bag. They have handling devices such as slings, loops, eyes, or frame attachments. They may also have inner liners.

All containers must have the strength to contain their intended lading. The paper containers must have enough water resistance to prevent their failure when loaded and exposed to moisture. Further, all the flexible container must resist aging and degradation by ultraviolet light. When filled, their height to width ratio can not exceed 2:1 (i.e., not more than twice as high as it is wide).

Composite Containers

In the case of composite intermediate bulk containers, the first letter indicates the receptacle's material of construction. The second letter indicates the outer container's material of construction. For example, a 31HA is a rigid, liquid service composite container. It has an inner plastic receptacle and an outer steel container.

Finally, rigid plastic, composite and flexible intermediate bulk containers follow the letter(s) with another single digit (Table 4-14). The single digit identifies the exact specification the container meets.

Table 4-14   The Exact Specification for Intermediate Bulk Containers

Type	Description
11H ? 1*	For solids, rigid plastic inner receptacle loaded or discharged by gravity
11H ? 2*	For solids, flexible plastic inner receptacle loaded or discharge by gravity
21H ? 1*	For solids rigid plastic inner receptacle loaded or discharged under pressure
21H ? 2*	For solids, flexible plastic inner receptacle loaded or discharge under pressure
31H ? 1*	For liquids, rigid plastic receptacle
31H ? 2*	For liquids, flexible plastic receptacle

*? indicates the location where the appropriate letter for the outer container's material of construction is inserted. For example, 11HA1 is a composite container with a rigid plastic inner receptacle with an outer steel container and is designed for solids that use gravity for loading or discharge.

Complete Container Markings

The coding systems identified so far are only part of the markings required for intermediate bulk containers. The following list identifies the required information in the order that it must appear.

1. The letters "UN" or the UN circle symbol (as found on nonbulk packaging)
2. The container's code number designation
3. The code letter designating the Packing Group of the container as follows:
   • X—Packing Group I, II, and III test requirements
   • Y—Packing Group II and III test requirements
   • Z—only Packing Group III test requirements

4. The month and year of manufacturing
5. The country authorizing the container
6. The name or symbol assigned to the manufacturer
7. The stacking test load in kilograms, 0 indicates the container is not designed for stacking
8. The maximum gross mass (weight) of the container in kilograms

The following helps to clarify these markings.

UN 13H3/Z/ 04 95 /USA/DML/0/1500

UN (United Nations) **13H3** (woven plastic with liner)/**Z** (packing group III)/
04 95 (mftr. 4/95)/**DML** (by DML)/**0** (no stacking)/**1500** (max. load 1500 kg)

This marking indicates that the container's design is based on United Nations criteria. The container is a flexible plastic container intended for solids and discharges by gravity. It meets the Packing Group III testing requirements and is authorized by the United States. The date of manufacture is April of 1995 by DML. Stacking is not allowed, and the container is designed for a maximum load of 1,500 kg or approximately 3,300 lb.

UN 31HA1/Y/ 11 94/USA/+ZT135/10500/1200

31 (liquid) HA1 (composite rigid plastic in steel)/**Y** (packing group II & III)/
11 94 (mftr. date)/**USA** (authorization)/ + **ZT135** (testing lab)/**10500** (stacking to max. weight of 10,500 kg)/**1200** (individual container capacity 1,200 kg)

This marking indicates the container is a composite container designed for liquids, with a rigid plastic receptacle and an outer steel body. It meets Packing Groups II and III, was manufactured in November of 1994, and was authorized by the United States. A third party testing laboratory tested the container to assure compliance with DOT requirements. Stacking is allowed to a maximum weight of 10,500 kg, and each container has a capacity of 1,200 kg.

Additional Markings

Besides the markings already described, each container must have the following markings readily located in the same area:

Rigid Plastic and Composite Containers:
1. Rated capacity in liters of water
2. Tare mass in kilograms
3. Gauge test pressure
4. Date of last leak testing (month and year), if applicable
5. Date of last inspection (month and year)

Metal Containers:

1. Rated capacity in liters of water
2. Tare mass in kilograms
3. Date of last leak testing (month and year), if applicable
4. Date of last inspection (month and year)
5. Maximum loading/discharge pressure, in kPa, if applicable
6. Body material and minimum thickness in mm
7. Serial number assigned by the manufacturer

Product Discharges

An additional point must be made about the Spec 31 (liquid product) containers and their discharges. All have some type of spigot located at the bottom of the container. The spigots are of two primary designs (Figure 4-8).

Some containers have a spigot nested in a relatively large recess located at the bottom and side edge of the container. Other containers have spigots that extend from the corner of the container, run parallel to the sidewall, and are recessed in channels to protect them from damage. Both nested and parallel spigots have unique quirks in design and operation.

In the case of nonregulated materials, the spigot may have a rotary valve that allows loading and discharge (Figures 4-9 and 4-10). These rotary valves have opened by themselves during transport. It is often difficult to determine what direction of rotation closes the valve. Careful examination of the container in the immediate area of the discharge may provide directions for opening and closing.

In the case of regulated materials, the spigot may also be nested in a bottom recess and have a quarter-turn valve. As with many quarter-turn valves, when the valve handle is perpendicular to the spigot piping, the valve is closed. When the valve handle is parallel to the piping, the valve is open.

However, when the spigot runs parallel to the side wall, the valve is located near the end of the spigot piping. In many instances, this arrangement also uses a quarter-

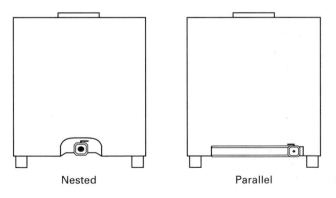

Nested   Parallel   **Figure 4-8** Spigot Location Options

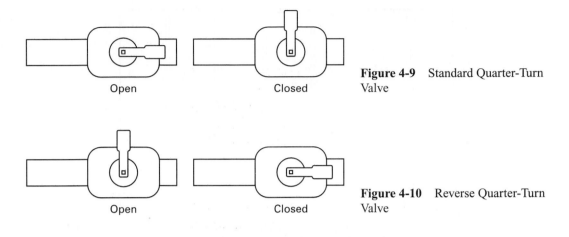

Open   Closed

Figure 4-9   Standard Quarter-Turn Valve

Open   Closed

Figure 4-10   Reverse Quarter-Turn Valve

turn valve. Unfortunately, the quarter-turn valve handle orientation is reversed. In other words, the open and closed handle positions are opposite to those of a standard quarter-turn valve.

The reason is simple. If the handle were perpendicular when the valve was closed, the handle would extend beyond the protection of the container's wall. As such, there would be a great potential for damage. In this case, the valve is closed when the handle is parallel with the piping and is open when it is perpendicular. Personnel must examine the side of the container in the area of the spigot to locate a graphic that shows the open and closed positions of the valve and handle before considering operation of the valve.

INTERMODAL CONTAINERS AND TANKS

Intermodal Containers

Intermodal containers are bulk, box containers (as opposed to tanks) consisting of metal cargo boxes that routinely measure approximately 8 feet square and range up to 45 feet in length. As with all intermodal containers and tanks, their mode of transportation can range from water to rail to highway transport.

These containers are very popular general commodity packaging because of the ability to load them with a wide variety and mix of individual packages. Since these box containers have no wheels of their own, they stack and load easily. In areas where there is sufficient bridge clearance, these containers are found stacked two high. Furthermore, this makes them ideal for intermodal applications since mechanized equipment can shift them from water to rail to highway transportation readily. As a result of this ease in handling, they are quite economical. In highway transport, a trailer frame or a flat bed hold these portable containers. In rail transport, a flat car holds the container and is called a Container On Flat Car (COFC) (Figure 4-11).

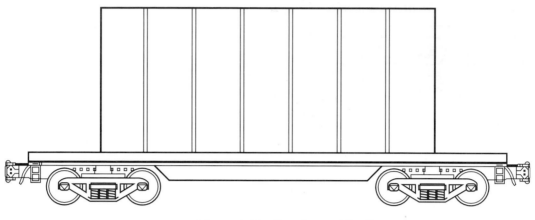

Figure 4-11   Container on Flat Car

Many of the newer flats used for this type of intermodal container are not typical flat cars. Rather, they are series of five flats that articulate over a shared truck assembly. This configuration is viable because the containers are of relatively low weight in comparison to a box, tank, or similar type of traditional car. The elimination of multiple truck sets helps to minimize friction caused by the excess wheels.

Intermodal Tanks

Other members of the portable tank family find widespread use in intermodal applications. As a result, they are known as intermodal tanks. The construction of intermodal tanks most commonly involves cylindrical bulk tanks with an exterior metal box frame. Some less common types include rectangular tanks, high pressure tube modules, and multi-compartmented tanks attached to a frame. Regardless of the specific tank type, the frames serve many purposes. They provide protection for the tanks, space saving loading and stacking capabilities, and easy handling for various types of loading equipment (e.g., fork trucks, straddle loaders, and gantries). Most commonly the frames measure 8 foot by 8 foot 6 inches on end and come in various lengths.

Materials of Construction

The vast majority (approximately 90%) of these containers use stainless steel as their material of construction. The remaining tanks are constructed of mild steel. The specific minimum thickness of the tanks depends upon the material of construction involved as seen in the table below.

Material of construction	Nonregulated materials minimum thickness	Regulated material minimum thickness
Stainless steel	Approximately $\frac{1}{8}''$	Approximately $\frac{3}{16}''$
Mild steel	$\frac{1}{4}''$	$\frac{3}{8}''$

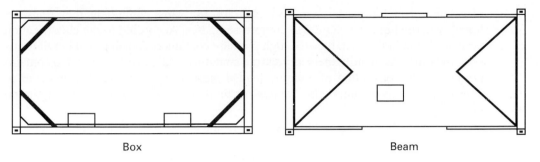

Box   Beam

Figure 4-12   Typical Frames

Frame Types

There are two types of frames used in the construction of intermodal tanks (Figure 4-12). The first type is the box type. Box type frames encircle the entire perimeter of the tank. As such, they help provide structural integrity to the tank as well. The second type is the beam type. Beam type frames only encircle the ends of the tank. In this case, the tank uses integral frame construction techniques to provide its own structural integrity. The design of the frames meets many container handling needs including load supports, base mounts, lift locations, and tie-down systems. It should be obvious that this arrangement is quite efficient for maximum use of space onboard ships as well as on rail and highway vehicles.

Tank Types

There are four primary types of intermodal tanks. They include Specification 51, Specification 60, IM 101, and IM 102. Their materials of construction, MAWP, minimum thickness, and so on vary greatly so it is helpful to examine each individually.

Specification 51

The DOT requirements for Spec 51 containers (Figure 4-13) are found in 178.245. Spec 51 tanks are steel portable tanks having a minimum water capacity of greater than 1,000 pounds, or approximately 120 gallons. Further, they must have an MAWP

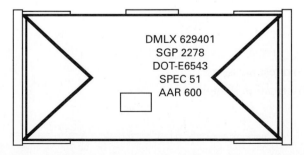

DMLX 629401
SGP 2278
DOT-E6543
SPEC 51
AAR 600

Figure 4-13   Typical Spec 51

of not less than 100 psi but not more than 500 psi. Most commonly, they transport highly volatile liquids or liquefied compressed gases. According to the classification systems found in this text, they are high pressure containers. When found in rail transport, the railroads identify them as pressure containers. All newly constructed containers require the protection of a spring-loaded pressure relief valve. The material of construction is most commonly stainless or mild steel having a 3/16 inch minimum thickness.

All openings, except liquid level gauges and safety devices, must be grouped in one location at the top or at one end of the tank. Each tank designed to transport a compressed gas that is uninsulated must have a reflective surface such as paint or bright metal. All tank outlets and inlets must have markings that indicate if they end in the vapor or liquid space when the tank is full. Further, all valves, fitting, accessories, safety devices, gaging devices, and so on must have protection against mechanical damage in the form of a protective device or housing. All outlets from the tank must have excess flow valves, and fill or discharge lines must have shutoff valves located as close as possible to the tank.

Spec 51 containers, when used in international transportation, are sometimes called IMO Type 5 tank containers. If the international containers are not constructed to DOT specification, they must have an exemption marking displayed on them.

Each tank must have a name/certification plate soldered, brazed or welded to one head. The plate must have the following information:

1. Manufacturer's name and serial number
2. Tank serial number
3. DOT specification number
4. Water capacity in pounds
5. Tare weight in pounds
6. Design specific gravity
7. Original test date
8. Tank retest pressure in psig and when

Additionally, they will have the following information on the ASME data plate:

1. Date of manufacture
2. Material of construction
3. Nominal capacity in gallons
4. ASME code symbol
5. Pressures including MAWP and test pressure
6. Insulation and heating coil information if applicable
7. Air connections if applicable

Specification 60

Spec 60 container requirements are found in 178.255. They must also meet the requirements found in 173.32, (qualification, maintenance and use of portable tanks other than Specification IM portable tanks) and 173.315, (compressed gases in cargo tanks and portable tanks).

They are very similar to a standard Spec 57 tote, except their construction must comply with the ASME Code, and have steel as their material of construction. The steel requires post weld heat treatment. To protect the tank from the lading, this specification allows the use of a liner within the tank. All valves, fittings, safety devices, gauging devices, accessories, and so on must have a protective housing equipped with a cover plate for protection.

Each tank must have markings according to ASME Code. Additionally, the tank marking must be by permanent stamping near the center of one head or on a brazed or welded plate with the following information:

1. Manufacturer's name and serial number
2. DOT specification
3. Nominal capacity in gallons
4. Tare weight in pounds
5. Date of manufacture

IM Portable Steel Tanks

The most common types (over 90%) of intermodal tanks are IM 101 or IM 102 specification tanks. The IM 101 tanks must have a maximum allowable working pressure of at least 25.4 psig (1.75 bar) but no greater than 100 psig (6.8 bar). The IM 102 tanks must have a MAWP of at least 14.5 psig (1 bar) but no greater than 25.4 psig (1.75 bar). By the classification system used in this text, they are low pressure containers. Since the railroads use the same classification system for intermodal tanks as they do for tank cars, IM 101 and 102 tanks are classified as nonpressure tank containers when in rail transport.

Again, each tank must have all fittings, gages, outlets, and so on protected from mechanical damage. Each outlet must have an excess flow valve with inlets and discharges equipped with shutoff valves.

All tanks with capacities of 500 gallons (1,893 liters) or less must have a primary frangible disk (rupture disk) or spring loaded (self closing) pressure relief valve. Tanks with capacities greater than 500 gallons require a spring-loaded primary pressure relief valve and may have a spring-loaded, frangible, or fusible secondary pressure relief valve. Any pressure relief device must have appropriate venting capacity based upon the exposed area of the tank. If the tank is exposed to a vacuum, it must have vacuum protection as well.

Each tank must have an identification plate permanently attached to the tank in an accessible location and in readable fashion. The plate must contain at least the following information:

1. DOT specification number
2. Country of manufacture
3. Manufacturer's name
4. Date of manufacturer
5. Manufacturer's serial number
6. Identification of DOT approval agency and approval number
7. MAWP in bars or psig
8. Test pressure in bars or psig
9. Total water capacity in liters or gallons
10. Maximum allowable gross weight in kilograms or pounds
11. Equivalent minimum shell thickness in mm or inches
12. Tank material and specification number
13. Metallurgical design temperature range in degrees Celsius or Fahrenheit

Additionally, the following information must appear on the identification plate when applicable.

1. Lining material
2. Heating coil MAWP in bar and psig
3. Corrosion allowance in mm or inches

When any of these tanks are intended for international transportation, it must have a safety approval plate also attached to it. These specific requirements are found in 49 CFR 451.21 through 451.25. Additionally, this section does not limit the display of any other pertinent information on the identification plate.

After construction, each tank must undergo a series of tests including hydrostatic, internal coils, pressure and vacuum relief, and inertia testing. Additionally, they must have periodic hydrostatic testing.

Intermodal Tank Markings

Intermodal tanks must display a series of markings similar to those found on rail tank cars (Figure 4-14). These markings provide shippers, haulers, and emergency responders a mechanism to identify the specific container, its construction specifications, its owner, and its hazardous contents if so placard. The most common location of these markings is on the right hand side of the tank when facing the tank. Again, this marking system closely resembles the markings found on rail tank cars. The markings include the following:

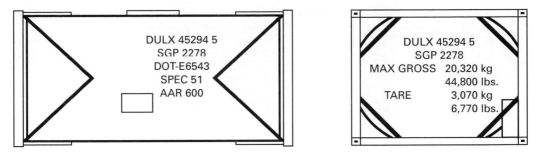

Figure 4-14   Intermodal Tank Markings

1. Tank initials and number (the equivalent to tank car of reporting marks and numbers)

2. Country and size/type markings

3. Specification marks and DOT exemption marks (if they apply)

4. AAR 600 marks (if they apply)

5. Identification plate

6. Tank and valve test dates

The ends of the intermodal tanks also have informational markings similar to rail tank cars. In this case, they contain the tank's initials and numbers and the country size/type marks. Below these two markings is the maximum allowable gross weight, generally in kilograms and pounds. Next comes the tare weight, also in kilograms and pounds. The tare weight is the weight of the empty container. When the tare weight is subtracted from the gross weight on the shipping papers, the total weight of lading is identified. The total possible weight of lading is identified by subtracting the tare weight from the maximum gross weight.

Tank Initials and Numbers

All intermodal tanks require registration with the International Container Bureau, located in France, that identifies specific marking requirements. These markings follow a system almost identical to the tank car system. In this case, they identify the owner and the specific tank's identification number (DULX 45294 5). As with tank cars, these marks act in the same manner as a standard Vehicle Identification Number (VIN).

Country and Size/Type Markings

The country and size/type markings come next. The particular marking, SGP 2278, includes three individual bits of information. SGP is the country of origin, 22 is the size of the container and 78 is the type code or pressure ratings. Tables 4-15, 4-16, and 4-17 provide the meaning of each code.

Table 4-15   Common Size Codes

Code	Length	Height
20	20 feet	8'
22	20 to 40 feet	8'6"
24	20 to 40 feet	>8'6"

Table 4-16   Common Country Codes

Code	Country	Code	Country
BM (BER)	Bermuda	LIB	Liberia
CH (CHS)	Switzerland	NLX	Netherlands
DE	Germany	NZX	New Zealand
DKX	Denmark	PA (PNM)	Panama
FR (FXX)	France	PIX	Philippines
GB	Great Britain	PRC	Peoples Republic of China
HKXX	Hong Kong	RCX	Peoples Republic of China—Taiwan
ILX	Israel	SGP	Singapore
IXX	Italy	SXX	Sweden
JP (JXX)	Japan	US (USA)	United States
KR	Korea		

Table 4-17   Type Codes Listing Maximum Allowable Working Pressure

	Nonhazardous lading	
Code	Test pressure in bars (1 bar is to equal 14.5 psig.)	Test pressure in PSIG
70	<0.44	<6.4
71	0.44 to 1.47	6.4 to 21.3
72	1.47 to 2.94	21.3 to 42.6
73	Spare	
	Hazardous lading	
74	<1.47	<21.3
75	1.47 to 2.58	21.3 to 37.4
76	2.58 to 2.94	37.4 to 42.6
77	2.94 to 3.93	42.6 to 57.0
78	>3.93	>57.0
79	Spare	

Specification Marks and DOT Exemptions

The next set of markings includes the DOT specification to which the container is built. As such, it will read Spec 51, IM 101, or IM 102. However, there are quite a number of portable tanks (both bulk and nonbulk), tank cars, cargo tanks, and so on that do not meet DOT specification yet can legally transport hazardous materials if they receive special DOT approval. In this case, DOT examines the specific container design and samples of the container to assure they are safe for use in transportation. When such assurances are available, DOT provides an exemption and assigns it a number. This exemption number must be clearly marked on any container constructed in this fashion. The exemption numbers always follow the same format. They start with "DOT-," which is followed by a capital E and several numbers (i.e., DOT-E2894).

AAR 600 Marking

The Association of American Railroads (AAR) has a series of specifications for all types of rail vehicles and containers. Most commonly known of these are the tank car and intermodal tank specifications. AAR Section 600 specifications indicate that the tank is acceptable for rail transport. AAR 600 indicates that the tank can transport regulated substances. On the other hand AAR 600NR indicates the tank can transport only non-regulated substances.

GATHERING INFORMATION—SECTION III
CONTAINER INFORMATION
Module III—Highway Transportation Containers

Although the intermodal containers are used in the transportation of commodities, they are also used in production and storage. Transportation containers or packaging, unlike intermodal tanks or containers, have the sole purpose of transporting cargo from one location to another. The most common types are used in four of the five transportation modes: highway, rail, water, or air. The fifth mode, pipeline, does not require containers because the pipeline is the container. Transportation containers can be for either single mode or intermodal use.

HIGHWAY CONTAINERS

Highway containers are the part of the vehicle that hold the product being transported. Highway containers include vans, box trailers, flat beds, cargo tanks, tube trailers, portable tanks, intermodal containers, and intermodal trailers.

To fully understand the relationship of these containers with the DOT regulations it is important to understand the basic distinction between these different containers. Specifically, a vehicle is considered packaging when it acts to hold the lading in the same fashion as any other container. On the other hand, a vehicle is a vehicle when it holds other packaging that actually contains the lading. This means a cargo tank is packaging, while a box trailer is not. Thus, a cargo tank's construction and performance must meet the specifications established in DOT regulations. On the other hand, since a box trailer is a vehicle, its construction and performance are not subject to the same regulations.

There are many types of highway vehicles. New designs are introduced continually. For example, previously it was rather simple to identify different types of cargo tanks by examining their silhouettes. Today, manufacturers construct cargo tanks that meet multiple classifications as well as exhibit silhouettes not normally associated with a general class. As such, the information provided in this text and in the supplemental information at the end of the text describes general configurations for which there are exceptions.

General Commodity Transports

Many highway transport vehicles are called general commodity units because they carry packaging of all shapes, sizes, and types. They can carry bulk and nonbulk containers or tanks. Among other things, they can also carry cardboard boxes, wooden crates, compressed gas cylinders, cryogenic cylinders, portable containers, and intermodal tanks. They include straight or step vans, box trailers, flat beds, and stake bodies. The enclosed units, such as box trailers, vans, and flat bed portable containers present unique problems in that their contents are not readily visible. One of the most common members of this group is the standard box trailer.

Box Trailers

Box trailers (Figure 4-15) are the most common highway transport vehicles and rather simple to understand. They are simply large, open boxes attached to frame rails. The large open area within allows the stowage of innumerable types, shapes, sizes, and forms of lading. As long as it fits within the box, almost anything can be transported in these units.

This particular situation gives rise to some of the problems when these units are encountered during an incident. First, the contents are almost always a mix of various and sundry goods and containers. It is rare to encounter a box with only one cargo. Second, the enclosure of the box is very effective at blocking a visual reconnaissance and routinely requires entry operations. Third, the haphazard stowage of various ladings often makes it difficult to fully recon its contents without "running a gauntlet." Finally, a fully loaded box provides little room for leak control and other operational activities. Although operations involving boxes are routine, there are a few special cases that require discussion.

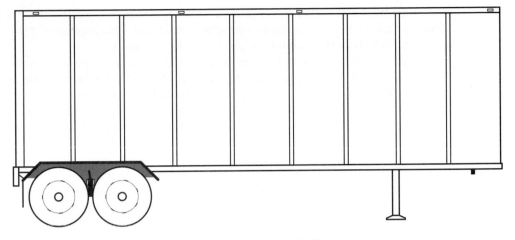

Figure 4-15   Box Trailer

Box trailers have become more involved in intermodal transport in various configurations. Standard boxes are loaded onto specially designed flat cars for shipment by rail. This configuration is the trailer on flat car (TOFC) (Figure 4-16).

The latest wrinkle is called a road railer which uses trailers similar to intermodal box trailers but does away with the flat car. In this case, the box looks like any other box with the exception of a wide gap between the rear tandem axles. The gap provides enough room to locate a standard rail wheel-and-axle assembly. The other configuration has the rear tandems shifted forward, allowing enough space to attach the rail wheels and axle at the very end of the trailer. When placed in rail service, the trailer has a flange that extends from the front of the frame. This flange attaches to the rear of the trailer in front. With the rear rail wheels lowered or attached, the assembly is ready for the attachment of the next trailer. It is not uncommon to find trailer after trailer attached and transported in this fashion.

Figure 4-16   Trailer on Flat Car (TOFC)

Both TOFCs and road railers present responders with major difficulties in identifying the exact trailers involved in a derailment situation. This is especially true if the derailment results in an accordion or similar stacks of trailers.

Refrigerated boxes used for hazardous materials transport pose a unique situation. The reason is that materials that require transport in refrigerated vehicles are usually unstable because they are sensitive to heat. As such, they may be capable of violent decomposition or explosion should the cooling capabilities of the trailer be lost.

Moving companies have long used box trailers for moving residential contents. However, over the past decade or so, some moving companies have entered the general commodities transport business. To the surprise of emergency responders, moving vans do transport hazardous materials. Remember that movers also handle offices and laboratories. Further, in some cases, these vans contain not just standard household wares but also the family car and large quantities of household chemicals.

Dry Bulk Carriers

Dry bulk carriers (Figure 4-17), as the name implies, carry dry bulk materials. Most commonly they carry solid granular, pellet, or powdered materials such as flour, cement, powdered plastic resins, and agricultural feed. However, they also carry hazardous materials. Obviously, the materials must be dry solids in a powdered or granular form. Some of the more common hazardous materials transported in this fashion are sodium hydroxide pellets, potassium hydroxide, and oxidizers such as the nitrates or sulfates.

Even if the lading of a dry bulk carrier is not hazardous, it can be hazardous substances or even an environmental hazard. For example, cement and lime present hazards due to their caustic nature when they contact water or moist tissues. Even flour can be quite hazardous due to the potential for dust explosions if it becomes airborne. This becomes an issue with dry bulk carriers in that they generally offload products by using a pneumatic system. Air from a compressor on the trailer or from a house air supply at the onload and offload sites is pumped into the carrier to entrain the lading. As such, a perfect situation exists for the creation of an airborne release of the finely

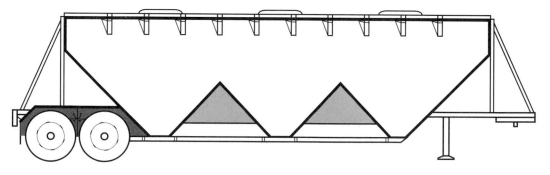

Figure 4-17   Dry Bulk Carriers

divided particulate being transported. However, the pressurized air needed for loading and unloading is normally vented before transport.

Tube Trailers

Tube trailers (Figure 4-18) are the large trailers that contain a series of large, horizontal compressed gas cylinders used as highway transportation vehicles. Although all of these units transport hazardous materials listed in the DOT divisions 2.1, 2.2, or 2.3, the entire unit is not covered by a specific DOT class (just like box trailers, flat beds, etc.). However, the individual "tubes" must meet the appropriate specifications for compressed gas cylinders based upon the type of gas they transport.

The reason for this approach is rather simple. Each tube is really nothing more than an oversized compressed gas cylinder. These large cylinders are attached securely to the frame and structural members of the trailer and then piped together in a manifold system. These units are nothing more complex than a glorified cascade system. Most commonly, the tube trailer is transported to a user facility and left on-site to provide a relatively high-pressure, low-volume gas supply. This method of supplying high pressure gas is quite economical for the smaller users because it does not require the installation of fixed cryogenic containers or pumping systems.

Although the pressure within each cylinder is often quite high, these units generally withstand accidents very well. Just as with standard compressed gas cylinders, the likelihood of a tube or trailer failure is relatively low. Sturdy construction methods and materials are the reason for their reliability. Also, rather than valves protected by a simple cap as in standard cylinders, the tubes have all valving completely enclosed in heavy cabinetry. Additionally, the tubes and cabinetry are fastened securely to the trailer's structural members.

When involved in traffic accidents, these units have received severe trauma yet not lost any product or structural integrity. Most commonly, product loss is due to failure in the pressure relief system. Each tube vents separately through its own relief device. In the case of flammable gases, each vent must have a stack that extends higher

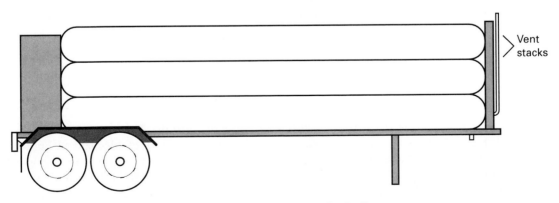

Figure 4-18   Tube Trailer

than any tube on the unit. When there is a large number of smaller diameter tubes on the unit, this results in a whole series of vent stacks, most commonly across the front of the trailer.

Due to the similarity to compressed gas cylinders, it is important to keep in mind the behavior of both during an emergency. In both cases, loss of product is most commonly the result of some type of valving problem. These problems include incomplete closure of valves, damage or wear on valve seats, corrosion of valve assemblies, and so on. The probability of actual container failure is quite low. However, when container failure does occur, the resulting energy release and associated damage are usually severe.

Highway Cargo Tanks

There are five DOT classes of highway cargo tanks. Before the passage of the Hazardous Materials Transportation Uniform Safety Act of 1990 (HMTUSA), these classes were known as Motor Carrier (MC) 306, 307, 312, 331, and 338. With the passage of HMTUSA (now simply called the Hazardous Materials Transportation Act), production of the MC 306, 307, and 312 classes after October 1993 is prohibited. The replacements for these three MC classes are the DOT classes 406, 407, and 412. All existing MC class cargo tanks used for the transportation of hazardous materials require retrofitting to meet all the new requirements of the DOT class vehicles.

The intent of these changes in the specifications is to help these units maintain their integrity through normal usage and during accidents. To accomplish these upgrades, the manufacturing of new units requires stronger and often heavier construction materials. Openings such as hatches, manways, valving, and piping generally must be stronger and function more effectively during accidents. Additionally, all new and existing cargo tanks must undergo more stringent and more frequent inspections and testing. All units must also have leak and pressure tests instead of simple visual inspections.

Unlike the other MC classes, the MC 331 and MC 338 classes and their specifications remain primarily as they were, with the exception that they must also meet more stringent testing and inspection requirements. In this text, all cargo tank designations follow the new DOT classes where they apply.

A discussion of the general requirements that apply to DOT 406, 407, and 412 cargo tanks follows. This information is found in 49 CFR Section 178.345-1 through 178.345-15. These general requirements identify the construction and performance criteria for all the DOT cargo tanks.

1. Materials of Construction and Their Minimum Thickness.   Section 178.345-2 identifies specific sections of the American Society of Mechanical Engineers' (ASME) Code of materials acceptable for use in the construction of these containers. This section goes on to identify the specific criteria used to identify minimum thicknesses of container heads and shells as well as corrosion and abrasion protection for the tank. Finally, this section addresses tank lining criteria.

2. Structural Integrity Criteria.   Section 178.345-3 identifies various considerations and calculations needed to assure that the tank design can withstand the stresses acting upon the container when it is loaded. Torsional, compressive, shear, and tensile stresses caused by various loading and transportation activities are all considerations.

3. Joint Welding.   Section 178.345-4 indicates that all welds that join tank shell, heads, baffles, baffle attaching rings, and bulkheads must meet ASME Code procedures and should be accessible for inspection where practical.

4. Manhole Assemblies.   Section 178.345-5, indicates that each cargo tank with a capacity of more than 400 gallons must have at least one manhole (manway) at least 15 inches in diameter. The manholes must withstand a pressure of at least 36 psig without affecting its structural integrity or ability to maintain the product. Further, each manhole must have a fastener that will not open when the tank is pressurized, during normal transportation, or during a roll-over.

5. Support and Anchoring Cargo Tanks to Nonintegral Frame Assemblies.   Section 178.345-6 indicates that each cargo tank with a nonintegral frame system (separate frame assembly) must be secured to the frame so there is no movement between the tank and the frame under normal transportation conditions. All attachments must be readily accessible for inspection and maintenance. If the frame is integral, full stress calculations must be performed and addressed.

6. Circumferential Reinforcements.   Section 178.345-7 indicates that any tank with a shell thickness of less than 3/8 inch requires circumferential (around the shell of the tank) reinforcement with bulkheads, baffles, ring stiffeners, or a combination of such. The greatest distance allowed between the reinforcements is 60 inches. All the reinforcements must meet specific criteria listed. Stiffening rings that prevent visual inspection are prohibited.

7. Accident Damage Protection.   Section 178.345-8 indicates that all the cargo tanks must minimize the potential for loss of lading due to an accident. Most domes, sumps, or washout cover plates that extend beyond the shell of the tank must be the same strength and thickness as the shell. If they extend more than 4 inches beyond the shell on the bottom third of the tank, or more than 2 inches beyond the shell on the upper two-thirds of the tank shell, they require accident damage protection. Such protection includes additional strengthening of the devices. Any piping, outlets, valves, closures, or other devices that, if damaged, could release product require protection. Further, any piping that extends beyond the accident damage protection requires a stop valve and a sacrificial device such as a shear section.

Bottom damage protection, consisting of a protective enclosure (cage or cabinet), must withstand a force of 155,000 pounds from any side of the device and protect piping or other components that could release lading. Any discharge opening protected with an internal self-closing stop-valve (internal valve or emergency shutoff) need not have the enclosure but requires a sacrificial device such as a shear section in its place. The shear section must be outboard from the internal valve and must fail at no more than 70% of the force that would damage the valve.

Rollover protection for all manholes, filling or inspection openings, valves, fittings, pressure relief devices, vapor recovery stop valves, or other lading retaining fittings on the upper two-thirds of the tank is required. A rollover protection device provides this protection. These devices must withstand specified stresses equal to twice the weight of the loaded cargo tank. If the device allows the accumulation of liquid on top of the tank, there must be a drain. The drain must direct the liquid away from any structural component of the tank.

Rear-end protection is required for all cargo tanks and their piping. The protection device (bumper) must deflect forward 6 inches without contacting any part of the cargo tank that contains lading during transport. Further, the protection device must extend at least 4 inches below the lower surface of any lading-holding portions of the tank.

8. Pumps, Piping, Hoses and Their Connections.   Section 178.345-9 indicates that systems involved in loading and unloading operations must prevent the overpressurization of the tank during either process. Hoses, piping, stop-valves, lading retention fittings, and closures require burst pressures of greater than 100 psig or four times the MAWP. Hose couplings require burst pressures greater than 120 psig or 4.8 times the MAWP.

All parts of the system require construction and protection that prevent leakage of lading. No nonmetallic piping, valving, fittings, or other connections are allowable unless they are outboard of the lading retention system.

9. Pressure Relief Systems.   Section 178.345-10 indicates that each cargo tank must have a pressure and vacuum relief system as identified for each tank specification. Each system must act to prevent tank rupture or collapse due to over-pressurization or vacuum that result from loading or unloading, or heating and cooling of lading.

Each tank must have a primary and secondary relief system. The primary system should consist of one or more reclosing pressure relief valves (spring-loaded relief valve). This system must actuate at no less than 120% and no more than 132% of the MAWP. Additionally, the valve must reclose at no less than 108% of the MAWP and remain closed at the lower pressure.

The secondary system consists of another pressure relief valve in parallel with the primary system. Nonreclosing relief devices (e.g., rupture disks and fusible plugs) are not authorized unless they are in series with a reclosing valve. The secondary system setting must be no less than 120% of the MAWP.

Both the primary and secondary systems together must have enough venting capacity to prevent the tank from exceeding its MAWP.

Each system must withstand a surge pressure 30 psi above its designed activation pressure. When sustained at that pressure for 60 milliseconds, it cannot release more than 1 gallon of lading and must reclose to a leak-tight condition. After August 31, 1995, each system must withstand a surge pressure of 30 psi above the designed activation pressure for 60 milliseconds and loose no lading.

The construction of all systems must prevent the unauthorized adjustment of their settings. Their installation must shield them, while allowing them to drain, without impairing their functioning. The manufacturer must certify the set pressure and flow rate of all pressure relief devices. Finally, each device must have permanent markings that indicate the manufacturer's name, model number, set pressure in psig, and rated flow capacity in standard cubic feet per hour (SCFH).

10. Tank Outlets. Section 178.345-11 indicates that tank outlets are any openings in the cargo tank wall used exclusively for loading or unloading lading. They do not include manhole covers, vents, vapor recovery, or similar devices.

Each outlet must have an internal self-closing stop valve or an external stop valve located as close as practicable to the tank wall. All of these self-closing valves must close within 30 seconds of actuation of the emergency shutoff mechanism. These self-closing stop valves must function even if their actuation systems receive damage or are sheared off. To be functional, the valve must remain securely closed and capable of retaining lading.

Each self-closing valve must include a remote emergency actuation device (emergency shutoff valve) located at least 10 feet from the outlet if the tank's length allows. Alternatively, the remote shutoff location can be at the end of the tank furthest from the outlet. The actuation mechanism can be either manual or mechanical but must be corrosion resistant and function in all types of environments and weather conditions.

Additionally, cargo tanks used for the transportation of flammable, pyrophoric, oxidizing, or poison 6.1 liquids must have emergency shutoffs that include thermal actuation mechanisms. The location of the thermal mechanism must be as close as practicable to the primary outlet(s) and actuate at temperatures at or below 250°F.

When piping extends beyond the self-closing valve, that piping requires the addition of another stop-valve located at the end of the connection. Further, any outlet not designed for loading or unloading requires the installation of a stop-valve or other leak-tight closure located as close as practicable to the tank outlet. If there is piping attached to this fitting, it requires a second closure at the end of the piping.

11. Gauging Devices. Section 178.345-12 indicates that each cargo tank, except those designed for filling by weight, require the use of some type of gauging device. The gauging device indicates the maximum liquid level to within 0.5%. No glass gauges are permitted.

12. Pressure and Leak Testing. Section 178.345-13 indicates that all cargo tanks must undergo pressure and leak testing as specified for the type of cargo tank. Pressure testing produces maximum stress in the container to assure there are no structural defects in the tank and that it can withstand the rigors of transportation. Leak testing procedures are less rigorous and produce lower levels of stress in the container to assure the container's ability to maintain its lading.

Each test involves either hydrostatic (i.e., pressurized liquid, usually water) or pneumatic (i.e., pressurized gas, usually air) methods. Of the two methods, hydrostatic is considered less risky and does not require the same degree of safeguards.

For pressure testing, each tank or tank compartment must be tested. In multi-tank units, the adjacent tanks must be empty and at atmospheric pressure. All outlets except pressure relief devices and loading/unloading vents that actuate below the test pressure, must be in place during the test. If the pressure relief or venting devices are not removed during the test, they must be made inoperative by clamps, plugs, or other devices. Upon completion of testing, the clamps or plugs must be removed.

During hydrostatic pressure testing, the tank is filled with water or a similar liquid and then pressurized specific to the tank. This pressure is held for at least 10 minutes during which time the tank is inspected for leakage, bulging, or other defects.

During pneumatic pressure testing, the tank is filled with air or an inert gas and then pressurized in steps. When the test pressure is reached, it is held for at least 5 minutes. After that, the pressure is reduced and held at the inspection pressure for the duration of the inspection. The entire cargo tank surface requires inspection for leakage or other defects. All joints and fittings receive a coating of soapy water or similar approaches to assure there are no leaks (indicated by bubbles) or other defects.

For leak testing, each cargo tank must have all its accessories in place and in operable condition. The tank requires testing at no less than 80% of the tank's MAWP, while maintaining the pressure for at least 5 minutes.

Any cargo tank that leaks, bulges or shows any other signs of a defect must be rejected. The rejected tank must undergo repairs and then be retested using the same test method before it can return to service.

Not only must new cargo tanks undergo such testing and inspections, so must all units that transport regulated materials. Before HMTUSA, the inspections primarily involved visual examination of the tank. Today however, visual inspection is only part of the process. When the inspection tests are performed, the date of the inspection and the code for the specific type must be marked directly on the cargo tank. There are six possible types of inspections and their marking code (Table 4-18). The inspection markings for a cargo tank that received an external visual, internal visual, and a pressure test in July of 1999 would read "7-99 VIP."

13. Markings.   Section 178.345-14 indicates that the manufacturer of each cargo tank must certify that it meets the appropriate design, construction, and testing

Table 4-18 Inspection Marking Codes

Inspection and test codes	
Type inspection	Code
External visual inspection	V
Internal visual inspection	I
Pressure test	P
Lining test	L
Leakage test	K
Thickness test	T

requirements established by DOT or ASME. Part of the certification process requires the manufacturer to permanently mount a name plate and a specification plate to each tank or its support structure. Plates must be attached to the left front side of the tank by either brazing or welding around their entire perimeter.

The name plate must list the following information:

1. The specific DOT cargo tank specification (e.g., DOT 406)
2. Original test date, month, and year
3. Tank MAWP in psig
4. Tank test pressure (Test P) in psig
5. Tank design temperature range in degrees Fahrenheit
6. Nominal capacity (Water cap.) in gallons
7. Maximum design density of lading in pounds per gallon
8. Material of construction specification number (shell)
9. Material of construction specification number (head)
10. Weld material
11. Minimum thickness of the shell in inches (if minimum thickness is different for different locations on the shell, each location's thickness must be listed)
12. Minimum thickness (head)
13. Manufacture's thickness in inches for the shell, top, sides, and bottom (if corrosion protection thickness is provided)
14. Manufacture's thickness in inches for head (if corrosion protection thickness is provided)
15. Exposure surface area in square inches

The specification plate must list the following information:

1. The cargo tank motor vehicle manufacturer (CTMV mfr.)
2. CTMV certification date
3. Cargo tank manufacturer
4. Cargo tank date of manufacture, month, and year
5. Maximum weight of lading in pounds
6. Maximum loading rate in gallons per minute at maximum loading pressure
7. Maximum unloading rate in gallons per minute at maximum unloading pressure
8. Lining material, if applicable
9. Heating system design pressure in psig
10. Heating system design temperature in degrees Fahrenheit

Multicargo tank vehicles where there are no voids between the tanks can combine the information found on the name and specification plates onto a single plate.

14. Certifications.   Finally, section 178.345-15 indicates that the manufacturer must provide certification documents to the owner of the vehicle. The documents must contain the manufacturer's registration number, the design certifying engineer, and the registered inspector. Further, the documents must indicate that the cargo tank meets all the design specification requirements of the tank.

DOT 406 Cargo Tanks

DOT 406 cargo tank specifications are found in #49 CFR part 178.346. Under this text's classification system they are atmospheric containers and used most commonly to transport liquids under ambient conditions. Most commonly they transport liquid fuels classified as either flammable or combustible liquids. Vacuum loading capabilities are prohibited for these units. DOT specifies that a 406 must have a MAWP of no less than 2.65 psig and no more than 4 psig.

Remember that psig is the gauge pressure, or the pressure within a container above atmospheric pressure. That means with the manhole open, the gauge reads "0". The gauge reads "0" even though it is actually exposed to atmospheric pressure is approximately 14.7 psi.

Most commonly, 406 cargo tanks have an elliptical (oval) cross section and a smooth shell (Figure 4-19). However, there is no mandate for either. Presently, there are more and more units constructed with round cross sections as well as some with elliptical cross sections and external ring stiffeners. The external rings are part of the vapor recovery system.

The best known example of the 406 cargo tank is the standard gasoline or other liquid fuel transport vehicle. The second most common example is the standard fuel oil home delivery truck. The vast majority of over-the-road liquid fuel haulers have capacities ranging from approximately 9,000 gallons to 14,000 gallons. The most common material of construction for these large capacity tanks is AL to minimize weight, although MS, SS, and HSLA are also allowable. The home delivery units generally have a capacity range of approximately 2,600 gallons to 5,000 gallons. About 50% of these home delivery units use AL in their construction with the majority of the remaining 50% using MS.

Most commonly, 406 cargo tanks are multicompartmented (i.e., having multiple individual cargo spaces). Each compartment is separated by a bulkhead. Bulkheads are liquid-tight metal sheets that are welded in place. They act to separate the entire tank

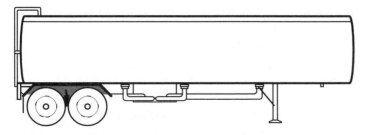
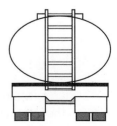

Figure 4-19   DOT 406 Cargo Tank

into smaller, individual cargo spaces. Baffles prevent liquid surge within each compartment. Baffles are non-liquid-tight, metal sheets welded within the compartment. Generally each baffle has an 8 to 12 inch hole in the middle of the sheet and a 4 to 6 inch hole at the top and bottom. These holes allow for rapid on- and off-loading yet drastically reduce liquid sloshing during transport. The easiest way to identify the number of compartments in a cargo tank is to count the number of manholes and outlet valves on the tank. For example, if there are four manholes and valves, there are generally four compartments.

An important design consideration for 406 tanks is the incorporation of truss type construction techniques. In the 406, the top rollover protection forms the top flange of truss; the baffles, bulkheads, and side sheets of the shell form the webbing; and the belly sheets or frame form the bottom flange.

These facts are vital to consider if the unit has rolled over. When the tank is on its side, it is just like a bar joist (a type of truss) on its side in two respects. First, it is incapable of supporting its design load. Any attempt to lift a loaded tank, unless lifted simultaneously along its entire length, can easily result in tank failure. Second, just as with a truss, a compromise (e.g., crack or break) of any part of the webbing, can lead to compromise of the truss and its complete failure. Since the location of baffles and bulkheads is within the tank, their inspection for damage is impossible. For both of these reasons, the righting of a loaded 406 tanker is generally inappropriate. The only righting methods shown to be relatively safe under these conditions is the coordinated use of multiple low pressure, highlift bags.

Each of these units must have the following:

1. A name plate and specification plate
2. Internal self-closing stop-valves (internal belly valve) to stop product discharge
3. One or more pressure relief valves set at no less than 3.3 psig
4. One or more vacuum relief devices that will limit off loading vacuum to a maximum of 1 psig
5. Circumferential reinforcement including bulkheads, baffles, and/or ring stiffeners
6. Accident damage protection including bottom damage protection, rollover protection, and rear-end protection
7. An emergency shutoff valve located at least 10 feet from the loading/unloading outlet where vehicle length allows, or at the end of the cargo tank furthest away from the outlet
8. If designed to haul flammable, pyrophoric, oxidizing or poison liquids, a thermally actuated emergency shutoff system
9. Compliant manholes (manways)

DOT 407 Cargo Tanks

DOT 407 cargo tank specifications are found in #49 CFR part 178.347. Under this text's classification system they are low pressure containers, and are most commonly used to transport liquids or solids under ambient conditions. Liquid examples include flammable or combustible solvents and fuels, mild corrosives, and poisons. Solid examples include molten sulfur and molten asphalt. DOT specifies that a 407 must have a MAWP of no less than 25 psig and that the unit may have equipment for external loading by vacuum. When designed for vacuum loading, the tank must withstand a negative pressure (referred to as external pressure) of at least 15 psi. They must also have a circular cross-section.

Commonly 407 cargo tanks have external ring stiffeners and a relatively large diameter (7 to 8 feet), and can be found as single or multiple compartment tanks (Figure 4-20). The number of manholes and rollover damage protection units seen on the top of the unit help identify the total number of compartments.

Additionally, there are many units equipped with thermal insulation (Figure 4-21). The insulation provides the 407 with a cross-section that most commonly resembles a horseshoe. The ring stiffeners are now under the fiber insulation and light external metal shell and thus not visible. Further, some of these units have heating coils. As their name implies, heating coils provide a heat source that allows the cargo tank to transport viscous liquids such as asphalt or molten solids such as sulfur. If not for these coils, the liquids would be too viscous and the molten solids would solidify, preventing their unloading.

The vast majority of these units have capacities ranging from approximately 4,500 gallons to 8,000 gallons. The most common materials of construction for these

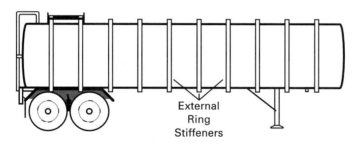

Figure 4-20   Uninsulated DOT 407 Cargo Tank

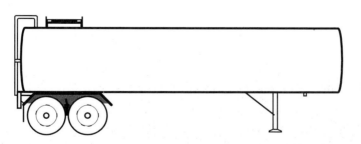
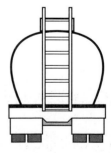

Figure 4-21   Insulated DOT 407 Cargo Tank

large capacity tanks are austenitic SS or MS, although HSLA and AL are becoming increasingly popular. Additionally, variable and dual specification cargo tanks are becoming more popular due to their versatility. It is not uncommon to find tanks constructed that can meet both the 406 and 407 specifications or an old 307 reworked to meet the new 406 specifications.

Each of these units must have the following:

1. A name plate and specification plate
2. Internal self-closing stop-valves or an external stop-valve located as close as practicable to the tank wall to stop product discharge
3. One or more pressure relief valves or systems set at no more than 132% of the MAWP which must reset at not less than 108% of the MAWP
4. Circumferential reinforcement including bulkheads, baffles, and/or ring stiffeners
5. Accident damage protection including bottom damage protection, rollover protection, and rear-end protection
6. An emergency shutoff valve located at least 10 feet from the loading/unloading outlet where vehicle length allows, or at the end of the cargo tank furthest away from the outlet
7. If designed to haul flammable, pyrophoric, oxidizing, or poison liquids, a thermally actuated emergency shutoff system
8. Compliant manholes (manways) that are capable of withstanding internal fluid pressure of 40 psig or the test pressure of the tank, whichever is greater

DOT 412 Cargo Tanks

DOT 412 cargo tank specifications are found in #49 CFR part 178.348. Under this text's classification system they are low pressure containers and are used most commonly to transport corrosive or poisonous liquids, generally with high densities. DOT specifies that a 412 must have a MAWP of no less than 5 psig. In actual practice the vast majority have MAWP of 15 psig or greater. If designed for vacuum loading, the MAWP must be 25 psig internal pressure and 15 psig external pressure (the result of a vacuum). Units with a MAWP greater than 15 psig must have a circular cross-section.

Typically, construction of these units involves the use of external ring stiffeners. They usually have relatively small diameters (4 to 6 feet) due to the density of the lading they transport. In some cases they may have insulation and thus have the horseshoe shape typical of insulated cargo tanks (Figure 4-22).

The most common example of the 412 cargo tank is the standard acid tanker. Such units generally have one compartment and capacities ranging from approximately 2,500 gallons to 5,000 gallons. Lading capacity is relatively low due to the high density of these products. Many of the acids weigh from 12 pounds to over 16 pounds per gallon.

The materials of construction for these cargo tanks generally are resistant to the attack of the lading. SS is the most resistant and a very commonly used material of construction. However, the use of AL, MS, or HSLA is allowable for various ladings if the

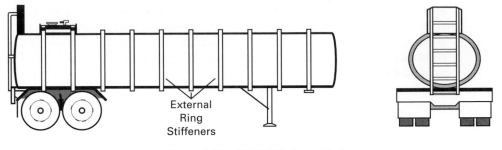

Figure 4-22   DOT 412 Cargo Tank

tank has a compatible liner or cladding installed. It is noteworthy that certain lading require even SS tanks be lined or clad.

Each of these units must have the following:

1. A name plate and a specification plate
2. Internal self-closing stop-valves or an external stop-valve located as close as practicable to the tank wall to stop product discharge
3. One or more pressure relief valves or systems set at no more than 132% of the MAWP which must reset at not less than 108% of the MAWP
4. Circumferential reinforcement including bulkheads, baffles, and/or ring stiffeners
5. Accident damage protection including bottom damage protection, rollover protection, and rear-end protection
6. An emergency shutoff valve located at least 10 feet from the loading/unloading outlet where vehicle length allows, or at the end of the cargo tank furthest away from the outlet
7. If designed to haul flammable, pyrophoric, oxidizing, or poison liquids, a thermally activated emergency shutoff system
8. Compliant manholes (manways)

MC 331 Cargo Tanks

MC 331 cargo tank specifications are found in #49 CFR part 178.337. Under this text's classification system they are high pressure containers and are most commonly used to transport liquefied compressed gases. These gases include various Division 2.1, 2.2, and 2.3 gases such as liquefied petroleum gas (LPG), propane, butane, anhydrous ammonia, anhydrous hydrogen chloride, methyl chloride, sulfur dioxide, and chlorine. The design pressure (identical to MAWP) cannot be less than 100 psig or greater than 500 psig.

The materials of construction allowable for these units are steel or aluminum. If made of aluminum, construction of the tank must include insulation as well as compatibility with the intended lading. The most common type of steel used in these cargo tanks is QT HSLA because of its ability to meet impact test requirements. The minimum shell and head thickness must be 0.187 inches for steel and 0.270 inches for aluminum. The shell and head require additional thickness for the transportation of corrosive materials such as sulfur dioxide and chlorine.

The tank (Figure 4-23) requires the protection of pressure relief valves located on the top or heads of the tank. If insulated, the insulation requires coverage with a reflective, bright, nontarnishing metal such as AL or SS. If noninsulated, the tank requires painting with white, aluminum, or a similar reflective color on the upper two-thirds. Cargo tanks designed to transport chlorine, the refrigerated liquids carbon dioxide or nitrous oxide, require insulation. In the case of chlorine, the insulation must be four inches thick and composed of cork or polyurethane foam.

Tanks designed for chlorine or anhydrous ammonia service must be heat treated after welding to improve the strength of the weld heat affected area (see Section V). Manufacturers must follow ASME code requirements in the heat treating. The tank metal temperature for anhydrous ammonia units must reach at least 1,050°F.

All product loading and unloading outlets require protection with some type of internal product flow valve and must be closed with a plug, cap, or bolted flange. The options for protecting product discharge openings include either excess flow valves or internal self-closing stop valves. The options for protecting product inlet valves include either check valves (directional flow valves) or internal self-closing stop valves. Each internal self-closing, stop valve, or excess flow valve must automatically close if any of its attachments shear off, or if any attached hoses or piping separate.

Further, the location of each of these valves either must be fully internal to the tank or in a welded nozzle that is an integral part of the tank. Tanks intended for the transportation of flammable liquids, flammable gases, hydrogen chloride (refrigerated liquid), or anhydrous ammonia require the use of remote controlled internal self-closing stop valves (e.g., emergency shutoff valves).

On tanks with a water capacity over 3,500 gallons, each internal self-closing stop valve must have two thermal and mechanical remote emergency shutoff valves compatible with the product and unaffected by environmental or weather conditions. The required location of the emergency shutoff valves is diagonal to the ends of the unit. If the loading/unloading location is not in the general vicinity of these shutoffs, the regulations require an additional fusible element that actuates in case of fire.

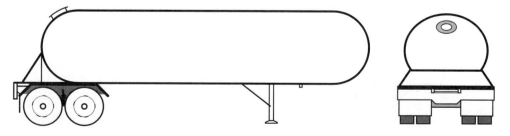

Figure 4-23   MC 331 Cargo Tank

On tanks with a capacity of 3,500 gallons or less, the regulations require the installation of at least one remote emergency shutoff at the spot farthest from the loading/ unloading location. The loading/unloading area is where hoses or hose reels connect to the permanent metal piping.

Each loading and unloading line must have a manual stop-valve as close as practicable to the tank. If an internal self-closing stop valve is used, the manual stop valve location may be on the piping before any hose connection.

Each of these units must have the following:

1. A name plate and specification plate
2. One or more pressure relief valves set to prevent a pressure of more than 120% of the MAWP
3. Accident damage protection including bottom damage protection, rollover protection and rear-end protection
4. Compliant manholes (manways)

MC 338 Cargo Tanks

The final DOT cargo tank is the MC 338 cargo tank. This unit's specifications are found in #49 CFR part 178.338. Under this text's classification system they are high pressure containers, and are most commonly used to transport cryogenic liquids (i.e., refrigerated liquids). Cryogenic liquids are liquids that have a boiling point lower than −130°F.

Each of these tanks (Figure 4-24) must include an inner tank (vessel) enclosed within an outer jacket. There must be some form of insulation present between the tank and jacket. This insulation can be in the form of an insulating material or a vacuum-insulating system. These units are often compared to giant thermos bottles. The most common materials of construction for these tanks are low temperature steels (commonly SS) or AL. The allowable materials of construction for the jackets are SS, MS, or AL.

All materials of construction must meet ASME code requirements or American Society for Testing and Materials (ASTM) specifications. The shell or head can have a minimum thickness of 0.187 inch for steel and 0.270 inch for AL. For tanks constructed

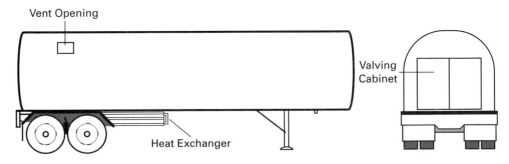

Figure 4-24   DOT MC 338 Cargo Tank

of steel and vacuum insulated, or double walled with a load bearing jacket, the minimum shell and head thickness can be 0.110 inch.

Each tank is designed for a specific cryogenic liquid and is based upon the boiling point of the liquid and its chemical properties (e.g., corrosivity, flammability, and oxidizing potential).

For tanks designed for oxygen or flammable cryogenic liquid services, each tank must have one or more pressure relief devices (spring-loaded) and a secondary system of one or more frangible disks (rupture disk) or pressure relief valves. Other cryogenic liquids only require spring-loaded pressure relief devices or combinations of spring-loaded and frangible disks. The relief devices for oxygen and flammable cryogenic liquid services must actuate at no more than 110% of the MAWP. All others must actuate at no more than 150% of MAWP.

Beside the mandated pressure relief systems, tanks (other than for carbon monoxide) can, and most commonly do, have optional pressure relief systems. These optional systems most commonly actuate well below the MAWP. This setting is often in the range of 14 to 25 psig. Additionally, they may have frangible disks set no lower than 150% and no greater than 200% of the MAWP.

Vacuum-insulated tanks required protection by a suitable device to relieve internal pressure. This relief device must function at a pressure below the design pressure of the jacket or 25 psig, whichever is less.

The MC 338 does not require the use of excess flow valves. Each liquid load or unload line must have a shutoff valve as close to the tank as practicable. Those units designed for flammable lading service must have a remote control shutoff valve. If there is no vacuum jacket, this valve must be internal. If there is a vacuum jacket, this valve must be as close as practicable to the tank.

On units with capacities greater than 3,500 water gallons, each remote control shutoff valve (emergency shutoff) must be diagonally at opposite ends of the unit. This valve must actuate mechanically and thermally at a temperature no greater than 250°F. Tanks with capacities of 3,500 gallons or less, must have one emergency shutoff on the side opposite the loading/unloading location. Again, the shutoff must be both mechanical and thermal.

Each unit must have a pressure gauge located at the front end of the tank. This pressure gauge must be positioned so the driver can read the gauge in the rear view mirror of the tractor. The gauge must have a reference mark to indicate the MAWP, the pressure relief valve setting, or the pressure control valve setting, whichever is lower.

All loading and unloading lines must have permanent markings that indicate whether they are liquid or vapor lines. To determine if a line is for liquid or vapor, the unit must be filled to its maximum permitted fill density.

Each of these units must also have the following:

1. Name plate and specification plate
2. Accident damage protection including bottom damage protection, rollover protection, and rear-end protection
3. Compliant manholes (manways)

This concludes our discussion of highway containers. In the next section, we will discuss rail containers.

GATHERING INFORMATION—SECTION III
CONTAINER INFORMATION
Module IV—Rail Containers

Rail transportation containers come in the same general types as highway containers. Their specific design and construction find their basis in the physical and chemical properties of their ladings. However, there are several major differences.

First, and most obviously, rail containers generally have significantly larger capacities than do highway vehicles. Most are capable of hauling weights ranging from about 50 to 130 tons. Before 1970, water gallon capacities ranged from about 4,000 to 45,000 gallons (in jumbo tank cars). However, since that time, the maximum capacity has decreased to 34,500 gallons. In any event, most rail containers present the potential for a major release when involved in an accident.

Second, since a train, by its nature, consists of multiple containers, an accident can easily involve many containers at one time. It is not uncommon for trains to contain over 100 cars. Low speed accidents usually result in simple derailments or minor rollovers that normally keep cars in their original alignment. As a result, it is usually not difficult to identify cars from their reporting marks and numbers, and the consist. In low speed derailments, it is possible for the rail line to rerail the cars by simply pulling the cars over a rerailer.

However, in high speed accidents, the cars very commonly "accordion" and pile on top of one another. In such situations, the last car in front of and the first car behind the derailment require identification. When these cars are identified, they are then located on the consist. All cars in between on the consist are part of the pile.

Third, with the exception of locomotives, the wheel assemblies attach to the cars by gravity and a king pin (Figures 4-25 and 4-26). This means that during rollover situations, the wheel assemblies, also known as trucks, separate from the car. The result can be a mass of truckless cars. To re-rail the cars, the trucks must first be re-railed and then the car lifted and placed back onto the truck.

Finally, the rail power unit, i.e., the locomotive or engine, has its own unique and potentially substantial hazards. The power for all locomotives comes from large, powerful electric motors that normally operate on 600 V. In rail passenger service, overhead electric cables or an extra (third) rail located between the two wheel rails supply the electricity. Obviously, the cable or third rail pose an extreme electrical shock hazard when energized.

However, for yard or over-the-road engines, electricity comes from an onboard diesel-powered electric generator. A diesel engine drives the generator that provides electricity to a bank of storage batteries. Herein lie two primary hazards: fuel and elec-

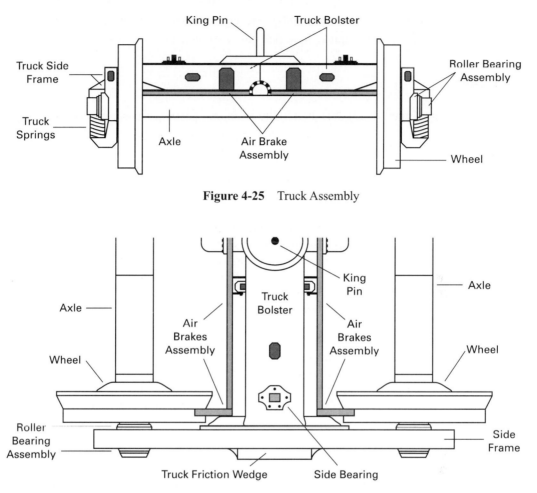

Figure 4-25   Truck Assembly

Figure 4-26   Top View of Truck Assembly

tricity. These engines have diesel fuel tanks with capacities ranging from 2,000 to 6,000 gallons. Further, the location of these tanks is under the engine and between the wheels. Also since these units have banks of electrical storage batteries, there is a constant electrical hazard even when the diesel is not operating.

The following is a list of the general types of rail containers and their primary uses:

1. Box cars haul small to large containers and crates.
2. Gondola cars haul loose, heavy, dry bulk materials that can be exposed to the elements.
3. Hopper cars haul dry bulk materials of relatively fine size and those that require protection from the elements.

4. Flat cars haul any type of large packaging or irregular-shaped objects.

5. Tank cars haul liquids, liquefied gases, molten solids, and so on.

BOX CARS

Box cars (Figure 4-27), like their highway counterparts, are a primary workhorse of rail transportation. These cargo boxes can hold almost any type of container or commodity. However, unlike box trailers that often contain mixed loads, box cars most commonly carry only one product at a time. There is a wide range of possible contents including household appliances, drums of chemicals, palletized paper bags, bales of cotton, crates of machinery, and similar cargo.

One rather common hazard associated with box cars is fire. When the tightly packed contents of a box car ignite, the resulting fire can be extremely hot, intense, and difficult to fight. Further, due to the tight seals on most box cars, backdraft explosion is always possible, especially if exterior thermal discoloration is present. Once identification of the car's contents occurs and it is determined that fire extinguishment is appropriate, an indirect attack is generally the safest and most effective method. Such an indirect attack most commonly involves the use of penetrating nozzles used high on the sides or top of the car.

Another major type of box car is the refrigerated unit or "reefer." These units carry perishable goods such as fresh produce. All new reefers have efficient insulation, most commonly foam. For short duration trips, they may be precooled with liquid nitrogen before loading. Others have their own diesel-powered refrigeration systems. These diesel units can have fuel storage tanks with capacities up to 500 gallons.

FLAT CARS

Most commonly, flat cars transport cars, irregular freight crates, intermodal tanks, trailers on flat car (TOFC), or containers on flat car (COFC). Obviously, the intermodal tanks, TOFCs and COFCs have the potential for containing hazardous materials.

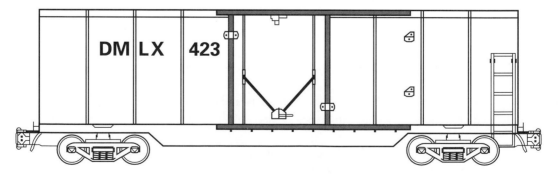

Figure 4-27 Standard Box Car

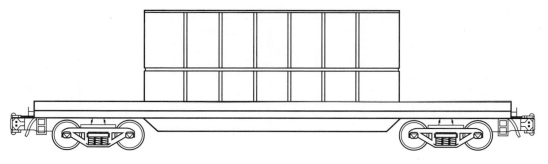

Figure 4-28   Double High Container on Flat Car (COFC)

Some interesting wrinkles have been added to COFCs over the past few years. First, many are shipped in the western and midwestern parts of this country with two containers stacked on top of each other. This configuration is known as "double highs" (Figure 4-28). Second, a new design of the flat car has also developed. In this particular case, a single flat car is composed of as many as five, articulating flats attached to each other. Each end of the flat has a set of trucks.

However, at the juncture of each articulating section, one set of trucks is shared by the "B-end" of the first and the "A-end" of the next flat. The result is rather strange looking. Also, each section may or may not have visible identification numbers. If they are visible, they often have the number followed by the letter that corresponds to its location within the system (e.g., GATX 1234A, GATX 1234B, and GATX 1234C).

GONDOLA CARS

Gondola cars (Figure 4-29) have low sides and open tops. They are used to transport heavy, loose, bulk materials, including coal, scrap iron, raw glass, and iron pellets. Since they have open tops, their lading is open and exposed to the elements. For speedy unloading, a rotating rail platform can invert these cars.

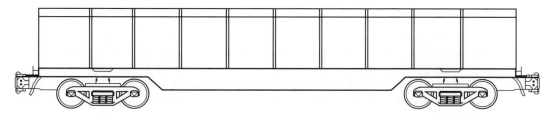

Figure 4-29   Gondola Car

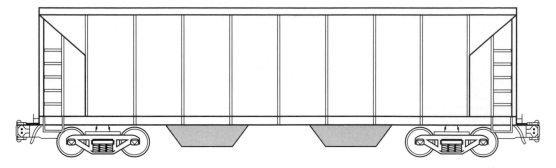

Figure 4-30   Hopper Car

HOPPER CARS

Hopper cars (Figure 4-30) are full height cars that contain sloped floors for bottom un-loading. The sloped floors may or may not continue below the bottom line of the car. When they do continue, they have the appearance of a funnel at the bottom of the car. Most commonly hoppers transport dry bulk materials such as wood chips, cement, pel-letized chemicals (e.g., sodium hydroxide), fertilizers, grains, and plastic resins.

One version of these cars hauls wood chips or other lightweight lading. These cars have open tops that may be covered by a tarp or similar material. As a result, they are sometimes called open hoppers. For unloading, these cars can be equipped with cou-plers that rotate. In this case, the car is positioned over a rotating section of track where it is clamped down. At that point, the track rotates and the lading simply falls from the open top of the car.

PNEUMATIC DRY BULK CARS

Pneumatic dry bulk cars (Figure 4-31) are similar to hoppers in that they are closed units with sloping floors. However, unlike hoppers, dry bulks have rounded as opposed to flat sides. As a result, their appearance can be confused with that of standard tank cars. Under some classification systems, the terminology, "special tank cars" is used to describe these pneumatically loaded/unloaded dry bulks.

Since the lading of these cars is generally fine particulate, powder, pellet, or similar physical form, pneumatic loading and unloading fixtures are common. Gener-ally, loading takes place through a series of hatches located on top of the hopper. By mixing the lading with pressurized air, the product moves like a liquid. The name for this type of loading process is *fluid bed product handling*. The reason for this name is that the solid material behaves like a liquid. During unloading, pressurized air moves through piping attached to the funnels. As lading pours from the funnel, a fluid bed results.

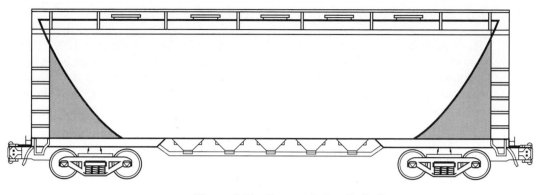

Figure 4-31   Pneumatic Dry Bulk Car

Even though fluid bed product handling is a highly safe and efficient way to move finely divided solids, it represents several potential hazards even when no hazardous substances are involved. First, the process inherently involves a finely divided solid mixed with a gas, most commonly, air. Such conditions are ideal for the development of a dust explosion inside or outside the system if the solid is combustible. Second, the movement of the solid through piping entrapped in the air stream often generates massive static electrical charges. If there is a problem with the bonding or grounding apparatus within the system, a static discharge spark could lead to a dust explosion.

TANK CARS

According to the Association of American Railroads, as of 1994 there were approximately 213,000 rail tank cars in the North American fleet. There are two primary categories used by the railroads to identify tank cars: pressure and nonpressure cars. Some also use the term specialized cars to describe a very small number of other tanks cars.

DOT uses these five categories, pressure cars, nonpressure cars, multi-unit tank cars, cryogenic liquid cars, and seamless steel tanks. The multi-unit, cryogen, and seamless steel cars are also sometimes called specialized cars. Of all these car types, approximately 76% are nonpressure cars, approximately 23% are pressure cars, and less than 1% are specialized cars.

The criterion by which the railroads differentiate between pressure and nonpressure cars is their test pressure. Nonpressure cars have tank test pressures of 100 psig or less, while pressure cars have tank test pressures greater than 100 psig. Specialized tank cars vary significantly. Using the pressure rating system explained earlier in this unit, nonpressure cars would fall into the category of low pressure containers, and the pressure cars would fall into the category of high pressure containers. For clarity, this section will follow the standard DOT terminology.

DOT Specification Numbers and Car Stencils

The regulations governing railroad tank cars are found in 49 CFR Part 179. All tank cars designed to transport hazardous materials and meet DOT regulations use a series of number and letter combinations to identify their specification. When stenciled on a tank car, these numbers and letters are part of the specification marks. The following are three examples of this type of marking:

1. DOT 105A300W
2. DOT 103W
3. DOT 111A100W4

Each of the letters and digits provides specific information based upon the specific class of the car to which it applies. The class of a car is a general type of car design. The primary DOT classes are:

DOT 103	DOT 104	DOT 105	DOT 106	DOT 107	DOT 109
DOT 110	DOT 111	DOT 112	DOT 113	DOT 114	DOT 115

Within a given class, there are multiple individual car specifications that are possible. The exact specifications required for a given car vary depending upon its intended use, the properties of the intended lading, allowable construction, and attachment options (e.g., top and bottom shelf couplers, thermal protection, and bottom outlets).

The letters "DOT" indicate that the car meets a specific DOT class. The next three numbers indicate the exact class of the car (e.g., *DOT* 103, *DOT* 112). In many instances, a second number follows the first number with a letter as separation. These letters are important in DOT Class 105, 111, 112, 113, and 114 cars. The explanation of each class follows this section.

The second number, if present, indicates the tank's test pressure (the pressure at which the tank undergoes hydrostatic testing). For example, a DOT 111A*100* has a test pressure of 100 psig.

The specification marking ends with a single letter or multiple letters and a single digit. When the specification ends with "W," it indicates the car is relatively new and constructed with the use of fusion welding. Fusion welding is the modern welding technique that uses an electrode rod to melt the base metal and the rod, thus joining the two pieces of shell plate.

When the specification ends with "F," it indicates the car is older and constructed with the use of forge welding. Forge welding is the old welding technique involving the heat softening (red hot) and then pounding of the pieces until they join. Since these cars are old, they have undergone modifications to meet newer DOT specifications. When there are multiple letters followed by a number, these markings indicate materials of construction and the presence of fittings, linings, and other protective measures.

As one faces the side of a tank car, the specification marks are always displayed on the right end. The minimum data found as part of the specification marks include the following:

1. The DOT class specification
2. The pressure setting of the relief device on the car
3. The test date of the device and the next test date
4. The test date of the car and the next test date

There can be additional information included as part of the specification marks. This additional data can include the following:

1. AAR or ICC specification
2. Proper shipping names or limitation stencil
3. Inhalation hazard stencil
4. Lining information stencil

AAR or ICC Specifications.   There are two other specification systems that exist for tank cars. However, cars that meet only these specifications can transport only nonregulated materials. These include the old Interstate Commerce Commission (ICC) and the Association of American Railroads (AAR) specifications. The specification marks of the cars start with AAR or ICC instead of DOT and generally the class numbers are in the 200s. In a few cases, the specification marks may indicate compliance with DOT and AAR, with the AAR marks directly below the DOT marks.

Other Stencils.   The proper shipping names of approximately 70 commodities must be stenciled on the tank car. In other instances, non-mandatory stencils indicate allowable or non-allowable lading (e.g., "Not For Flammable Liquids" or "For Sulfuric Acid Only").

Commodities identified as poisons by inhalation in Table 172.101, Special Provisions Column (7), [also see 173.116(a) and 173.133(a)] must stencil the term "Inhalation Hazard" on both sides of the car, in letters at least 3.9 inches in height. Most commonly, the location of this stencil is directly above the DOT car specification and accompanies the proper shipping name.

Finally, some car owners include information about linings as well. This information often includes the type of lining, special handling needs, date of installation, testing dates, and so on. Although this information is not mandated, it is obviously of value in the proper usage and handling of the car.

Reporting Marks

On the opposite end of the side from the specification marks will be the reporting marks (Figure 4-32). The reporting marks consist of the initials of the company that owns the car (maximum of four letters). The letter X indicates that the car's owner is someone other than the railroad. A series of numbers (maximum of six) follows these letters to

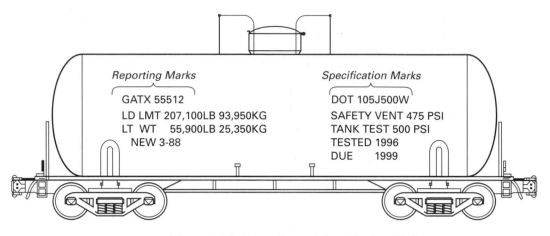

Figure 4-32   Reporting and Specification Marks

identify the exact car. The latest wrinkle in reporting marks is their conversion into bar codes. These bar codes are then read by a computer system at the classification yard and used to properly route the cars. The reporting marks are located on both sides and each end of the car. Bar codes are in the area of the specification marks.

Included with the reporting marks are capacity marks. Capacity marks indicate the weights of the car and its volumetric capacity. Found on either side of the car is the maximum load limit (LD LMT on the graphic). This weight indicates the gross vehicle weight of the car. Under this marking is the light or tare weight (LT WT on the graphic). The light weight is the weight of the empty car. As in the case of the intermodal tanks, when the light weight is subtracted from the load limit, it indicates the maximum weight of the lading.

Found on either end of the car is the volumetric capacity in both gallons and liters. Such information along with the weight can be invaluable when personnel try to estimate the amount of the release, the potential spread of lading, size of dike areas, capacity of retentions, and so on.

Since the reporting marks help with car identification, they also can help identify its contents by locating it on the consist. If the consist is unavailable, the contents of the car can be identified by the computer database of the carrier (i.e., railroad) or possibly by the owner.

New Versus Existing Tank Cars

The primary tank car specifications identified in this text are from 1997. However, there are many tank cars constructed before this time that are still in use. As such, their original construction is often not compliant with today's more stringent requirements. Because of the lessons learned from routine use, accidents, and improved technologies, DOT often requires car owners to upgrade or modify their cars. These upgrades help establish the new standard of compliance.

For example, a situation known as coupler override was a primary contributor to BLEVEs and other types of tank car breaches. In an override, the coupler of one car literally slides above or below the coupler of the next car. As a result, a coupler hits the end of the car (the head) causing potentially severe damage that could lead to catastrophic failure of the tank.

To prevent such occurrences (Figure 4-33), DOT now requires all new tank cars be equipped with top and bottom shelf couplers (49 CFR 179.14). DOT also requires specific existing tank cars in Classes 105, 111, 112 and 114 be equipped with head puncture protection (e.g., head shields). Further, DOT requires designated existing cars that carry certain flammable liquids and liquefied compressed gases be equipped with thermal protection. These modifications or retrofit programs have been highly successful and have led to the dramatic decrease in BLEVEs over the past decade or so.

General Terminology

As with the other modes of transportation, rail has certain unique terminology that it uses to describe appurtenances on its containers. Two terms that are often confused are insulation and thermal protection. The difference between the two is not the mechanical methods used but rather the intended protection provided. Insulation acts to prevent normal atmospheric temperatures from adversely affecting the contents of the car. The insulation requires the protection of a metal jacket of not less than 11 gauge. Thermal protection acts to protect the car from the heat that results from a fire. This heat can come from an engulfing pool fire or direct flame impingement. Generally, insulation will also perform the function of thermal protection. Obviously, a tank car protected with insulation or thermal protection is less subject to the effects of a fire situation.

Openings on the bottoms of tank cars present potential problems if there is an accident, because they are potential avenues through which product can escape. As a result, it is important to understand the uses and differences between the two primary types of

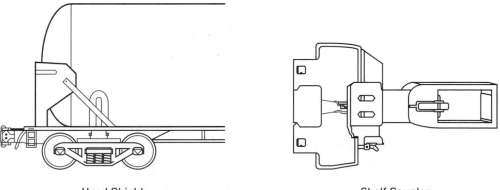

Head Shield   Shelf Coupler

Figure 4-33   Accident Damage Protection

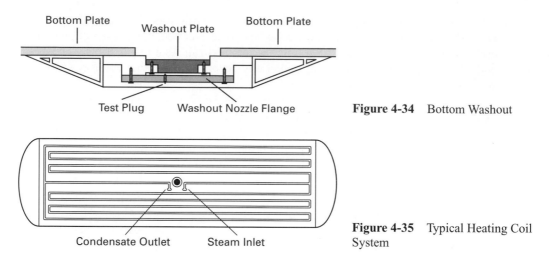

Figure 4-34   Bottom Washout

Figure 4-35   Typical Heating Coil System

bottom openings. Bottom outlets are valve controlled openings in the bottom of a tank car that are designed for the loading or unloading of the car. Bottom washouts (Figure 4-34) are openings in the bottom of a tank car that allow condensate, water, or other cleaning and purging substances to be removed. An analogy is the drain plug in a sink. Generally, washouts have a plate bolted to a flange, equipped with some type of gasket material to assure a seal.

Heating coils (Figure 4-35) are pipes located on the interior or exterior of the tank. A heating medium such as steam or hot water flows through the piping to heat the lading in the car. Such lading includes extremely thick and viscous liquids like molasses or asphalt, or solids such as sulfur or phosphorous. The location of the steam inlet and outlet is normally on the bottom, often on either side of the bottom outlet.

NONPRESSURE TANK CARS

The three primary types of nonpressure cars are the DOT 103 Class, DOT 104 Class, DOT 111, and the DOT 115 Class cars. Class 103 and 104 make up about 11% of the nonpressure car fleet, Class 115 about 1%, Class 111 cars the remaining 88%. Nonpressure cars transport substances that are normally liquids under ambient conditions, highly viscous liquids, molten solids, and slurries (viscous mixture of powder and liquid). These products include the following:

1. Food grade liquids (e.g., corn syrup, molasses, vegetable oil)
2. Acids and bases
3. Molten sulfur, phosphorous, sodium, and so on
4. Phosphorous/water slurry
5. Flammable and combustible liquids
6. Poisons
7. Oxidizers

All the nonpressure cars have manways (hatches), loading/unloading systems, and pressure relief devices. Insulation is optional for most nonpressure cars with only four out of twenty-four specifications requiring insulation. With few exceptions, all valving, relief devices, and manways are visible on the top portion of the car. Further, Class 103, 104, and 111 can have up to six individual compartments, each with its own loading and unloading fittings.

Depending upon the exact specification, some of these tanks have either top, bottom, or both top and bottom loading and unloading systems. For top side applications, the outlet may or may not have a built-in valve. Rather they may have an outlet that ends with a blank cover plate attached to the outlet's flange. During unloading, the blank cover plate is removed and a valve or an unloading hose is bolted in place. In other instances, the outlet is secured by a pipe nipple. Generally, when there is a nipple, there is also a valve.

Commonly, the location of top loading or unloading outlets is in some type of enclosure. Sometimes the enclosure resembles a mail box with hinges that opens from the side. In other cases the enclosure is simply a rounded cylinder, while in still other cases, the covering is nonexistent. In such a situation the top side inlets and outlets are attached directly to the cover of the manway.

Top unloading systems have a pickup tube that runs to within an inch or so of the bottom of the car. Since these are nonpressure cars, this type of unloading requires some kind of energy to force the lading out of the tank. Most commonly, the energy takes the form of pressure from a house air system. As a result, top unloading systems normally have an air coupling to pressurize the tank for unloading.

Bottom unloading systems have outlets equipped with valves. The location of the valve is at the outlet. However, there are two optional types of valves (Figure 4-36). The first type is a simple ball valve located in the outlet. In this case, the valve handle

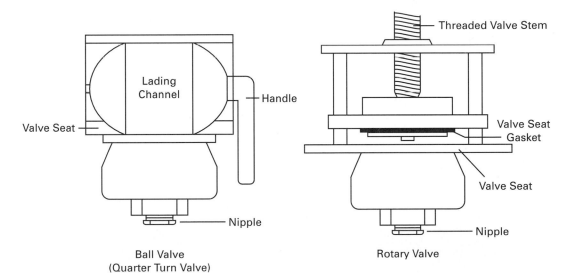

Figure 4-36   Ball and Rotary Outlet Valves

protrudes from the side of the outlet and functions as a quarter turn valve. Recall that quarter turn valves are open when their handle is in line with the piping. By simply rotating the valve handle to the perpendicular (turning it 90° or a quarter turn), the valve is closed.

The second type of valve is a rotary valve, sometimes called a "top operated bottom valve." This type of valve is similar to a garden hose valve. In this case, the valve has a threaded stem. At one end is the handle; at the other is a valve seat gasket. Both are perpendicular to the stem. Since the valve is at the bottom of the tank, the stem runs from the top to the bottom of the tank. At the other end of the stem, located at the top of the car, is a cover called a stuffing box. The stuffing box is the handle for the valve when it is removed and inverted.

DOT CLASS 103 AND 104 CARS

Class 103 (Figure 4-37) and 104 cars are considered together because they are very similar. There are nine 103 specifications and one 104 specification. All ten of these tank car specifications must have an expansion dome. An expansion dome is a relatively large cylindrical structure located on top of the car and designed to allow for thermal expansion of the lading. The capacity of the expansion dome must be 1–2% of the tank's capacity. The materials of construction for these cars include steel and steel alloys, aluminum alloy, or nickel. The test pressure for all of these cars is 60 psig.

All tank cars, pressure as well as nonpressure, require protection against excessive internal pressure. Tank car pressure relief devices fall into three primary categories: pressure relief valves, fusible plugs, and pressure relief vents. Pressure relief valves (called valves in 49 CFR) are spring-loaded devices where a spring pushes the valve against its seat. As the pressure within the tank increases, it overcomes the force of the spring and the valve leaves its seat. At that point, pressure escapes from the tank. When

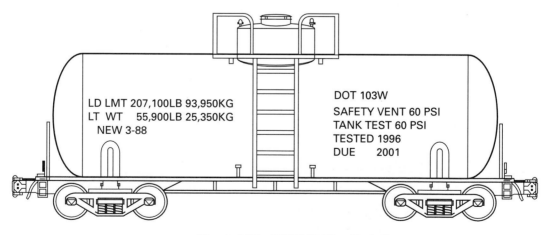

Figure 4-37   DOT 103 Class Tank Car

Table 4-19 Minimum Plate Thickness for 103 and 104 Tank Cars*

Inside tank diameter-ins.	Bottom sheet	Shell sheet	Expansion dome sheet	2:1 ellipsoid heads	3:1 ellipsoid heads	Int. compart-ment heads
60 and under	7/16″	1/4″	5/16″	7/16″	1/2″	5/16″
Over 60 to 78	7/16″	5/16″	5/16″	7/16″	1/2″	5/16″
Over 78 to 96	1/2″	3/8″	5/16″	7/16″	1/2″	3/8″
Over 96 to 112	1/2″	7/16″	5/16″	7/16″	9/16″	7/16″
Over 112 to 122	1/2″	1/2″	5/16″	1/2″	5/8″	1/2″

*NOTE: All the specifications listed are from 1996 standards.

enough pressure escapes, the spring reseats the valve. This is why this type of relief valves is referred to as self closing.

Fusible plugs (as well as fusible links) are weak elements within the container, designed to fail when heated to a specified temperature. Their construction generally involves either a heat sensitive plastic or lead-like alloy that will melt at relatively low temperatures (well below other parts of the tank). When the container is exposed to heat, the plugs melt, thus controlling internal tank pressure. It is important to note that once activated, a fusible plug cannot close itself.

Finally, pressure relief vents are also weak elements within the container, designed to fail mechanically at a predesignated pressure. Their construction generally involves lead or plastic membranes placed over a specially designed opening in the tank. When the pressure exceeds the rated pressure of the disk, it tears and fails, thus controlling internal tank pressure. These vents are also called frangible disks or rupture disks. It is important to note that once activated, pressure relief vents also cannot close themselves.

To reiterate, all cars must have either a pressure relief valve or vent. All relief valves (e.g., spring loaded, self closing valves) must start to discharge at 35 psig and relief vents (e.g., frangible or rupture disks) must rupture at 60 psig. The only exception is the 103CW tanks which must have a relief valve only with no option for a vent. Gaging devices are optional on all Class 103 and 104 tank cars.

Class 103 and 104 cars commonly haul hazardous materials. The most common types of lading include corrosives, some flammable or poisonous liquids, and oxidizers. It is allowable for Class 103 cars to transport non hazardous materials, but it is not as common. The minimum plate thicknesses are listed in Tables 4-19, 4-20, and 4-21.

DOT 111 CLASS CARS

The 111 Class cars (Figure 4-38 shown on page 136) make up approximately 87% of the nonpressure cars in the North American fleet. As such, there are over 20 different individual specification options for this class. Their materials of construction include

Table 4-20   103 Class Tank Cars

Requirements	103W	103AW	103ALW	103A-ALW	103ANW
Materials of construction	Steel	Steel	AL alloy	AL alloy	Nickel
Insulation	Optional	Optional	Optional	Optional	Optional
Bursting pressure	240 psi	240 psi	240 psi	240 psi	240 psi
Min. thickness shell head	Based upon tank diameter (see table)	Based upon tank diameter (see table)	1/2 inch 1/2 inch	1/2 inch 1/2 inch	Based upon tank diameter (see table)
Min. expansion dome capacity	2% minimum	1% minimum	2% minimum	1% minimum	1% minimum
Test pressure	60 psi	60 psi	60 psi	60 psi	60 psi
Safety relief device	Valve or vent	Vent, valve if lading requires	Valve or vent	Valve or vent	Vent, valve if lading requires
Top load/unload	Optional	Required, valves optional	Optional	Required, valves optional	Required, valves optional
Bottom outlet	Optional	Prohibited	Optional	Prohibited	Prohibited
Bottom washout	Optional	Optional	Optional	Optional	Optional
Manway closure	No-open when under pressure	X	No-open under pressure	X	Cover made of nickel
Postweld heat treatment	Required	Required	Prohibited	Prohibited	Not required
Special require/ references	X	X	X	X	X

aluminum alloy, steel, or steel alloy. Some must have insulation while others must have a lining. Some may have bottom outlets while others may not. Those with bottom outlets can have a stuffing box, which is simply a valve stem extending from the top of the car to the bottom outlet. Unlike 103 or 104 Class cars, 111s have no expansion domes.

As the numbers indicate, 111 Class tank cars are the workhorse of the tank cars. They commonly transport hazardous as well as nonhazardous materials. These cars also carry many food-grade substances such as molasses, corn syrup, vegetable oil, fruit juice, whiskey, and caramel. Hazardous materials include liquid corrosive (e.g., acid and bases), flammable and poisonous liquids (e.g., ethyl ether and benzene), oxidizers (e.g., hydrogen peroxide and fuming nitric acid), and flammable solids (e.g., phosphorous and sodium). All must have a 2% expansion capacity.

There are also some older 111 Class cars that have been fitted with various protection systems including jacketed thermal protection, puncture resistant tank head shields (head shields), and top and bottom shelf couplers. These units are designated as DOT

Table 4-21 103 and 104 Class Tank Cars

Requirements	103BW	103CW	103DW	103EW	104W
Materials of construction	Steel	Alloy steel	Alloy steel	Alloy steel	Steel
Insulation	Optional	Optional	Optional	Optional	Optional
Bursting pressure	240 psi	240 psi	240 psi	240 psi	240 psi
Min. thickness shell head	Based on tank diameter (see table)	Based on tank diameter (see table)	Based on tank diameter (see table)	Based on tank diameter (see table)	Based on tank diameter (see table)
Min. expansion dome capacity	2% minimum	1% minimum	2% minimum	1% minimum	1% minimum
Test pressure	60 psi	60 psi	60 psi	60 psi	60 psi
Safety relief device	Valve or vent	Valve	Valve or vent	Valve or vent	Valve or vent
Top load/ unloading	Required, valves optional	Required, valves optional	Optional	Required, valves optional	Optional
Bottom outlet	Prohibited	Prohibited	Optional	Prohibited	Optional
Manway closure	Metal with acid protection	Metal ring and cover	Metal/no-open under pressure	Metal ring and cover	No-open under pressure
Bottom washout	Prohibited	Prohibited	Optional	Optional	Optional
Postweld heat treatment	Required	Required	Required	Required	Required
Special require/ references	Rubber lining required	All fitting, etc. AAR Spec.	All fitting, etc., AAR Spec.	All fitting, etc. AAR Specs.	X

111J instead of DOT 111A. Tables 4-22 and 4-23 provide information on 14 present 111 Class specification tank cars. Note that all specifications are 1996 49 CFR requirements.

One of the older classes not identified in the tables is the DOT 111J***W4. The 111J***W4 class tank cars are equipped with thermal protection, top and bottom shelf couplers, and head shields. Bottom outlets and washouts are prohibited.

Some older DOT Class 105 cars have been modified to be DOT Class 111 cars. They include the following:

1. *111A60F1:* steel, a modified DOT 105A300, 105A490, or 105A500 tank (bottom outlets and washouts optional)
2. *111A100F1:* steel, a modified DOT 105A300, 105A490, or 105A500 tank (bottom outlets and washouts optional)
3. *111A100F2:* steel, a modified DOT 105A300, 105A490, or 105A500 tank (bottom outlets and washout prohibited)

Table 4-22 111 Class Tank Cars

Requirements	111A60ALW1	111A60ALW2	111A60W1	111A60W2	111A60W5	111A60W7	111A100ALW1
Mat. of const.	AL alloy	AL alloy	Steel	Steel	Steel	Alloy steel	AL alloy
Insulation	Optional	Optional	Required	Required	Optional	Optional	Optional
Burst pressure	240 psi	240 psi	240 psi	240 psi	240 psi	240 psi	500 psi
Min. thick. shell	1/2"	1/2"	7/16"	7/16"	7/16"	7/16"	5/8"
head	1/2"	1/2"	7/16"	7/16"	7/16"	7/16"	5/8"
Test pressure	60 psi	60 psi	60 psi	60 psi	60 psi	60 psi	100 psi
Relief device	Valve/vent	Valve/vent	Valve/vent	Vent	Valve/vent	Valve/vent	Valve/vent
Valve activates	35 psi	35 psi	35 psi	35 psi	35 psi	35 psi	75 psi
Vent burst	60 psi	60 psi	60 psi	60 psi	60 psi	60 psi	100 psi
Top load/unload	Optional	Required, valve opt.	Optional	Required, valve opt.	Required, valve opt.	Required, valve opt.	Optional
Bottom outlet	Optional	Prohibited	Optional	Prohibited	Prohibited	Prohibited	Optional
Bot. washout	Optional	Optional	Optional	Optional	Prohibited	Prohibited	Optional
Manway closure	No open under pres.	X	No open under pres.	X	Metal ring and cover	Metal ring and cover	No open under pres.
Postweld heat treatment	Prohibited	Prohibited	Required	Required	Required	Required	Prohibited
Special references	X	X	X	X	Rubber lined	Fitting, etc. AAR Spec.	X

Table 4-23 111 Class Tank Cars

Requirements	111A100ALW2	111A100W1	111A100W2	111A100W3	111A100W4	111A100W5	111A100W6
Mat. of const.	AL alloy	Steel	Steel	Steel	Steel	Steel	Alloy steel
Insulation	Optional	Optional	Optional	Required	Required	Optional	Optional
Burst pressure	500 psi	500 psi	500 psi	500 psi	500 psi	500 psi	500 psi
Min. thick. shell	5/8"	7/16"	7/16"	7/16"	7/16"	7/16"	7/16"
head	5/8"	7/16"	7/16"	7/16"	7/16"	7/16"	7/16"
Test pressure	100 psi	100 psi	100 psi	100 psi	100 psi	100 psi	100 psi
Relief device	Valve/vent	Valve/vent	Vent	Valve/vent	Valve	Valve/vent	Valve/vent
Valve activates	75 psi	75 psi	75 psi	75 psi	75 psi	35 psi	75 psi
Vent burst	100 psi	100 psi	100 psi	100 psi	100 psi	100 psi	100 psi
Top load/unload	Required, valves opt.	Optional	Required, valves opt.	Optional, if used, valves required	Required, valves required	Required, valves opt.	Optional, if used, valves required
Bottom outlet	Prohibited	Optional	Prohibited	Optional	Prohibited	Prohibited	Optional
Bot. washout	Optional	Optional	Optional	Optional	Prohibited	Prohibited	Optional
Manway closure	X	No open under pres.	X	No open under pres.	No open under pres.	Metal with protective covering	No open under pressure/metal ring & cover
Postweld heat treatment	Prohibited	Required	Required	Required	Required	Required	Required
Special requirements/references	X	X	X	X	Marked with water weight capacity	Rubber lined	Fitting, etc. AAR Spec.

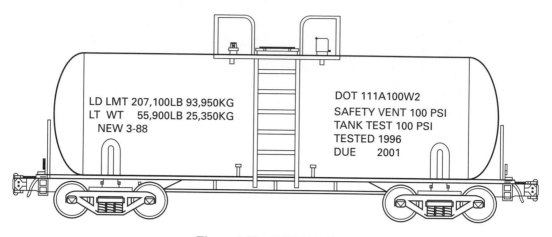

Figure 4-38   DOT 111 Class Tank Car

DOT 115 CLASS CARS

The DOT 115 Class cars are specially designed tank cars. Another name for them is "tank within a tank" because they consist of an inner tank, a support system for that tank, and an outer shell. The space between the inner and outer shell must contain approved insulation. They must have a manway on the tank's top and can have multiple compartments. The materials of construction include steel, steel alloy, or aluminum alloy. The inner tank and the outer shell must have fusion welded joints.

All 115 cars must have stencils that comply with AAR specifications. The stencils must be located near the center of the car and indicate the safe upper temperature limit, when applicable, for the inner tank, insulation, and support system. The stencil shall have letters at least 1.5 inches high. Additionally, cars that transport chloroprene must have "CHLOROPRENE" stenciled on both sides in 4-inch letters. The three 115 Class specifications are listed in Table 4-24.

Most commonly, 115 Class cars transport high-hazard and dual-hazard liquids (e.g., inhibited chloroprene, chlorosilane, ethylene dichloride, ethyl mercaptan, and epichlorohydrin). Many of the listed materials are highly reactive, toxic, and have relatively high vapor pressures. All manway construction must prevent their opening when there is pressure within the tank.

TANK TRAIN SYSTEM

The Tank Train System (Figure 4-39) is the trade name for a unique tank car configuration and loading/unloading system developed by the General American Transportation Corporation. It consists of up to 100 tank cars (generally in the DOT 111 Class) that have been interconnected to allow the loading and unloading of all the cars from

Table 4-24   115 Class Specifications

Requirements	115A60W1	115A60ALW	115A60W6
Inner container mat.	Steel	AL alloy	Alloy steel
Burst pressure	240 psi	240 psi	240 psi
Min. plate thickness	1/8″	3/16″	1/8″
Test pressure	60 psi	60 psi	60 psi
Safety relief device	Valve or vent	Valve or vent	Valve or vent
Valve activation	35 psi	35 psi	35 psi
Vent rupture	45 psi	45 psi	45 psi
Gaging device	Required	Required	Required
Top load/unload	Optional	Optional	Optional
Bottom outlet	Optional, valved	Optional, valved	Optional, valved
Bottom washout	Optional	Optional	Optional

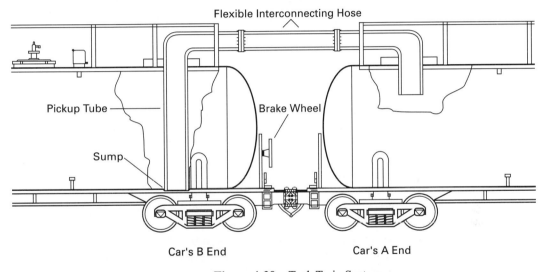

Figure 4-39   Tank Train System

one location. The manufacturer claims a two-member crew can unload up to 90 cars in about three hours by using the basic principles of a siphon. An incident involving such a system could potentially result in a major product release.

PRESSURE CARS

There are four Classes of pressure tank cars: 105, 109, 112, and 114. By far, the least common is 109 Class. All four classes transport liquefied compressed gases and certain volatile liquids. These volatile liquids include substances such as fuming sulfuric acid with 30% or greater sulfur trioxide content, and fuming nitric acids. Also certain substances that present inhalation hazards are transported in these types of cars.

Visual recognition of a pressure car is rather simple. Unlike nonpressure cars, pressure cars have all of their valves, fittings, and relief devices located within a protective housing on top of the car. This protective housing is sometimes called a bonnet. The bonnet most commonly is located at the center of the car and is surrounded by the work platform. This design provides maximum protection for the car's fittings during routine use or during an accident.

Within the bonnet are pressure relief valves, loading and unloading valves, gaging devices, thermometer well, and sampling ports. Relief devices in pressure cars are all pressure relief valves (spring loaded).

There are normally three or four loading and unloading valves in the bonnet, depending on the specific lading intended for use. These valves are called angle valves because they start by coming straight up out of the tank and then make a 90° turn. They may have rotary or quarter turn valves to control the flow of lading.

Valves that run in line with the length of the car are liquid lines because their pickup tubes go to the bottom of the car. There will always be two liquid lines. Valves that run perpendicular to the length of the car are vapor lines because their pickup tubes stop in the vapor space at the top of the car. There will be either one or two vapor lines. If there is only one vapor line, generally a gaging device will be located where the second valve would be located.

All pressure cars and some nonpressure cars have several sampling and measuring devices available. These devices include gauging devices, thermometer wells, and sampling tubes.

Gauging devices are openings into the tank that allow the measurement of product levels. This measurement process is most important during loading operations because the tank must have a specified amount of unfilled space to allow the lading to expand during transport. The common name for this remaining space is *outage*.

There are various types of gauging devices in use, but they all function to identify the product level within the tank. Open gauging devices have direct openings into the tank while closed devices have indirect, mechanical systems for the measurements. Open gauging devices use a graduated slip tube that is similar to a straw. As the tube is lowered into the tank, it spits liquid when the liquid level is reached. The graduation marks on the tube help identify the quantity of lading in the tank. Closed gauging devices have a magnetic float that stays on the surface of the lading and transmits the liquid level to a gauge on top of the car.

Thermometer wells are closed tubes filled with an antifreeze. A thermometer can be placed in the well to take the temperature of the lading. By knowing the temperature of the lading, the pressure can be determined.

Sample tubes are direct openings into the car through which product samples can be taken without having to open any other valves. The sampling tube normally has a small shutoff valve and a pipe nipple through which the product flows.

The allowable materials of construction for 105 and 109 Class cars is either aluminum alloy or steel. The only allowable materials of construction for 112 and 114 Class cars is steel. The 109 and 114 Classes may have optional bottom washouts. Only

114 Class may have bottom outlets. Neither 105 nor 112 Class cars can have bottom washouts or outlets.

DOT 105 CLASS CARS

The allowable materials of construction for 105 Class cars (Figure 4-40) are either aluminum alloy or steel. A typical example of a 105 Class is a chlorine tank car. All specifications for this Class require the use of insulation and prohibit bottom outlets or washouts. As a result, all of these cars have top loading and unloading only. Tank test pressures range from 100 to 600 psi, depending upon the exact specification.

Specification DOT 105A200F is a group of conversion cars. In this case, they are converted from existing forge welded cars. Specifically, their previous specifications were 105A300, 105A400, 105A500, or 105A600. Their joints must meet specific weld requirements and rivets cannot be used as anchors. These tanks must meet all other requirements for Class 105. Finally, they are the only Class 105 cars not required to have a circular cross section. There are ten specifications within Class 105. The specific information about their requirements is found in Tables 4-25 and 4-26. Note that all specifications are 1996 49 CFR requirements.

There are additional modifications to older cars. The following is a key to the additional protective measures taken. The location of the letter indicating the modification is between the class number and the test pressure (e.g., DOT 105J600 W). The following list provides the meaning of each letter.

- "**A**" means the car is equipped with top and bottom shelf couplers.
- "**J**" means the car is equipped with jacketed thermal protection, tank head puncture resistance, and top and bottom shelf couplers.
- "**S**" means the car is equipped with a tank head puncture protection and top and bottom shelf couplers.

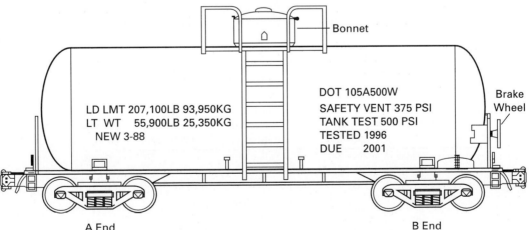

Figure 4-40   Class 105 Tank Car

Table 4-25   Class 105 Tank Cars

Requirements	105A100ALW	105A100W	105A200ALW	105A200F	105A200W
Materials of construction	AL alloy	Steel	AL alloy	Steel	Steel
Insulation	Required	Required	Required	Required	Required
Burst pressure	500 psi	500 psi	500 psi	Conversion	500 psi
Min. plate thickness—all	5/8″	9/16″	5/8″	9/16″	9/16″
Test pressure	100 psi	100 psi	200 psi	Conversion	200 psi
Relief valve activation	75 psi	75 psi	150 psi	150 psi	150 psi
Valve vapor tight pressure	60 psi	60 psi	120 psi	120 psi	120 psi
Manway cover thickness—min.	2 1/2″	2 1/2″	2 1/2″	2 1/4″	2 1/4″
Special requirements	Based on intended lading	Based on intended lading	Based on intended lading	Based on intended lading	Based on intended lading

Table 4-26   105 Class Tank Cars

Requirements	105A300ALW	105A300W	105A400W	105A500W	105A600W
Materials of construction	AL alloy	Steel	Steel	Steel	Steel
Insulation	Required	Required	Required	Required	Required
Burst pressure	750 psi	750 psi	1,000 psi	1,250 psi	1,500 psi
Min. plate thickness—all	5/8″	11/16″	11/16″	11/16″	11/16″
Test pressure	300 psi	300 psi	400 psi	500 psi	600 psi
Relief valve activation	225 psi	225 psi	300 psi	375 psi	450 psi
Valve vapor tight pressure	180 psi	180 psi	240 psi	300 psi	360 psi
Manway cover thickness—min.	2 5/8″	2 1/4″	2 1/4″	2 1/4″	2 1/4″
Special requirements	Based on intended lading	Based on intended lading	Based on intended lading	Based on intended lading	Based on intended lading

DOT 109, 112, and 114 CLASS CARS

Class 109, 112, and 114 cars are very similar to the Class 105 cars except they do not require the application of insulation. Many have thermal protection in the form of an insulation layer and outer shell or sprayed-on thermal materials. All Class 112J, 114J, and 114T cars must have thermal protection that meets performance requirements. Also, all Class 112 (Figure 4-41) and 114 cars (except those used for hydrogen fluoride transport) must have a minimum of the upper two-thirds of the car painted white if they have no insulation.

There are additional modifications that can be found in the 112 and 114 specification cars. The following is the key to the additional protective measures taken. The location of the letter indicating the modification is between the class number and the test pressure (e.g., DOT 112**J**600W). The following list provides the meaning of each letter:

- "**A**" means the car has top and bottom shelf couplers.
- "**J**" means the car has jacketed thermal protection, tank head puncture resistance, and top and bottom shelf couplers.
- "**S**" means the car has tank head puncture protection as well as top and bottom shelf couplers.
- "**T**" means the car has nonjacketed thermal protection system, top and bottom shelf couplers, and head shields.

There are four 109 Class tank specifications, four 112 Class tank specifications, and two 114 Class tank specifications. Many of the requirements are similar to those found in the 105 Class tanks. Tables 4-27 and 4-28 provide information on the specifications of all three Classes. All specifications are 1996 49 CFR requirements.

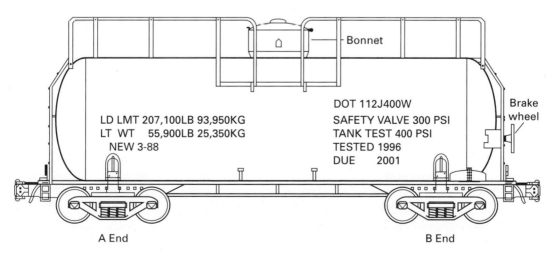

Figure 4-41   112 Class Tank Car

Table 4-27 Class 109 and 112 Tank Cars

Requirements	109A100ALW	109A200ALW	109A300ALW	109A300W	112A200W
Materials of construction	AL alloy	AL alloy	AL alloy	Steel	Steel
Insulation	Optional	Optional	Optional	Optional	Optional
Burst pressure	500 psi	500 psi	750 psi	750 psi	500 psi
Min. plate thickness	5/8″	5/8″	5/8″	11/16″	19/16″
Test pressure	100 psi	200 psi	300 psi	300 psi	200 psi
Relief valve activation	75 psi	150 psi	225 psi	225 psi	150 psi
Valve vapor tight pressure	60 psi	120 psi	180 psi	180 psi	120 psi
Manway cover thickness—min.	2 1/2″	2 1/2″	2 5/8″	2 1/4″	2 1/4″
Bot. washout	Optional	Optional	Optional	Optional	Prohibited
Bot. outlet	Prohibited	Prohibited	Prohibited	Prohibited	Prohibited
Special requirements	X	X	X	X	Based on intended lading

Table 4-28 112 and 114 Class Tank Cars

Requirements	112A340W	112A400W	112A500W	114A340W	114A400W
Materials of construction	Steel	Steel	Steel	Steel	Steel
Insulation	Optional	Optional	Optional	Optional	Optional
Burst pressure	850 psi	1,000 psi	1,250 psi	850 psi	1,000 psi
Min. plate thickness	11/16″	11/16″	11/16″	11/16″	11/16″
Test pressure	340 psi	400 psi	500 psi	340 psi	400 psi
Relief valve activation	225 psi	300 psi	375 psi	225 psi	200 psi
Valve vapor tight pressure	204 psi	240 psi	300 psi	204 psi	240 psi
Manway cover thickness—min.	2 1/4″	2 1/4″	2 1/4″	[1]	[1]
Bot. washout	Prohibited	Prohibited	Prohibited	Optional	Optional
Bot. outlet	Prohibited	Prohibited	Prohibited	Optional	Optional
Special requirements	Based on intended lading	Based on intended lading	Based on intended lading	Based on intended lading	Based on intended lading

[1]See AAR specifications for tank cars; also can be internal self-energizing and can be outside of bonnet if there are no attachments.

Class 114 cars are the only members of this group that may have a cross section that is not circular. However, these cars still must have an approved cross section that is of sufficient thickness and reinforcement to withstand its designed internal pressure.

Since these cars are pressure cars, all of their venting, loading and unloading valves, measuring, and sampling devices require the protection of a bonnet. Also, all of these cars are equipped with pressure relief valves instead of vents. The Class 114 cars may have cleanouts located on each head.

DOT 113 CARS—CRYOGENIC CARS

There are a relatively small number of cryogenic rail tank cars: DOT Class 113 (Figure 4-42). As with almost all types of cryogenic containers, the construction of these tank cars is a container within a container separated by an insulation layer. The insulation layer incorporates a vacuum. Both the inner and outer tanks require the protection of pressure relief devices. Most commonly, the location of all loading/unloading piping and valves is in a compartment above the trucks.

Primarily, these tank cars transport cryogenic ethane, ethylene, methane, natural gas, or hydrogen. Most common Class 113 tanks have a standard configuration except they are quite long and have a large diameter. Additionally, there are no bonnets, valving, manways, etc. visible on the top or bottom. One rather unusual and uncommon configuration of this car is the tank located within a box car. These cars are known as boxed tanks or "XT" boxed tanks.

Many of the construction requirements are very similar to those of the MC 338 cargo tanks since they both carry cryogenic liquids. Some of the requirements follow.

All of these tanks must have some type of insulation system. Such systems incorporate a vacuum with some type of insulating material. The entire insulation system requires performance testing to provide appropriate thermal protection for the lading. Additionally, the insulation space must have a pressure measuring device with an easily accessible connection or permanently mounted device. The device must be

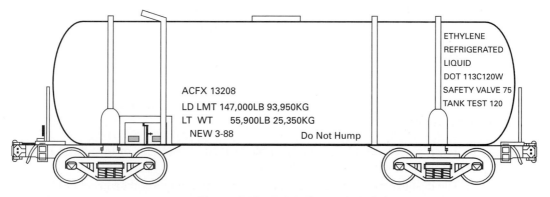

Figure 4-42   DOT Class 113 Tank Car

readily visible to the operator. All joints for the inner tank and shell must be fusion welded and be postweld heat treated. Tanks can have fully compliant stiffening rings as dictated by design.

Piping for the inner tank includes product loading and unloading lines and blow down lines. The outlets of all liquid and vapor lines must direct all accidental discharges away from the outer jacket, car structures, trucks, or safety appliances. All loading and unloading lines require two manually operated shutoff valves. The first is located as close as practical to the jacket with a secondary closure that is liquid- and vapor-tight. The secondary closure must relieve any trapped pressure before it can open. On DOT 113A60W tank cars, the piping that runs between the inner tank and discharge requires the use of a vacuum jacket. Further, the shutoff valve also requires vacuum jacketing. All valves must contain materials compatible with the product and approved for that use. All liquid lines require the use of extended stems.

Each car must have a vapor blow down line. Vapor blow down lines attach to the vapor space of the tank and allow the reduction of pressure during loading and unloading operations. During such times, the vapor line is part of a closed system, preventing its normal functioning. These tanks can have approved pressure building systems as well.

Each tank must have an approved liquid-level gauge or a fixed-length dip tube with a manual shutoff valve and a vapor gauge. The vapor gauge indicates the vapor pressure within the tank and must be visible to the operator.

Each tank must have multiple pressure relief devices for its piping and tank. These devices include valves and vents. The inner tank's relief devices attach to the vapor lines and remain at ambient temperatures prior to their operation. Each inner tank must also have at least one pressure relief valve and one pressure relief vent. The outer jacket must have a pressure relief system (valve or vent) that will prevent the pressure to exceed 16 psig (or the external pressure for which the tank is designed, whichever is less). Also, any isolated piping (cut off from the relief system) requires pressure relief as well.

Each valve, gage, closure, and pressure relief device must have a protective housing. The housing protects the device from solar heating, transportation debris, adverse weather, and normal operations of the tank. All housed devices must have easy access for operation and inspection. Because operation of these devices requires the use of heavy gloves, the housing must have sufficient space.

Each device found in the protective housing must have a plate that identifies its type and operation. These operating instructions include a diagram of the tank, its piping system, gages, control valves, and pressure relief devices clearly identified and located.

The stenciling requirements are somewhat different for these tanks. Each must have the following:

1. The replacement date of the frangible disk and the initials of the person making the replacement
2. The design service temperature and maximum lading weight adjacent to the hazardous materials stencil

3. The water capacity in pounds
4. The statement, "Do Not Hump or Cut Off While In Motion," on both sides of the car

Class 113 cars have specified loading and shipping temperatures. Identification of these temperatures is by the letter found between the DOT Class and the test pressure (e.g., 113A60W). The following codes are used:

1. "**A**" minus 423°F, for liquid hydrogen service
2. "**C**" minus 260°F, for liquid methane or liquid natural gas, ethane, or liquid ethylene

There are two specifications for 113 tank cars. They are listed in Table 4-29.

Table 4-29   Class 113 Tank Cars

Requirements	113A60W	113C120W
Design service temperature	−423°F	−269°F
Material of construction— inner tank	Stainless steel, ASTM Spec. A240, Type 304 or 304L	Stainless steel, ASTM Spec. A240, Type 304 or 304L
Burst pressure	240 psi	300 psi
Min. plate thickness shell head	3/16″ 3/16″	3/16″ 3/16″
Test pressure	60 psi	120 psi
Safety vent burst pressure max.	60 psi	120 psi
Safety valve activation pres.	30 psi	75 psi
Alternate safety valve activation	X	90 psi
Pressure control valve activation	17 psi	Not required
Transfer line insulation	Jacketing required between outer tank jacket and shutoff valve	Not required

DOT 107 CARS

The description of Class 107 cars (Figure 4-43) is, "seamless steel tank car tanks." These tank cars are analogous to highway tube trailers. Specifically, these Class 107 cars consist of a series of compressed gas cylinders attached to a modified flat car. Also, since the cylinders are not the vehicle, but simply attached to the vehicle, the regulations only address the cylinders.

The specific construction of each cylinder is very similar to ordinary compressed gas cylinders. These cylinders are either hollow forged or drawn in one piece and then the ends drawn down. The minimum wall thickness is 1/4 inch. At each end of the

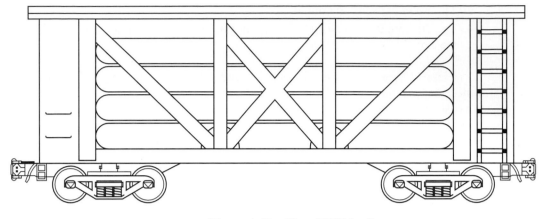

Figure 4-43   Class 107 Tube Car

cylinder, a protective housing encloses all safety devices and loading and unloading valves. Each cylinder must have an approved safety relief device. Cylinders that transport flammable gases must have the devices equipped with an approved ignition device.

DOT 106 AND 110 CARS

DOT Class 106 and 110 cars are very similar to Class 107 except that they have steel, one-ton cylinders mounted on a flat car instead of large compressed gas cylinders. The Class 106 specifications for the cylinders are the specifications used for the construction of the standard one-ton cylinder. The Class 106 tanks (Figure 4-44) have cylinders with concave heads (heads that curve inward) while the Class 110 tanks have convex heads (heads that curve outward). In both Classes, the cars are designed so the individual tanks can be removed for filling or use. Neither of these Classes are very common in most parts of the country.

None of the cylinders can have insulation. Fusion and forge welds are acceptable. All cylinders must have valving and relief devices located on one head of the cylinder. All cylinders must have relief valves, vents or fusible plugs unless prohibited. Table 4-30 provides information on the two Class 106 specifications and the four Class 110.

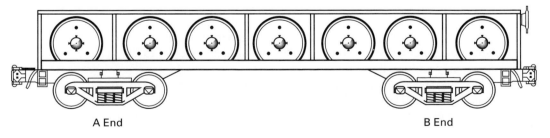

A End   B End

Figure 4-44   Multi-Unit Tank Car—Class 106

Table 4-30   Classes 106 and 110

Requirements	106A500-X	106A800-X	110A500-W	110A600-W	110A800-W	110A1000-W
Burst pressure	none specified	none specified	1,250 psi	1,500 psi	2,000 psi	2,500 psi
Min. shell thick.	13/32″	11/16″	11/32″	3/8″	15/32″	19/32″
Test pressure	500 psi	800 psi	500 psi	600 psi	800 psi	1,000 psi
Start to discharge	375 psi	600 psi	375 psi	450 psi	600 psi	700 psi
Vapor tight min.	300 psi	480 psi	300 psi	360 psi	480 psi	650 psi

GATHERING INFORMATION—SECTION III
CONTAINER INFORMATION
Module V—Air Containers

Air transportation, including that of hazardous materials, also is regulated by DOT. The specific requirements for general aviation are found in CFR 14, while requirements for hazardous material transportation are found in CFR 49. The Federal Aviation Administration (FAA) is responsible for enforcement of both regulations. Transportation of hazardous materials by the military onboard military aircraft is governed by the Department of Defense (DOD).

It is obvious that not all aircraft covered by these various agencies transport hazardous materials, but all of them do carry hazardous materials, at least in the form of fuel. To better understand the potential problems associated with aircraft, it is helpful to identify the various types of aircraft that can be encountered, their fuels, and the probability that they carry other hazardous materials.

Like other modes of transportation, there are many ways to classify specific types of aircraft based on a variety of criteria. The first way classifies aircraft by their power plant: reciprocating, turbopropeller (i.e., turbo prop or prop jet), and jet. The second way classifies aircraft by the type of wings they have: either fixed wing or rotary wing (helicopters). The third way classifies aircraft by their designed use: passenger, work, or cargo. The last way classifies aircraft based upon the owner and operator of the aircraft: private (sometimes called civil aviation), commercial, and military.

Although all four systems are useful when identifying the potential hazards associated with a given aircraft, we will concentrate first on the type of power plant. Later in this section, we will examine their wing type, designed use, and finally owner/operators.

POWER PLANTS

Reciprocating engine power plants produce thrust through a propeller attached to an internal combustion engine. Pistons rotate a drive shaft attached to the propeller. As the propeller turns, it produces thrust by the air forced out behind the aircraft. This thrust moves the aircraft forward. Most commonly, these planes are privately owned civilian aircraft or, in some cases, small commuter craft. They use a type of fuel known as aviation gasoline or *Avgas*. Aviation gasoline is similar to standard high-octane gasoline.

It has a flash point of $-50°F$, an ignition temperature of over $800°F$, and a flammable range of 1.3 to about 7%. Obviously these properties indicate there is the potential for severe fire hazards should this fuel be involved in an incident.

Turbo prop jet power plants produce thrust through the rotation of a propeller attached to the shaft of a jet turbine. The engine draws air into the turbine where it is compressed, and fuel is added. The fuel and air mixture is then ignited. As the fuel burns, it produces combustion gases and heat. The heat in turn causes expansion of the gases within the turbine. This expansion produces the energy that causes the turbine to rotate. Due to the design of this type of turbine, there is only a limited amount of thrust generated as these gases exit through the exhaust port. Generation of the primary thrust is through the propeller attached to the rotating turbine shaft. Turbo jets find common use in commuter and corporate aircraft. These turbines use a jet fuel commonly called *Jet A*. Jet A has a flash point ranging from about $95°F$ to about $100°F$ and an ignition temperature of approximately $475°F$.

Jet power plants produce thrust from the exhaust gases produced by the turbine. New generation jet turbines incorporate a fan in the rear section of the turbine to further boost thrust. The turbine functions in the same fashion as turbo jet prop, except there is no propeller. Both private and commercial jets use Jet A as fuel, while military jets use JP-4 as fuel. JP-4 is a combination of kerosene and gasoline. It has a flash point in the range of $-10°F$ to $35°F$ and an ignition temperature of $468°F$. As with Avgas, JP-4 presents an increased fire hazard due to its high volatility, which is reflected by its low flash point.

Type fuel	Flash point	Ignition temperature
Avgas	$\sim-50°F$	$>800°F$
Jet A	$\sim95°F$ to $100°F$	$475°F$
JP-4	$-10°F$ to $35°F$	$468°F$

Most commonly, all modern fixed and rotary wing aircraft are powered by one of the two forms of turbines. Again, reciprocating engines are generally found in older, privately owned craft.

Obviously, all of these aircraft carry the fuel needed for their power plants to function. As such, each presents a potential hazardous material problem based upon its fuel load. Larger aircraft carry thousands of gallons of product. Since weight is a key safety consideration to all aircraft, fuel loads are specified in pounds as opposed to gallons. As a rule, aircraft fuels weigh about 6 pounds per gallon. In other words, to convert pounds of fuel to the approximate number of gallons, simply divide the weight by 6.

At airports, incidental and sometimes rather large spills are all too common occurrences. Normally, airport Crash Fire Rescue (CFR) or fire department personnel safely and effectively handle such problems. (It is important to note that airports with five or more daily passenger flight departures per day are required to have on-site CFR services.) Further, many newer, large airports have fixed systems designed to limit the spread of released fuel and to aid in its recovery. However, airports without CFR capabilities need municipal response agency involvement. Even airports with CFR agencies may require assistance handling large releases.

When a crash occurs, fuel is one of the primary and often overriding operational consideration. The simple fact is that fire normally accompanies most crashes, unless the crash is a result of the plane running out of fuel. Commonly, large flammable liquid fires result. These fires produce intense heat that often consumes much of the craft and its contents. If there is unignited fuel, possible ignition represents a real and present danger.

The fuel storage cells for smaller aircraft are commonly found in the wings. However, larger aircraft may contain fuel in the main fuselage and the tail as well. Rear-engine aircraft may have fuel lines running from the wings and fuselage back to the engines. In this configuration, their fuel lines can be as large as four inches in diameter.

Another hazard that all larger aircraft have is their hydraulic system. Hydraulic systems operate the control surfaces (flaps, rudder, etc.). Hydraulic fuel lines move fuel from the tanks to the engines. In either case, pressures in the range of 3,000 psi are not uncommon.

DESIGNED USE

The transportation of hazardous materials as cargo or for delivery from aircraft (e.g., crop dusting) is a common practice. Depending upon the specific type and use of the craft, the hazardous materials that may be transported range from explosives (e.g., military craft), to radioactives (e.g., cargo craft), to biohazards (e.g., passenger aircraft), to almost any other type. As such, one of the most important considerations involving the potential hazards associated with the aircraft is its intended use.

The primary use categories are passenger aircraft (including private pleasure and commercial passenger craft), work aircraft (including public service, privately owned work, and military tactical craft), and cargo aircraft (including commercial and military craft). It is helpful to consider each type individually.

Pleasure and Work Aircraft

Out of all these groups, the private pleasure craft and public service work craft (e.g., air ambulance, police and fire craft, and news/traffic craft) present the lowest probability of chemical hazards. However, pleasure craft are used for smuggling activities ranging from drugs to endangered species. In such cases some relatively high hazards may be present.

Passenger Aircraft

Passenger aircraft generally present limited probability and relatively low danger due to the high degree of regulation. These regulations restrict the transportation of many substances onboard passenger aircraft. For example, explosives, many corrosives, liquefied compressed gases, and poison gases are forbidden from transportation by passenger aircraft. Of the substances allowable for such transportation, restriction on their quantities exists. Generally, quantities are limited to a maximum of 60 liters (approximately 15 gallons) of liquid and 100 kilograms (about 220 pounds) of solids.

However, it is important to note that quite a few rather harrowing experiences involving hazardous materials have occurred on passenger aircraft. Most commonly these situations result when a hazardous material is undeclared (not reported to the airline) and thus shipped as ordinary cargo. There have also been situations where declared (reported to the airline) and properly stowed hazardous substances have escaped their containers and caused in-flight incidents.

Work Aircraft

Work aircraft include those intended for use as sky cranes and in crop dusting and spraying operations. Sky cranes generally handle cargo that does not include hazardous materials. However, crop dusters and sprayers carry relatively large quantities of pesticides for aerial application. Further, this type of craft operates at low altitude and has been involved in many crashes. Contamination of victims, responders, and the environment is an ever present danger. Fire as a result of spilled fuel often acts to complicate operations and to spread contamination. They could also be used by terrorists.

Cargo Aircraft

Commercial cargo aircraft also commonly transport hazardous materials. Aircraft operated by companies such as Federal Express, UPS, and other rapid delivery companies are the most common example of this type of aircraft. These aircraft range in size from standard to jumbo jets where their entire cabin space is designed to accept aircraft cargo containers as opposed to passengers. These large metal or plastic cargo containers are sometimes called igloos, cans, or half cans. Many carriers color code these cans red and equip them with a fixed suppression system if they are intended to carry hazardous materials. These suppression systems are manually activated by the flight engineer. All hazardous substances are located immediately behind the cockpit area in the number one position of these aircraft. This allows for crew access and inspection during their pre- and in-flight checks so they can readily identify any problems.

Military Aircraft

Obviously, military cargo aircraft transport hazardous substances, but they may also transport military personnel. In some instances, these aircraft transport both cargo and personnel. The potential hazardous materials run the gambit from munitions to pesticides to vehicles.

Military work aircraft include fighters, bombers, and tankers. The fighters and bombers are high performance, highly sophisticated aircraft. Not only do these craft contain munitions that include bullets, rockets, cannon, or bombs, many also contain liquid oxygen (LOX) for breathing during high altitude operations. In some cases (e.g., an F-16) a mixture containing 70% hydrazine and 30% water is used to boost engine performance. Further hazards include ejection seats that use explosive charges to propel the cockpit occupants away from the aircraft.

Military tankers are used for in-flight refueling operations. In essence, these tankers are the equivalent of a Boeing 707 that contains a huge bladder tank aft of the cockpit. Obviously, huge quantities of fuel are found in these tanks.

AIRCRAFT SHIPPING PAPERS

Just as in highway and rail transportation, air transportation also requires the use of shipping papers. The papers are known as airbills. The airbill for any given flight is supposed to be located in the cockpit. In some cases, additional copies may be found on individual containers.

AIRPORT OPERATIONS

Whenever responders operate on airport facilities they must be familiar with airport requirements, including operational practices. These procedures are based on the unique nature of an airport and the aircraft themselves.

First and foremost, response personnel must understand the Air Operations Area (AOA) of an airport. The AOA includes taxiways and runways. Ground access to these areas is restricted and must only occur after clearance is provided by the control tower. Whenever driving on an airport facility, it is important to note that all visual warning signals must be in full operation. It is vital to understand that ground vehicles yield the right-of-way to aircraft. The lights lining both taxi way and runway areas provide cues to their use. Blue lights indicate a taxi way while white lights with yellow at the end indicate a runway. Neither area must be entered without approval from the control tower. As such, communication with the tower before any movement in the AOA is essential.

When operating around aircraft, it is also vital to be cognizant of the pilots' highly restricted field of vision. Primarily, the pilot's field of vision is restricted to the area in front of the wings. The sheer size of larger aircraft further hinders visibility. Additionally, all three types of engines produce high velocity air that extends to the rear of the aircraft. This air can easily blow debris considerable distances. Due to both of these facts, one should always approach aircraft from the front, in plain view of the crew.

AIRPORT EMERGENCY OPERATIONS PLANS

As would be expected, all certified airports must have emergency operations plans. These plans must be tested on an annual basis through the use of table-top exercises. Every three years, a full scale exercise must also be held. Thus, it is important that all potential airport emergency responders be knowledgeable of the facility and of their role in any operation.

Beyond the specific airport emergency operations plan and exercise requirements, FAA uses an alert classification system (similar to the NFPA Incident Level System) to help identify the type of problem involved in an emergency.

Alert 1	Possible in-flight emergency. Stand-by.
Alert 2	Confirmed in-flight emergency. Stage at designated location.
Alert 3	Aircraft emergency (crash).

Airport and aircraft emergencies are not common occurrences in most jurisdictions. These emergencies are normally complex and require highly specialized preparation. More in-depth and facility specific training, education, and planning are needed to adequately prepare for and participate in such emergencies. Your local airport can assist in identifying appropriate training needs and sources.

GATHERING INFORMATION—SECTION III
CONTAINER INFORMATION
Module VI—Water Containers

Water is another mode available for the transportation of hazardous materials. Again, DOT regulates the transportation of hazardous materials using this mode of transportation. Potential problems with this mode can be significant due to three primary factors. First, all of these vessels have large cargo capacities. In the event of an incident, tremendous amounts of product can be involved. Second, with the exception of some barges, most vessels have relatively complex designs that are unfamiliar to most emergency responders. Finally, access to these vessels or their cargo is often quite difficult. The problem can occur when the vessel is at sea, in a harbor, or while traveling remote sections of rivers or lakes. Even when they are moored, access is restricted.

Obviously, incidents involving water vessels are potentially complex and difficult. This section is not designed to provide in-depth information about this topic, but rather it will provide an overview of general concepts of operations.

PRIMARY TYPES OF VESSELS

There are two primary types of vessels used to transport products by water, barges, and ships. The primary difference between the two types of vessels is their power plants.

Barges

Barges do not have an onboard power plant. Rather they rely on tug boats to push them through the water. Commonly, a group of barges is secured to each other while a tug pushes the group along. There are three primary types of barges: hoppers, tanks, and flat deck.

Hopper barges are similar to gondola or hopper rail cars. They are designed to handle bulk materials. Some hopper barges have no tops and haul materials such as coal, wood chips, garbage, and similar materials that are not weather sensitive. Covered hoppers haul materials such as fertilizers, flammable solids, and grain.

Tank barges transport liquid or liquefied gases. Due to this fact, these containers can be atmospheric, low pressure, or high pressure. The products transported by these barges may include hydrocarbons, corrosives, and chlorine. Due to their capacity, these barges can transport quantities many times greater than those transported by even rail tank cars.

Flat deck barges have heavy metal decks that allow the transportation of large, heavy objects. These objects can include cranes, derricks, rail cars and intermodal containers, and tanks. As a result, an assortment of other mode containers can be found on these barges.

Ships

Ships are vessels equipped with their own power plant. They come in all shapes, sizes, and configurations, and they are intended for all types of uses. This section focuses on merchant as opposed to work, military, or passenger vessels. In the merchant category, there are six primary types of vessels. They include container, general/multi-cargo, roll on/roll off (RO/RO), dry bulk, break bulk, and tankers.

Container.   Container ships are very popular because they transport containerized cargo. Specifically, containerized cargo is cargo shipped in any type of intermodal container. These intermodal containers include portable box containers, Spec 51 tanks, IM 101 and 102 portable tanks, and so on. Because these ships can transport over 1,000 individual containers, they are extremely cost effective and popular.

However, imagine the potential difficulties that could arise should there be a problem with one or more of the containers. In one incident, approximately 25 to 30 containers fell from the deck of a ship during a northeaster. The containers carried arsenic compounds in 55-gallon drums. Some of the containers failed, spreading drums over a large area of the sea floor. The ensuing clean-up was performed in several hundred feet of water and was extremely expensive.

Loading and unloading involve large derrick systems designed to lift the containers. On the ground, the containers are normally handled using straddle carriers or a specifically designed forklift type of equipment.

General/multi-cargo.   General cargo and multi-cargo ships carry a wide variety of goods, including paletized drums, boxes, containers, shrink wrapped loose goods (e.g., paper and plastic bags), and endless nonbulk containers. In some cases, these ships have refrigerated holds. General cargo ships locate most cargo in the holds while multi-cargo ships also use their decks for intermodal containers.

Roll on/roll off (RO/RO).   As the name implies, roll on/roll off ships carry cargo that have wheels. Cargo is driven or pushed onto the ship by way of ramps located on the sides or ends of the ship. RO/ROs are commonly used to transport cars and trains. The military uses this type of ship (i.e., LST) to haul tanks and other vehicles. The primary advantage of using RO/ROs is the ease and speed of loading and unloading.

Dry bulk.   As in highway and rail transportation, dry bulk ships haul bulk quantities of unpackaged solid materials. These materials include grain, fertilizer, coal, iron ore, and so on. Cargo is loaded and unloaded using various types of equipment, including fluid bed technologies, conveyor systems, or clamshell buckets.

Quite a few incidents have involved the spontaneous heating and ignition of cargo. To extinguish such fires, it is often necessary to completely unload the cargo to most safely and efficiently extinguish the fire. By unloading, the potential for steam explosions caused by water contacting the deep-seated burning mass is eliminated. In the case of coal, efforts to extinguish a fire in the hold have produced sulfuric type acids. These acids result from the steam hydrolyzing the sulfur oxides produced by the combustion. Once produced, these acids then attack the structural members of the ship causing structural failure. Such failures have led to several sinkings.

Break-bulk.   Break-bulk ships (breakable bulk) transport packaged, dry bulk cargo that is subject to breakage due to lightweight packaging. Most commonly such cargo includes bagged and paletized materials such as grains, coffee, fertilizer, cement, and various solid chemicals.

Tankers.   Just as in tank barges, tank ships carry products in the forms of liquids, liquefied compressed gases or cryogenic liquids. Under ambient conditions, these products are either liquids or gases. The capacity of large tankers is over one million barrels (42 million gallons). Most commonly, tankers transport hydrocarbon liquids including crude oil. Less commonly, tankers carry corrosives and even food grade materials (e.g., vegetable oil and syrups).

One of the most specialized tankers is the type designed to transport liquefied natural gas (LNG), a cryogenic liquid. LNG tankers are huge ships outfitted with large, hemispheric domes extending above the main deck. As with all cryogenic containers, these LNG tankers are the equivalent of huge thermos bottles. Because the product remains a liquid due to the absence of heat, LNG tankers do not require pressure to maintain the product in the liquid form.

SHIP DESIGN AND EMERGENCY CONSIDERATIONS

All of these various types of ships have several aspects in common. First, they all have some type of superstructure that extends above the main deck (the primary exterior deck). The superstructure houses the bridge from which navigation and general control of the vessel occurs.

Second, all large ships are subdivided into compartments. Compartments are structural members of the ship that help in identifying onboard locations. Horizontal floor structures located below the main deck simply are called decks. Those located above the main deck are called levels. Vertical wall structures that normally run from side to side are called frames. Small subdivisions between frames are called compartments. There are several systems used to identify a specific location within a ship based upon the deck or level, frame, and compartment numbers. However, these systems are quite confusing at times and often require significant assistance from the crew to identify exact locations.

Finally, all cargo vessels need some place to store cargo during transport. These areas are called holds. Holds are large, open areas located below the main deck. In some instances they are simply large, open-topped caverns used to hold dry bulk materials such as coal. In other instances, they are almost completely enclosed, with the exception of hatches, and used to hold huge volumes of liquid cargo.

SHIPPING PAPERS AND EMERGENCY PLANS

All vessels transporting hazardous materials are required to have shipping papers. In this case the shipping papers are called the Dangerous Cargo Manifest and are usually found with the captain or first mate. Another required document is known as the Station Bill. The station bill contains a listing of the duties and assigned emergency station of all crew members.

In addition to the dangerous cargo manifest and the station bill, ships are required to have a Fire Control Plan (FCP). The FCP is designed to provide land based firefighters with information needed to fight a shipboard fire. The FCP includes information regarding cargo locations, deck by deck prints, fixed suppression systems, standpipe systems, and so on. Further, the International Maritime Organization (IMO) mandates that the FCP be located in a red box near the gangway. The box must have a white front with a red silhouette of a ship on the cover. If it is not located in the immediate vicinity of the gangway, clearly marked signs must direct firefighters to its location. Normally, the station bill will also be located in this box.

Fixed suppression systems and firefighting equipment onboard ships can provide the same benefit or frustration as those found within buildings on land. When the systems and equipment are appropriate and well maintained, they provide responders with many advantages for dealing with an emergency. However, it is not uncommon for responders to find both fixed systems and equipment that are not functional.

One such example is a recent engine room fire onboard a cargo ship. In this case, the captain informed fire department personnel that the CO_2 system was activated. As personnel worked their way to the system panel and cylinder bank, they found that multiple failures in the piping system resulted in discharge of the CO_2 into the bank room. When additional activator handles were pulled, more CO_2 gushed into the area. In one instance, when the activator handle was pulled, approximately a 10-foot section of cable followed. Unfortunately, the cable was not attached to anything at all. In the same vein, house fire hoses were found to be rotted and unusable.

When an emergency occurs, poor radio communications can be a problem. Primarily this situation is a result of the massive amounts of steel used in large ships. Normal portable radio signals may not carry very far inside a ship (the same type of problem found in certain types of buildings). As a result, responders may be required to string hard-line communications systems or rely upon hard wire shipboard systems.

Obviously, shipboard operations are unique. Response agencies that may become involved with shipboard emergency-response activities need further training, education, and planning to effectively handle such incidents. Contact the Harbor Master or the United States Coast Guard for assistance in these matters.

GATHERING INFORMATION—SECTION III
CONTAINER INFORMATION
Module VII—Pipelines

Pipelines are not commonly considered to be containers or part of transportation. However, they are. DOT has a separate office, the Office of Pipeline Safety, that deals with pipelines. The National Transportation Safety Board investigates incidents involving pipelines, just as they investigate incidents involving all four of the other modes of transportation.

The simple reason for this inclusion is that huge quantities of hazardous materials are transported throughout most of this country by pipeline. Pipelines transport gaseous and liquid hydrocarbons across the country. They distribute natural gas to homes and industries in most cities. They transport specific products such as ethylene, butane, oxygen, liquid nitrogen, and numerous other products between various production facilities. In some instances, they even transport solids in the form of a slurry. Further, in many instances, they transport multiple products at the same time. They even involve seasonal storage of natural gas in old wells and salt domes.

In short, pipelines transport solids, liquids, and gases that are flammable, corrosive, poisonous, oxidizers, and so on. As such, it is important to understand some basic features of this mode of transportation.

PIPELINE FEATURES AND PROTECTION OPTIONS

Pipeline construction and materials must conform to ASME standards, the American Petroleum Institute (API) standards, and DOT regulations. The majority of these standards are designed to enhance the safety of pipeline operations by establishing minimum standards for materials of construction, installation, operating procedures, and line testing.

All pipelines use pressure to move product through the line. Pressure within the pipeline may be created by gravity in the form of head pressure. In other instances, pressure is created by the weight of a container as found in the old style telescoping tanks. However, most commonly, pressure is created by pumps. In long distance pipelines, booster pumps are strategically located along the length of the line.

Pipeline Corrosion Protection

Most pipelines are laid underground or even underwater. Because of mineral content, as well as varying levels of both oxygen and moisture found in the soil, pipes corrode. This form of corrosion is an electrochemical reaction similar to that found in a typical battery. For this reaction to occur, several components must be present. First, there must be an anode that is the source of the electrons. As a result, the anode looses electrons. Second, there must be a cathode that receives the electrons. As a result, the cathode has excess electrons. Third, there must be an electrolyte solution that incorporates dissolved mineral ions so it conducts electricity. Finally, there must be a completed circuit that includes a conductor and an electrolyte solution between the anode and cathode. When all of these components are present, a corrosion cell results.

There are several important points worth noting about corrosion cells. First, corrosion cells result when moisture content, soil oxygen levels, or soil composition change. Second, corrosion cells result when new pipe is placed into an existing pipeline. Third, corrosion cells result from minute variations in the composition within a pipe. As a result, many individual corrosion cells appear within any pipeline.

In essence, a corrosion cell develops when an electrical current (a flow of electrons) starts within the section of pipe due to the condition and composition of the pipe and surrounding electrolyte solution. Based on the variables in soil or pipe conditions and composition, some location on the pipe becomes the anode while another location becomes the cathode. Metal atoms in the anode release electrons to the surrounding metal and act to form the electrical current. As electrons leave the metal atoms in the anode, metal ions go into solution where they combine with non-metal ions (anions). The cathode receives the excess electrons released to the electrolyte solution where they often combine with hydrogen ions to form hydrogen gas. Since there is no ionization at the cathode, there is no corrosion of the pipe.

This process of metal ionization and dissolution at the anode is corrosion and its significance is the focal point of the entire corrosion problem. For example, a steel pipe that establishes a 1 ampere current will lose 20 pounds of iron in one year. This corrosion results in a decrease in the wall thickness of the pipe. A reduced wall thickness coupled with the pressure from within the pipe could easily result in a pipe failure. Until rather recently, corrosion was a primary cause of all types of pipeline failure. With a better understanding of corrosion cell behavior and protection methods, corrosion failure of large oil pipelines has decreased. However, it is still a major problem for water and natural gas delivery systems.*

To minimize the formation of corrosion cells, most large, long distance pipelines incorporate two primary methods of protection. First, the pipes have some type of protective coating. Second, they are provided with cathodic protection. The action of a protective coating is clear. The coating protects the pipe from exposure to moisture in the soil, and thus prevents contact to the electrolyte solution. Without the electrolyte solution, there can be no electrochemical reaction. However, failure in the coating results from many types of mechanical or sometimes thermal damage. As such, cathodic protection is also needed.

The specifics of cathodic protection are rather complex. However, in function, they all involve the intentional development of a corrosion cell where the pipe itself is always the cathode as opposed to the anode. As such, the pipe, behaving as a cathode,

---

*NOTE: This type of electrochemical reaction is the same one involved in the corrosion of dissimilar metals such as has been seen in fire apparatus bodies or valve assemblies.

has excess electrons so the metal atoms do not ionize and dissolve in the electrolyte solution.

One type of corrosion cell involves connecting the pipe to a source of direct current (DC) electricity such as a rectifier or solar panel. (A rectifier converts alternating current, or AC, to DC.) By designing the system appropriately, the pipe becomes the cathode and is connected to anodes by wires. Commonly, this type of anode is made of carbon or similar materials.

A second type of corrosion cell involves the use of a sacrificial anode. Most commonly, this option is used when a source of electricity is a problem, as well as in some older systems. Here, the pipe is wired to an anode composed of a more reactive metal such as magnesium or magnesium alloys. A major drawback to this type of protection is the need to routinely replace the corroded anodes, thus assuring that the system continues to function in its intended fashion.

Pipeline Mechanical Protection

As mentioned, pipeline materials and construction methods must meet the requirements established by ASME, DOT, and API. As such, specific attention is paid to preparation of the bed, including the removal of large stones and vegetation, depth of the pipe, and so on.

Wherever the pipeline intersects another mode of transportation such as a highway or rail line, special protection is required. Further, protection may be required when the line runs parallel to and within a specified distance from highways or rail lines. The first form of protection is the placement of pipeline markers. These markers are strategically placed to warn everyone of the presence of the pipeline. Each marker must identify the product or products transported through the line, the company that owns the line, and a contact telephone number.

Second, the pipeline is reinforced in some manner. In some instances, heavy gauge pipe is used in these locations. Alternatively, a reinforcement known as a stress casing is used. Stress casing is nothing more than an oversized pipe through which the pipeline passes. In this arrangement, the stress casing takes the brunt of any stresses and thus protects the product line. To allow for monitoring of the product line, the stress casing is normally equipped with vents.

TYPES OF PIPELINES

There are three primary types of pipelines: gathering lines, trunk lines, and distribution lines.

Gathering lines, as the name implies, gather raw products from well heads and transport them to trunk lines. Obviously, multiple gathering lines will supply a trunk line. Since these lines only transport raw product, they are not required to add odorants such as mercaptans (—SH or thiol) as part of their leak detection systems. This means that products, such as natural gas, having no inherent odor, require the use of air monitoring equipment to detect their presence.*

Trunk lines are normally large lines that deliver products such as crude oil to refineries, transport processed products to distribution centers, or transport natural gas cross country. In some instances, trunk lines transport products such as ethane, ethylene, propane, butane, oxygen, and nitrogen from a production facility such as a refinery or air separation unit to a secondary production facility. Since these lines are not intended to provide products directly to the consumer, they may not contain odorized product.

Trunk lines normally transport large volumes of product. To move these large quantities, the lines have diameters ranging from about 12 inches to over 36 inches. The internal pressure of these lines is routinely in the range of 600 psi to 1,600 psi. In some instances, the pressures are even higher. For cross country transport, these lines are 1,000 or more miles in length. Even with these large diameter lines, friction loss is a problem, especially over long distances.

Distribution lines provide direct service to the customer. The most common type of distribution line provides home owners, industry, schools, and others with natural gas for heating and cooking. Since the lines provide direct service for consumers, they most commonly contain odorants.

PUMPING AND LEAK DETECTION SYSTEMS

Whenever liquids or gases are transported through piping, friction loss becomes an issue. To counter this, these lines have booster pumps strategically located along their length. These booster pumps function in the same way as fire department engines in a water relay operation. As friction loss acts to slow the flow of product, these pumps boost both the volume and pressure of the product flowing through the pipe. Due to the large diameter of the lines, pump stations are normally only needed every hundred miles or so. Obviously, this indicates that large quantities of product are found in the piping between each pump station.

The vast majority of these systems are computer controlled. On the basis of data provided by sensors located along the pipeline, the flow of product is controlled. Should these sensors detect a major drop in pressure, the system shuts down. However, such systems are not a total guaranty that no major loss of product will occur should there be a problem. For example, if the pressure drop is small, such as may accompany a minor failure in the pipe, the system may not shut down. On the other hand, even if the pumps shut down, residual pressure will force additional product from the pipe. If the incident should occur in a low lying area located between pump stations, all the product within the surrounding pipe will exit at the failure site. It is not uncommon for a failure in a trunk line to result in the loss of hundreds of thousands of gallons of product.

*NOTE: The industry is starting to use new odorants that are not as toxic.

For the most efficient transportation of products, pipelines must carry multiple products at the same time. Traditionally, multiple products have been separated by devices known as pigs. Pigs are solid or inflatable devices, placed within the pipeline to separate different products. In some of the newer systems, an air bubble may simply be placed between the products. In still other systems no separation is used. This results in minimal commingling of product and is thus not a major drawback.

However, from the perspective of responders, this could be a major problem. For example, when there is a failure in the pipeline, one product (e.g., gasoline) could flow from the pipe. As product continues to flow, it is conceivable that the release could change from gasoline to some other product without warning. Obviously, should this situation occur, major operational difficulties could result.

Along with pressure and flow detectors along the pipeline, owners are required to periodically inspect their lines and the right-of-way. In some instances, inspections are performed from the air, while others are performed from the ground. During these inspections, personnel look for any indications of a slow leak or other problems. Some clues to the possible presence of a leak can include soil and plant discoloration, presence of dead animals or plants, standing pools of product, or a visible sheen on nearby bodies of water. These clues can also be used by response personnel to help identify locations of possible leaks.

Mapping and Digging Requirements

As part of the ongoing Pipeline Safety program, each state is required to establish a one telephone number excavation contact point. This one number system is designed to help contractors identify the location and types of pipelines that are potentially present where any excavation is anticipated. Additionally, these contact points can provide responders with information or actual maps of all pipelines within their response area. Such maps are invaluable for preplanning activities such as identification of pipeline locations, companies owning the pipelines, and so on.

As with the other modes of transportation, this brief discussion addresses only the general topics. For further information, check with any local pipeline companies, your state pipeline oversight organization, or contact DOT's Office of Pipeline Safety.

GATHERING INFORMATION—SECTION III
CONTAINER INFORMATION
Module VIII—Fixed Containers

Fixed containers are designed to hold products at one location. Their capacities can run from about 200 gallons to millions of gallons. They contain molten solids, liquids, or gases. Most commonly, fixed containers are broadly classified as above ground or underground tanks. Any tank that has 10% of its side wall located below ground level is

Vertical Tank   Horizontal Tank   Spherical Tank   **Figure 4-45**   Above Ground Tanks

classified as an underground tank. Within each classification, there are many individual types of container. Each type is identified by its structural design and configuration as well as by its design working pressures. This section will discuss each classification and type individually.

ABOVE GROUND TANKS

Above ground tanks are identified as either vertical, horizontal, or spherical (Figure 4-45). If the cylindrical portion of the tank is perpendicular to the horizon, it is a vertical tank. If the cylindrical portion of the shell is parallel to the horizon, it is a horizontal tank. Finally, a spherical tank is in the shape of a sphere. We will first discuss vertical tanks, horizontal tanks, and then spherical tanks.

VERTICAL TANKS

There are four primary types of vertical tanks: cone roof, open floating roof, closed floating roof, and dome roof tanks. There are also two other less common tanks known as telescoping tanks and lifter roof tanks. Each has a recognizable design and is built for specific types of products and environmental conditions. Before discussing the individual types of tanks, it is helpful to examine some options and features that may be found in all four types.

Fixed Suppression Systems for Vertical Tanks

Since most of these tanks are designed to hold flammable or combustible liquids, it is not uncommon for them to be equipped with some type of fixed suppression system. The most common fixed suppression system is the foam chamber (Figure 4-46).

Foam chambers are located at the top of the side wall of the tank. They aerate foam solution supplied through a foam piping system. There is an opening in the side wall through which the aerated foam enters the tank and flows down the inner surface. In some instances, the foam chamber is equipped with a flexible, fire resistant tube that unrolls when foam is applied. The advantage of the foam tube is that it deposits foam directly on the products' surface as opposed to flowing over the hot metal tank wall.

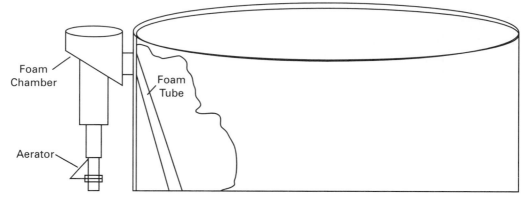

Figure 4-46   Typical Foam Chamber

Other tanks may be equipped with subsurface application systems (Figure 4-47). In this case, an existing product line or manifold is fitted with a high backpressure foam eduction system. This eduction system enables responders to generate foam solution and pump it into the line. The product line now transports the foam into the tank. A key consideration in this type of foam system is the size of the product line. Specific calculations must be performed to assure that the line has the capacity to flow the required gallons per minute of foam for effective foam operations.

Additionally, it is important to determine if the tank is equipped with a sump and, if so, its location in relation to the product line end. A sump is a depression in the bottom of the tank that holds water or similar product contaminants. If the sump should fill with water and the product discharge falls below the water level, subsurface foam will dissipate. A further discussion of foam applications and calculations is found in Chapter 7, Section 7, Fire Control.

Cone Roof Tanks

One of the most common types of vertical tank is the cone roof tank (sometimes called fixed roof tank). As the name implies, this type of tank has a conical roof attached directly to the side walls of the shell (Figure 4-48). The tank is then equipped with vents

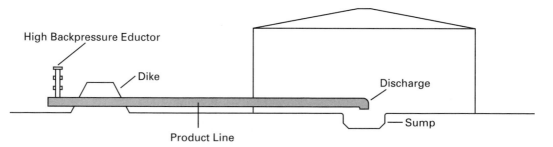

Figure 4-47   Subsurface Injection

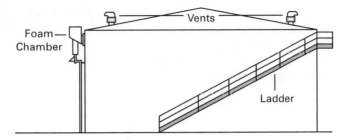

Figure 4-48 Cone Roof Tank

located on the roof to allow air in and out while the tank is being filled or emptied. These vents are fitted with a flame arresting system similar to a fine screen material. The flame arrestor acts to prevent an exterior fire from propagating back into the interior of the tank.

Along with the vent, each tank commonly is equipped with an exterior ladder and roof access port to the interior of the tank. Most commonly, larger diameter (greater than 50 feet) cone roof tanks are approximately 50 feet high.

The space above the liquid level in the interior of the tank is filled with product vapors. With vents on the roof, these vapors easily escape the confines of the tank. This situation allows product loss through evaporation. As a result, most facilities do not store highly volatile liquids in these tanks because of potential product loss and subsequent environmental contamination. This means that most of these tanks are filled with combustible rather than flammable liquids.

Since the space above the liquid contains vapor, it is possible that this space is within the flammable range. If a lightning strike or some other ignition source ignites these vapors, an explosion within the tank is possible. All modern tanks are designed with a weak side wall to roof weld (Figure 4-49), so the roof will separate from the walls if there is such an explosion.

In older tanks without weak shell to roof seams, an explosion within the tank can be devastating. Generally, these tanks are of small diameters (30 feet or less). Should ignition occur within the tank, the weakest section of the tank will fail. If this happens

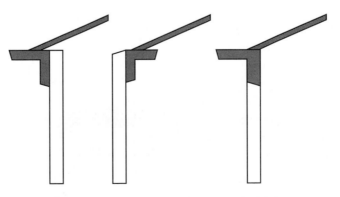

Figure 4-49 Typical Weak Shell to Roof Seam

to be the bottom to shell seam, the tank can take off like a rocket and trail burning product. The identification of all such tanks is critical during preplanning operations to assure this potential is identified.

Open Floating Roof Tanks

Open floating roof tanks (Figure 4-50) have no permanently attached roof. Rather, they are equipped with a roof that floats directly on the surface of the product. As product fills the tank, the roof rises and, as it leaves, the roof falls. Since the roof rests directly on the surface of the product, its surface area is decreased. Consequently, vaporization and its associated product loss are minimized. As a result, these tanks are ideally suited for the storage of highly volatile products such as flammable liquids.

There are two design options for the floating roofs of these tanks. The first type is the annular pontoon. The annular pontoon incorporates a single, steel bottom deck that covers the entire surface of the product (Figure 4-51). Next, there is an air space covered by a single, steel top deck. However, the top deck does not cover the entire surface of the bottom plate. Rather, there is a section in the middle where there is no air space or top deck. Commonly, this air space is likened to a donut. Due to this design, when one walks on the center of the roof, one is walking on a single metal deck that is floating directly on the surface of the product.

The second type of roof is the double-deck floating roof (Figure 4-52). As the name implies, this type of roof has a full top and bottom deck separated by an air space.

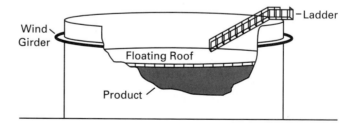

Figure 4-50   Open Floating Roof Tank

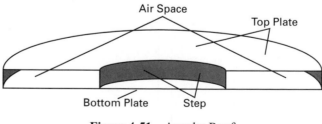

Figure 4-51   Annular Roof

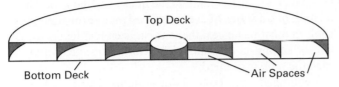

Top Deck

Bottom Deck   Air Spaces   **Figure 4-52** Double-Deck

Since both types of roofs are exposed to the elements, water can build up on their top surfaces. If there were no provisions for removing this water, the roof could easily become too heavy to support itself on top of the product. If this occurs, the roof will sink. To prevent such situations, each roof is equipped with some type of drain. The drain's inlet is located in the center of the roof. An articulating drain pipe runs from the roof inlet to a discharge located on the side of the tank. There is a valve on the outside of the tank that controls the flow from the line.

The drain has several articulating joints that allow it to move as the product level changes within the tank. In some instances, the seals at these joints have failed and allowed product to enter the drain line. Should this occur, product can escape from the valve located on the exterior of the tank. If the exterior valve is closed, product can then back-up through the drain and onto the floating roof. To prevent this from occurring, many facilities place a plug in the drain and keep the exterior valve shut. The only time they may open this valve is when precipitation is in the forecast.

Obviously, if the facility operator forgets to open the drain or exterior valve, the precipitation could cause the roof to sink. Also, the application of foam or cooling water could have the same effect.

Open floating roofs are also equipped with support legs (Figure 4-53). These legs are metal tubes designed to keep the roof from resting directly on the bottom of the tank when the liquid level is very low. Further, they provide a space ranging from 3 feet to

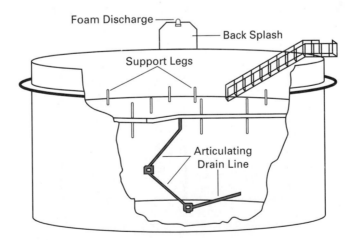

Foam Discharge

Back Splash

Support Legs

Articulating
Drain Line

Figure 4-53 Roof Drain and Legs

as much as 8 feet for personnel to enter, clean, or maintain the interior of the tank. These legs, when viewed from above, look like a series of evenly spaced pipes protruding several feet from the top of the roof. In most instances, there are at least two levels for leg adjustment. Generally, the shorter setting is maintained for normal usage, while the longer setting is employed when the tank is drained.

These tanks may also be equipped with fixed foam systems. Most commonly these systems include a foam pipe that extends to the top of the tank and over a back splash plate. There is an air aspirator located below the back splash. When located next to the ladder, this line may also be equipped with a shut off valve and an 1 1/2 inch hose connection to assist in firefighting operations. The discharge from the foam pipe is curved toward the back splash. This allows foam to be discharged against the back splash. In this configuration, the foam contacts the back splash and flows in a wide sheet down the inside of the tank wall and over the seal.

These roofs also have various openings that enter the air space in the roof. These openings allow the operators to inspect the air space for any type of leakage or damage. Further, these roofs are often equipped with ports that open directly to the product space. These openings allow sampling of the product as well as inspection of the tank.

Around the edge of each floating roof is an area known as the seal (Figure 4-54). The seal is a space between the roof and the side wall of the tank. This space allows the roof to float easily up and down as product enters or leaves the tank. There are three primary types of seals: flexure ring, tube seal, and leaf spring. All of these seals act to keep the roof at least 3 to 6 inches away from the tank shell as well as prevent excessive product vapor loss. Each is equipped with some type of weather shield to prevent precipitation from accumulating around the seal.

Roof seals do have their down side, however. They are the most common location of fires on this type of tank. Routinely, some ignition source, such as a lightning strike, ignites vapors above the seal. Since the roof is steel, the fire normally is confined to a relatively small section of the seal. As the fire continues to burn, the weather shield is damaged or destroyed, and the size of the fire increases. If firefighting operations are initiated rapidly, minimal damage results.

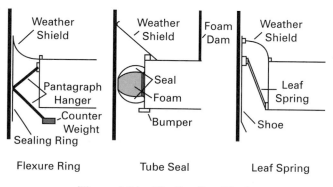

Figure 4-54   Floating Roof Seals

It is important to note that tube seals are generally the least susceptible to this type of fire. There are several reasons for this fact. First, since there is a tube located between the roof and the wall, there is no exposed product surface area (although there are still vapors). Second, tube seals normally employ metallic weather shields that are less susceptible to damage and destruction from a fire.

However, should the fire burn for an extended period of time, the entire seal area can become involved. The most effective method of extinguishing such seal fires is by accessing of the top of the tank and placing hand foam lines in operation from the ladder or the roof itself. Another option includes the use of fixed foam systems that incorporate foam dams. Foam dams are metal plates welded perpendicular to the top deck of the roof that extend several feet above the deck. These dams keep foam from spreading over the surface of the roof and greatly increase the speed and effectiveness of the fixed foam system.

Portable foam application from the ground or aerial appliances is generally ineffective. Further, such applications risk flooding the roof and causing it to sink. Should this occur, the fire will intensify and magnify since much, if not all, of the product's surface is able to burn.

Closed Floating Roof Tanks

Closed floating roof tanks (Figure 4-55) are hybrids between fixed roof tanks and open floaters. Closed floaters have a floating roof as described in the previous section. However, they are also equipped with a fixed roof. The advantage to this type of construction is that the fixed roof protects the floating roof from all types of precipitation. As a result, there is little problem with a potential weight buildup due to precipitation.

This type of tank is distinguishable from a standard fixed roof tank simply by examining the upper section of the tank wall. On this section, a series of vents is found. These vents allow the atmosphere above the floating roof to move in and out of the tank as it is filled or emptied. Normally, these vents are equipped with a flame arresting screen to prevent the flame propagation from the outside to the inside of the tank.

As with other tanks, these can be equipped with fixed foam systems. Such systems are similar to those found on cone roof tanks. They generally incorporate an

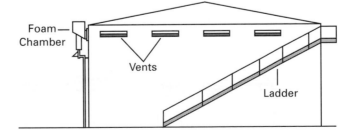

Figure 4-55   Closed Floating Roof Tank

external foam chamber and may have a foam tube located on the inside of the tank. In some instances, these tanks also have subsurface injection capabilities.

Fires inside these tanks, where the fixed roof is still intact, are extremely difficult to fight. This is especially true if the floating roof has sunk. Most commonly, the roof does not completely sink. Rather, it pivots, allowing only part of it to submerge. As a result, the other part of the roof extends above the surface of the product. This alignment hinders or prevents the spread of foam over the surface of the product from fixed systems, subsurface injection or portable application.

There is a special type of closed floating roof tank known as retrofit open floating roof tanks (Figure 4-56). As the name implies, originally these were constructed as open floating roof tanks. However, the operators decided to retrofit these tanks with covers. Visually, these tanks have the wind girder found in open floaters, but they also have a fixed roof attached.

There are two primary roof configurations found in these retrofits. The first, commonly called a geodesic dome, incorporates a series of triangles, pentagons or hexagons to form the roof. The second incorporates a flat or concave roof. Either type generally incorporates the addition of some type of venting immediately above the wall of the original tank. Very commonly, these retrofit roofs are made of aluminum. As a result, the roof is generally not a problem if there is fire. These tanks can be protected with the same types of fixed suppression systems as found in a standard closed floater.

Recently, a new system that incorporates a roof design similar to a traditional closed floater has been introduced. This retrofit is detected by the presence of the wind girder at the upper edge of the wall.

Physical Surroundings

All three of these flammable or combustible liquid storage tanks have requirements for their physical surroundings. First, all must have dikes built around them. Each dike must accommodate 110% of the capacity of the largest tank within the dike. If there are multiple tanks within a single dike, the volume of each tank must be factored into the calculations for the exact capacity and dimensions of the dike.

Each dike area must have some type of drain that will allow the removal of water created by precipitation. These drains are pipes that extend from the inside of the dike to the exterior. Each must have a shutoff valve. Normally, the valve is kept in the closed position to prevent any spilled product from exiting the dike. However, there have been instances where the valves were not closed and product has escaped.

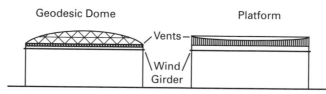

Figure 4-56   Retrofitted Open Floating Roof Tanks

During any emergency operation involving diked tanks, this valve must be checked to assure that it is closed. If there is a fire or heavy precipitation, control of this valve can be critical. For example, if the dike area fills with water from protection lines or precipitation, the tank or tanks within have a very bad habit of floating on the water. If this occurs, product lines feeding the tank can break and release product into the dikes.

Additionally, the tanks generally rise unevenly and thus may be damaged or spill their contents. Should cooling or suppression water build up in the dike, it may be necessary to open the drain. Obviously, if there is product floating on the water, the outflow must be monitored to prevent its escape. Also, it is essential to identify where the outflow goes to determine what additional control action may be needed.

Many dike areas contain additional equipment as well. Most commonly, this equipment includes above ground piping, valves, and product pumps. When preplanning these areas, it is essential to identify the location of such equipment, shutoffs, probable contents, and so on. Effective preplanning can be the key to handling incidents at this type of facility.

Failure Modes for Atmospheric Vertical Tanks

Fire is the most common problem associated with cone roof, open floating roof, and closed floating roof tanks. Overfills, lightning, and contractors account for most fire involving these tanks. Occasionally, a lightning strike will ignite flammable vapors within a cone roof or closed floating roof tank. The majority of these ignitions result in a partial or complete separation of the roof from the tank walls (except potentially in the case of weak wall-bottom weld tanks).

Overfills rarely result in a tank roof explosion, but commonly result in dike fires caused by vapors from the accumulated product. Such dike fires can result in the failure of aboveground piping, flanges, valves, pumps, and similar equipment located within the dike. These failures can be explosive if there is no venting of the pressure that builds up within the piping, valves, etc. Further, these failures routinely result in three-dimensional flammable liquid fires that are extremely difficult to control and extinguish. If valving at the base of the tank is not present or is not functional, these three-dimensional fires will continue until they can be extinguished or the tank loses all of its contents. Yet, extinguishment of these fires is essential to controlling the dike fire.

These tanks do not collapse as a result of any internal or external fire. Rather the walls tend to curl toward the inside of the tank as a result of their own weight. There is one documented exception to this behavior, however. In this particular incident, a three-dimensional fire occurred at the main product valve located at the bottom of the tank wall. As the fire continued to burn, it heated the steel shell and, in this case, the shell curled outward. Although there was no actual tank failure, the outward curl led to some very tense moments on the incident scene.

Another potential danger with these tanks involves the inherent properties and molecular weight of the specific product they contain. Specifically, tanks containing heavy crude oil or heavy waste oils are subject to two phenomena known as frothover and boilover. The high molecular weight and the presence of water are key factors in both cases. In either case, massive quantities of product are almost instantaneously ejected from the tank and cover huge areas around the tank. Both phenomena will be discussed at greater length in Chapter 7, Section 7.

Dome Roof Tanks

Unlike the previously discussed vertical tanks, dome roof tanks (Figure 4-57) are not atmospheric. Generally, they are classified as low pressure containers with maximum allowable working pressures of 100 psi or less. They may hold a range of products that include highly volatile liquids. Most commonly, they contain cryogenic liquids.

Many dome roof tanks are constructed with a relatively small diameter ranging from about 10 to 30 feet. However, bulk cryogenic liquid tanks such as those found at air separation plants can have diameters over 100 feet. If these tanks hold cryogenic liquids, their construction is similar to the MC 338 or any other cryogenic containers. Specifically, there is a product tank separated from an outer shell by a vacuum space and an insulating material.

When found at user sites, these cryogenic tanks normally have heat exchangers and in some instances pumps. The heat exchangers are metallic columns with multiple fins. They act to warm the product so it converts to a gas and pressurizes the container. This pressure provides the energy for product delivery on the site. At locations where the user needs the product in the gas state, the heat exchangers convert the liquid to gas, and then pumps build the pressure to desired levels.

Spherical tanks

Spherical tanks (Figure 4-58) are nothing more than large, spherical, high pressure tanks. They almost exclusively contain liquefied compressed gases such as LPG, propane, or butane. Routinely they have diameters of about 50 feet and are found in or near oil refineries. In some instances, they are co-located within the dike area of other vertical tanks. The only type of fixed system used for their protection is a deluge water system located in a ring around the top of the sphere.

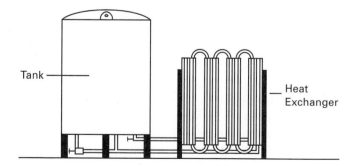

Figure 4-57   Cryogenic Dome Roof Tank

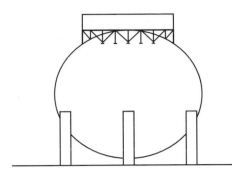

Figure 4-58   Typical Spherical Tank

Telescoping and Lifter Roof Tanks

Occasionally, two unique vertical tanks are encountered (Figure 4-59). Although these tanks are not very common, if they are present in a community, they require consideration. The two types are the telescoping tank and the lifter roof tank. In both cases, the tanks' walls rise and fall as product enters the tank. The weight of the tank provides a nominal pressure to products stored within.

In the telescoping tank, there are multiple sections of tank that nest within each other. Routinely, a steel superstructure of beams and girders surround these tanks. In this configuration, product entering the tank causes sections of the tank to telescope upward in a fashion similar to an extension ladder. In some instances, these tanks are hundreds of feet high and are used to store reserves of natural gas. In this case, the weight of the tank itself provides pressure for the gas.

In the lifter roof tank, a vapor space is maintained above a volatile, liquid hydrocarbon product. Due to the volatility of the product stored in these tanks, the vapor

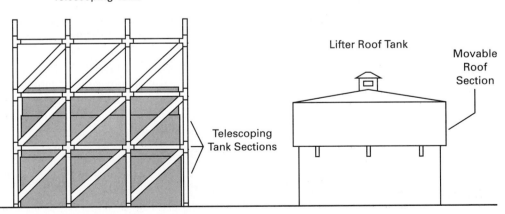

Figure 4-59   Two Vertical Tanks

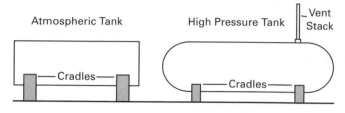

Figure 4-60   Typical Horizontal Tanks

space is normally too rich for ignition to occur. However, if this type of tank is emptied rapidly, outside air may enter through a breather valve. Should this occur, the vapor space may contain an ignitable atmosphere.

HORIZONTAL TANKS

Horizontal tanks fall into two categories: atmospheric and high pressure (Figure 4-60). Atmospheric horizontal tanks routinely hold hydrocarbons, hydrocarbon derivatives, corrosives, or similar liquids. High pressure horizontal tanks routinely contain lique-fied compressed gases including LPG, propane, butane, ammonia, chlorine, and so on. Their maximum capacity is in the range 30,000 gallons. However, the smaller sizes that range from 500 to 10,000 gallons are the most prevalent. They may be found individu-ally or in groups. Routinely they are supported by some type of cradle assembly. The cradle can be part of the tank or separate and constructed of steel or concrete.

If there is a surrounding fire, a steel cradle is a definite liability because the heat of the fire can cause it to lose its structural integrity. Should this occur, the weight of the tank can cause the cradle to collapse. Such an occurrence can have catastrophic ef-fects on the ability of the tank to maintain its integrity.

All horizontal tanks have pressure relief devices. In the case of liquefied com-pressed flammable gases, the vents must have stacks that extend well above the surface of the tank. This is to prevent flame impingement should discharging gas be ignited.

Because of all the recent problems with leakage of underground storage tanks, horizontal aboveground tanks are gaining in popularity. To meet safety and environ-mental needs and requirements, a new type of horizontal aboveground tank, involving the placement of a tank within a tank, has been introduced. The inner tank holds the product while the space between the tanks provides a secondary containment should there be some type of tank failure. Their behavior under fire or other incident condi-tions is unknown due to their recent introduction.

UNDERGROUND TANKS

When the term *underground tank* is used, people think of underground gasoline stor-age tanks. However, such tanks are only part of the picture. There are many other types of products and applications for these tanks. Again, any tank with 10% or more of its structure below ground is considered an underground tank. The products stored in these tanks range from fuel oil, to chemical intermediaries, to solvents, and even liquefied compressed gases. For example, the tank containing methylisocyanate (MIC) that was released in Bhopal, India, was an underground tank.

Until recently, the majority of underground tanks, whether atmospheric or high pressure, were constructed of steel. Unfortunately, these tanks have been prone to the same type of failures found in underground pipelines, namely corrosion cell deterioration. As a result, leaking underground storage tanks (LUST) have become a major source of environmental pollution.

To remedy this situation, laws and regulations require the replacement of these susceptible tanks, or the installation of monitoring and other safety equipment. There are three widely employed remedies for this problem. The first remedy involves replacing the steel tank with a fiberglass tank. The second remedy involves placing steel tanks within underground vaults of some type (Figure 4-61). The vault's design allows detection of any leakage, so environmental contamination does not result. The third option involves replacement of underground tanks with above ground tanks.

As mentioned, the most common failure mode for underground tanks is corrosion cell deterioration. When this occurs in an atmospheric tank containing liquids, the release often goes unnoticed until product shows up in drinking water, sewer systems, basements, or other locations. Under these circumstances, there is little a hazmat team can do other than monitor where product is detected, notify the appropriate authorities, take actions to minimize further spread of the product, and exercise necessary public protective actions.

In addition, there are several less common failure modes that must also be considered. First, tanks placed in porous or sandy soils have a bad habit of floating to the surface when ground water levels rise close to the surface. Such tank floating is rather common during severely wet periods or during flooding when the soil becomes saturated. Second, should the tank contain reactive substances that can decompose, polymerize or in some other way react violently, the internal chemical reaction can lead to

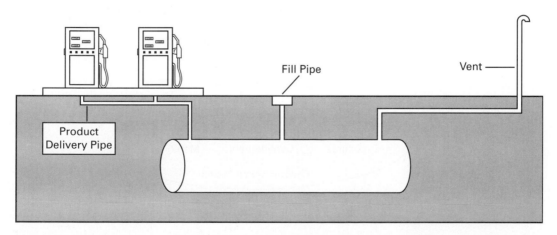

Figure 4-61   Typical Underground Storage Tank

explosive tank failure. Finally, and least commonly, high pressure tanks containing liquefied gases can fail explosively should corrosion cell deterioration weaken the tank.

GATHERING INFORMATION—SECTION IV
ENVIRONMENTAL INFORMATION

The final hazmat incident component to be considered is the environment. In this context, the environment is anyone or anything that is or may be affected by any situation or occurrence associated with the incident. Obviously, the potential number of considerations involving the environment is huge. For our purposes however, we consider nine broad environmental areas. These areas include occupancy/location, environmental media, time, exposures, meteorology, topography, geology, confinement, and conduits (Figure 4-62).

OCCUPANCY AND LOCATION

The occupancy and location of an incident are vital environmental considerations. Location considerations such as transportation corridors versus fixed facilities can provide specific operational advantages or hindrances.

There are different types of containers found in a highway incident versus a rail incident or a port for example. Transportation locations are generally more difficult to preplan as thoroughly as fixed facilities. Generally, vulnerable zones are identified as a half-mile strip on either side of the corridor. It is highly advantageous to preplan corridors to identify drainage patterns and retentions. Such information is generally available for recently constructed or reconstructed interstate highway corridors. For rail, airport, and water ways, topographic mapping may be the only information available. Pipeline maps correlated with topographic maps are extremely helpful.

Fixed facilities vary widely depending upon the specific occupancy of the facility. Incidents occur in occupancies ranging from residential locations, to storage facil-

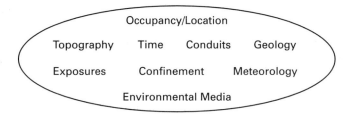

Figure 4-62   Environmental Considerations

ities, to production facilities, to SARA Title III planning facilities, and so on. By identifying the specific type of occupancy, responders can identify much information. Commonly, they can identify information about the types of products and containers potentially involved, as well as various design features of the facility.

For example, a bulk petroleum storage facility will have tanks, dikes, drainage systems, piping and pipelines, valving locations, and often loading racks. Refineries have a wide variety of hazardous substances and hazardous processes. A semiconductor production facility has high security, clean rooms, sophisticated detection and monitoring systems, gas cabinets, and so on. Thermal generating facilities have some type of fuel (either fossil or nuclear), high pressure and temperature systems, water treatment chemicals and systems, and so on. Illegal drug labs have the cooker (person making the drugs) and many different raw and intermediate chemicals. Furthermore they may have possible chemical or mechanical booby traps, mechanical systems, and modifications not designed for dwellings. Obviously, identification of the types of occupancies within a jurisdiction and the occupancy's associated characteristics and systems is a must for all hazmat responders.

ENVIRONMENTAL MEDIA

The first area of consideration is the environmental media into which the product has been or threatens to be released. There are four media into or onto which product release can occur. Those four media are air, water, surfaces, and subsurfaces (Figure 4-63). Air and water are self-explanations. However, it is helpful to examine surface and subsurface release media in a little more detail to fully understand when they apply.

Surface releases occur when a product encounters a solid material. Surfaces would include materials such as blacktop, macadam, concrete, gravel, soil, flooring, or a lab table. One key factor about surfaces is their relative absorbency and porosity. For example, a highly compact, nonabsorbent surface such as clay soil will tend to inhibit downward movement and percolation. However, a porous, absorbent surface such as a sandy or gravely soil will tend to rapidly draw and percolate a released liquid below its surface.

As the name implies, subsurface releases occur below the surface of the ground. Further, subsurface releases occur directly into the soil. Most commonly subsurface releases occur from pipelines, piping systems, underground storage tanks, or from a surface release involving a highly porous and absorbent soil. Product that has entered a storm or sanitary sewer system is not a subsurface release because the product has not been released directly into the soil.

Figure 4-63   Environmental Media

TIME

As with most all emergency response operations, time is an important consideration. Traditional time considerations include time of day, time of week, time of month, and time of year. As with almost all types of emergency operations, time considerations present major implications for hazmat response. Some of these include life safety hazards, access problems, product reactivity, container internal pressure, potential personnel demands, and similar operational activities and problems.

In transportation incidents, time variables such as rush hour will affect access to the scene, the number of potential victims on the roadway, the number of hazmat shipments, and so on. In fixed site incidents, time variables include things such as the number of occupants at the site, possibility of active chemical processing, alacrity of release detection, and so on.

EXPOSURES

The term exposure is used in the broadest possible sense. Specifically, exposures include the incident location, structures, persons, animals, biological habitats, ecosystems, or geologic structures and strata that have been or may be affected by the release of product. In other words, an exposure is anything that can be affected by the incident.

Depending upon the specific incident situation encountered, the identification of possible exposures may be extremely difficult. Further, it may require the assistance of environmental protection specialists. This is especially true for some biologic or geologic situations.

In the case of biologic exposures, even food materials such as molasses or milk can produce severe environmental impacts due to their high biologic oxygen demand (BOD). If such products enter a body of water, especially when there is little flow, they are capable of killing most forms of life found there. As microorganisms consume these products, they rapidly reproduce. The rampant reproduction produces huge numbers of the organisms that, in turn, consume huge amounts of oxygen. As a result, oxygen levels can become so depleted that aquatic organisms die from asphyxiation.

In the case of geologic exposures, groundwater contamination is always a potential. One key factor to consider is the proximity of the groundwater level to the surface. Another factor is the porosity of the soil. If the groundwater level is close to the surface, the soils are very porous, or there is direct access to the groundwater (e.g., a sink hole, quarry, or mine shaft), contamination is a very real possibility.

METEOROLOGY

Meteorologic conditions often play a tremendous role in the effective handling of many incidents. Considerations such as wind speed, wind direction, temperature, humidity, and barometric pressure, all impact the incident. Unfortunately, the implications and potential impacts of the weather are all too often unrecognized or underestimated.

Most commonly, responders identify basic and fundamental meteorologic conditions as the only ones needing consideration. For example, they know the wind may aid or hinder an operation, and its direction will help identify vulnerable zones, populations, and exposures. They realize that too much wind can blow fine powders and dusts over an extensive area, while too little wind will allow a vapor cloud to remain static.

Most responders are aware of some basic interactions between heat (quantified by temperature) and matter. Some understand that a direct relationship exists between temperature and chemical reactivity. Specifically, as the temperature of a product (or products) increases, so does its reactivity. Conversely, as the temperature of the product (or products) decreases, so does its reactivity. They further understand that temperature also directly affects the internal pressures found within any closed containers. They also understand that high humidity may cause water-reactive products to decompose or otherwise react.

However, there is much more to the realm of meteorology and understanding its impacts on an incident. For example, how will a plume behave when the wind speed is between 5 and 10 miles per hour, versus 15 to 20 miles per hour? Intuitively, most think that higher wind speeds result in more rapid mixing and dispersion. However, in most instances, higher wind speed (above 12 to 15 mph) will tend to cause a narrow plume that remains cohesive and extends over a greater distance down wind (Figure 4-64).

What about humidity? Most people are familiar with the basic implications that humidity has on the behavior of water-reactive substances. They recognize the potential negative impacts on vapor and plume dispersion and similar situations. But what about humidity-related properties such as the dew point? Dew point is the temperature at which water vapor from the air will start to condense. But what does it mean when the weather service advises that the dew point will be reached at a certain time during an incident? When the dew point is reached, the relative humidity will closely approach, if not reach, 100%. Obviously, this produces potential problems such as inhibited vapor and plume dispersion, a greater degree of hygroscopic wetting, and accelerated water reactivity.

Further, dew point can produce a significant impact on vapor buoyancy, more commonly known as vapor density. The standard vapor density of air is given as 1. This value is based on air having a combined molecular weight of 29. A quick way of determining approximate vapor density of a substance is to determine its molecular weight and compare it to that of air. If the substance's molecular weight is less than 29, it is lighter than air and will therefore rise. If its molecular weight is greater than 29, it is heavier than air and will therefore sink. However, when the dew point of the air is reached, water vapor starts to condense and form minute droplets. These droplets increase the density of the air.

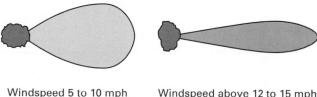

Windspeed 5 to 10 mph   Windspeed above 12 to 15 mph

Figure 4-64   Plume Dispersion

As a result, a substance that is normally heavier than air may become buoyant and rise above ground level (Figure 4-65). This is the reason so many "odor investigations" occur in the evening hours, after the dew point is reached. The vapor responsible for the odor is often present before this time but is much closer to the ground. When the dew point is reached, the vapor rises because the actual density of the air has increased. This makes the vapor somewhat more buoyant so it rises to nose level.

What about temperature? Again, most hazmat personnel are familiar with the more common impacts temperature has on an incident. They are less familiar with implication of thermal mass, rate of reaction and heat, topographic surface texture impacts on temperature, and so on.

Thermal mass is matter that holds heat (thermal energy). All matter has a tendency to hold heat. Generally, dense matter tends to hold heat better and longer while taking longer to warm up initially. This means there is a constant lag between the temperature of the matter and the surrounding ambient air temperature. Not only is there thermal lag, but there is often a substantial temperature difference as well. For example, on a warm summer day, it is not uncommon to find surface temperatures ranging from 135° to around 190°F, while the ambient air temperature may range only from 85° to 110°F. This means that a spill of a combustible liquid such as diesel fuel or kerosene can easily be heated to or above its flash point even though the ambient air temperature is well below that temperature. Obviously, this can have a substantial impact on the hazard of the incident.

Conversely, low temperatures can present many operational difficulties. For example, when temperatures are below freezing, the ground can freeze and eliminate a potential source of diking materials. Further, cold weather can lead to hypothermia, decontamination difficulties, equipment stiffening, embrittlement of containers, freezing of relief devices, and numerous other problems.

What about precipitation? Again, most response personnel readily identify rain as a potential problem during an incident. When the incident involves a water-reactive material, rain can obviously initiate reactions. When the incident involves liquid materials, rain can cause more rapid and greater spread of the substance.

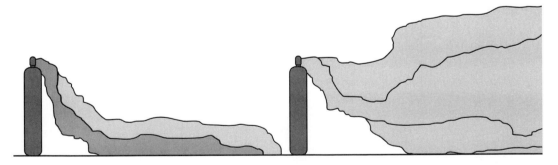

Above Dew Point   Below Dew Point

Figure 4-65   Behavior of a Substance with Vapor Density Greater than 1

However, few responders consider rain as a potential cause of an incident. In recent years, flooding associated with hurricanes and heavy rains have produced substantial hazardous materials incidents. For example, flooding in Texas led to the failure of multiple petroleum pipelines. These failures involved large releases of product and multiple fires. In other instances, flood waters have swept up a multitude of containers and deposited them very far from their sources.

If the containers are fixed, this movement generally involves the rupturing of attached piping systems and the release of product. If the containers are portable, they may impact fixed obstacles such as buildings or bridges and release their products. In other instances, as flood waters recede, numerous unmarked containers have been found throughout a large geographic area. Identification of products and responsible parties is tedious and time consuming. Residue of released products can produce widespread contamination. All in all, flooding presents a very difficult, frustrating, and time-consuming problem.

TOPOGRAPHY

Topography (i.e., the geologic lay of the land) will also play a major role in all aspects of the operation. As with other environmental factors, topography may be either a great ally or a great enemy. It will help determine the overall scene setup, the location of access points, decontamination points, command posts, and even the overall size of the operational zones. Topography will determine which way and how rapidly a spill will flow.

The United States Geologic Survey (USGS) has excellent maps, known as quadrangle maps, available. They show the topography, transportation corridors, fixed structures, bodies of water, and other topographic features. They are available through the USGS and the Superintendent of Documents in Washington, D.C. They are also available through various sporting goods and wilderness outfitters. One important note: one must know the quadrangle designation to purchase the right maps.

Topography's impact on plume spread is quite critical. Responders who use CAMEO, ALOHA, and other plume modeling systems are familiar with the terms *rural* and *urban*. However, in the case of plume modeling, these two terms do not simply reflect population density. An urban setting is any topography that will act to impede the spread of a plume in a fashion similar to a city setting. Hilly or mountainous terrains are examples of possible urban settings. Conversely, a rural setting is any setting that will allow maximum movement and dispersion of the plume. Flat terrain with single family dwellings is an example of a possible rural setting.

GEOLOGY

The geology surrounding the incident is yet another important environmental consideration. The specific makeup of the soils, underlying rock formations, surface water, and ground water are but a few critical considerations. For example, coarse sandy or gravely soils will tend to speed percolation of released liquids down through the soil. If the water table is close to the surface, such as in Florida, percolation can lead to rapid contamination of groundwater.

On the other hand, clay type soils will normally slow or possibly stop this downward movement. Even when the water table is close to the surface, the likelihood of major groundwater contamination is relatively low. However, this does not mean that groundwater contamination is not possible. Rather, this means that there is more time before contamination is a major result.

In other locations, bedrock is on the surface. In this case, percolation is not the initial concern, but the horizontal spread of a released product is. Since most bedrock has very little absorptive qualities, a liquid will rapidly move down hills over the surface. Such a situation can lead to rapid contamination of surface water such as streams, rivers, lakes, or ponds. In other areas, this bedrock layer may lead directly to wetlands, swamps, tidal zones, or the ocean. In any case, this product movement normally produces more response headaches.

DEGREE OF CONFINEMENT

The degree of confinement *existing* in the media must be considered. One of the first existing confinement considerations is whether the incident has occurred within a confined area (e.g., room, structure, or chemical reactor vessel) or outdoors. Probably the most familiar type of built-in confinement is the dike systems that surround above ground storage tanks. Other types of confinements include retention ponds, holding tanks, sumps, and similar systems. During an emergency response, confinements can dramatically limit the potential spread of released liquids. As a result, they are an ally of responders.

CONDUITS

Conduits are systems or contours that act to increase the spread of released product. Conduits include drainage ditches or swales, storm sewers or pipes, and sanitary sewers or laterals. Conduits normally complicate response efforts because they extend a released liquid, and often its vapors, over a greater area. Conduits often disappear underground and hide the actual direction of spread. Further, without good maps, it is often difficult to determine the full extent of the spread, outflows from the conduits, and the full extent of vapor dispersion through the system.

Obviously, environmental factors play a major role in an emergency response and a substantial amount of information must be gathered about them. As data about the product, container, and environment are gathered, they must be stored for use during the incident. Ideally, the data will be written down to provide a hard copy for anyone to use and double check. It is also very helpful to identify the sources of technical data in case they must be re-examined. Additionally, the writing down of the data and sources can be an extremely useful tool in the documentation process after the incident has been completed.

GATHERING INFORMATION—SECTION V
DAMAGE ASSESSMENT

Damage assessment is the final phase of gathering information. Specifically, it is the transitional phase between gathering information and estimating potential course and harm of the incident. Further, it incorporates functions found in both these steps of the GEDAPER process. Thus damage assessment forms the bridge between these first critical steps.

Traditionally, damage assessment strictly encompasses consideration of damage to the container and its potential outcomes. However, this myopic focus fails to identify that damage to the product or to the environment is sometimes equally as important as damage to the container. This is because damage to any of the incident components can have important impacts on the course and harm of the incident.

Damage assessment is the process of determining how incident occurrences have affected and will affect the product, container, and the environment. These occurrences include those already complete, those ongoing, and those that will occur during the control and cleanup phases of the incident. In other words, damage assessment involves determining the damage that has been done, is being done, and will be done. Through this determination, responders seek to identify the most probable impact the damage has or will have and its probable impact on the incident. As such, this process involves the gathering of information relating to damage, assessing that information, and then extrapolating its potential impact.

To understand the process more thoroughly, it is essential to understand the relationships between damage, energy, force, stressors, and their potential outcomes.

DAMAGE, ENERGY, AND FORCE

Damage is a result of the interaction of matter and energy, specifically when some type of energy acts upon matter. When this happens, it is generally called force. Granted, there are various and sundry forces (resulting from both potential and kinetic energy) that constantly act upon matter. Yet, these forces normally do not result in damage. The keys to whether damage occurs or not revolves around the amount of force exerted and the nature of matter so exposed.

For example, some unstable products, such as various organic peroxides and monomers, will be damaged by normal ambient temperatures in that they start to chemically decompose. The potential result of this decomposition is a violent chemical reaction or explosion. Containers can be damaged if they are exposed to the force known as pressure. Should an atmospheric container develop an internal pressure of 150 psi, it would most likely fail quite violently. Further, most ordinary structures will sustain damage or be destroyed by a relatively low, yet sudden, internal, positive pressure force.

In other words, as long as the matter is exposed to forces that fall within its inherent or design properties, there is no damage. However, expose matter to forces that exceed its inherent properties, and damage will result. The extent of the damage

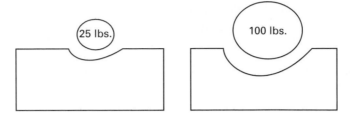

Figure 4-66   Degree of Damage versus Force Exerted

depends on the total amount and specific types of energy involved as well as on the properties of the matter. For example, more container damage will result when a container is exposed to an impact force of 100 psi as opposed to an impact force of 25 psi (Figure 4-66). On the other hand, more damage will occur to an aluminum container than to a high-strength low-alloy steel container when both are exposed to the same amount of force. Most commonly, these energies and forces are called stressors.

Stressors

There are three primary types of stressors that can act on the incident components (Figure 4-67). These stressors are mechanical, chemical, and thermal. Each of the stressors is composed of one or more types of energy that act on the matter through different types of forces.

Mechanical Stressors.   Mechanical stressors are those involving a physical force acting upon the container. Mechanical stressors include forces such as blunt force (e.g., impacts and falls), elongation (e.g., stretching and pulling), torsional (e.g., twisting), compression, pressure, and similar types of forces. Probably the two most severe and influential mechanical stressors in hazmat incidents are impact and pressure. As a result, each must be considered early in the assessment.

Impact (Figure 4-68) can be the result of the container banging into an object or some object banging into the container. In either case, at least one of the objects was in motion. This type of blunt force creates other forces as well. Impact can lead to elon-

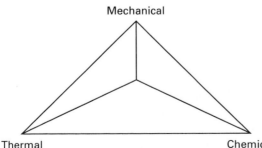

Figure 4-67   Primary Stressors

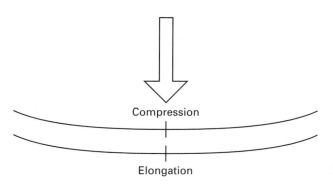

Figure 4-68   Impact Force

gation and compression at different locations in a container. As the container deflects, the edge of the wall facing the force tends to compress while the edge away from the force tends to elongate. Granted, if the force is strong enough, elongation will occur on both edges unless the impact force comes from the ends.

If the impact is accompanied by a twisting motion, torsional forces will also be encountered. This type of situation is quite common in any type of vehicle roll-over incident. Further, these torsional forces, especially when accompanied by compression and elongation, can lead to severe container damage.

Impact forces can also affect certain types of products as well. For example, older style acetylene cylinders contained a porous, solid media saturated with acetylene dissolved in acetone. Impact or other mechanical damage to the solid media would allow the acetylene to vaporize from the acetone. Since the pressure is generally above the self polymerization pressure of approximately 15.75 psi, the acetylene would polymerize. This polymerization reaction generates extreme heat and often causes the cylinder to explode. Obviously, shock sensitive substances would also be adversely affected by similar mechanical stressors.

Impact is also a problem when considering environment. Impact that results from accidents, explosions, and similar violent occurrences, can cause structural damage to buildings, highway structures, and other exposures. This type of damage can result in compromised structural integrity and the potential for structural collapse or failure. Any of these potential outcomes can also compromise the safety of response personnel and any victims.

Unlike impact, pressure is routinely found within the container. Pressure can be a result of normal shipping procedures or it can be induced by thermal stressors, chemical stressors, chemical reactions, or crushing impact forces. As the pressure pushes outward in all directions, it exerts elongation forces. These forces can lead to container failure, especially if they exceed the design limits of the container.

Further, these forces can lead to failure should the container be structurally compromised by other stressors. Since containers under pressure are continually exposed to this internal mechanical stressor, the potential for failure exists at almost any point during an incident. This is especially true if external stressors such as impact, elongation, compression, heat, or corrosion have substantially weakened the container.

Additionally, the higher the MAWP of the container, the greater the potential for such failures. This is because the ongoing stress produced by internal pressure can result in failure all by itself.

Chemical Stressors.   In the truest sense, chemical stressors involve products that are incompatible with their containers. Specifically, the product acts to attack, weaken, and possibly destroy the integrity of the container. Most commonly this attack occurs on the inside of the container. As a result, the damage to the container is undetectable through external container examination. Only after the chemical attack has reached the exterior of the container can it be seen. In the process, the chemical stressor can literally destroy the integrity of the container while providing little or no visual warning. Such chemical attack is most commonly the result of acids, oxidizers, bases, or a combination of these three. This is one reason why corrosives are involved in approximately 40% of all reported hazmat incidents.

In one incident involving two oxidizing acids, it took less than two hours for the incompatible corrosive mixture to eat through the bottom of the container. During an interior inspection after the incident, the liquid level was readily identifiable because over half of the container shell (cross-sectional thickness) was corroded away. Not only did the reaction eat away the shell, it produced gases and heat. Both the gases and heat produce the mechanical stressor (i.e., pressure) that also acted to cause failure.

It is worth noting that when acids react with a metal, two primary gases are produced. In the case of nitric acid, the most common gas produced is toxic, reddish-brown nitrogen dioxide. Most other acids will produce flammable hydrogen. In either case, the gas, heat, and pressure produced can lead to extremely violent failure of the container.

In some instances, chemical stressors can act on the exterior of the container. Such exposure can also compromise the structural integrity of small portable containers such as 55-gallon drums. But it is conceivable that much larger containers could be adversely affected if the exposure was long enough and the chemicals strong enough.

These external chemical stressors could be the result of exposure to a released corrosive solid, liquid, or gas. Severe corrosive attack is often seen affecting containers and structural members at facilities that process, use, or handle corrosives. This is especially true when the corrosives are gases, have high volatility (fumes), or are heated.

External chemical stressors are also associated with long-term container exposures to moist environmental conditions. These types of conditions exist with outside storage where there is exposure to the elements. In one instance, an employee went to move a 150-pound propane cylinder to a loading dock. As it was moved, the bottom failed. There was an instantaneous release of product, and the cylinder rocketed. The failure was so violent that the worker was fatally injured.

In any event, chemical stressors can have a profoundly negative impact on an incident. For example, normally acceptable tactics such as repositioning, plugging, or patching a 55-gallon drum can lead to complete container failure. Furthermore, in the presence of additional mechanical or thermal stressors, the container can simply fail while standing there with little or no outside help or advanced warning.

One misconception that requires attention is that self polymerization, decomposition, and incompatibility reactions between products are chemical stressors. They are not. Although each is a chemical reaction, it is not the reaction itself that leads to failure. Rather, they cause thermal and mechanical stressors that will affect the container. These reactions and some other types of chemical reactions are discussed in the next section.

Thermal Stressors.   Thermal stressors are those involving heat or its absence (cold) acting upon the container and, in some instances, the product. Heat transmission or absorption can be through radiation, conduction, convection, flame impingement, or gas law behaviors. However, the mechanisms through which thermal stressors affect containers and products are quite different and require separate examination.

In the case of the product, thermal stressors normally take the form of heat or elevated temperatures. It is important to note that elevated temperatures for various unstable products can range from 0° to 40°F. As the product warms, two types of chemical reactions can result: decomposition or polymerization. Those are dependent upon the specific chemical properties of the reactive substance involved.

Decomposition involves the chemical breakdown of the substance. Literally, the molecules of the substance break apart into small fragments. The process is similar to the pyrolysis of standard combustibles except that this decomposition is self-sustaining and self-accelerating. In any event, the process is exothermic and releases energy in the forms of heat, gases, and often flammable byproducts. The breakdown may start slowly and then rapidly accelerate as is common with organic peroxides. In other cases, the re-action is almost instantaneous, as is the case with various explosives. In either instance, massive amounts of energy can be released in milliseconds.

It is important to note that decomposition can also be induced through the intro-duction of contaminants or incompatible substances. In this particular case, although the decomposition is not induced by a thermal stressor, both thermal and mechanical stressors are most commonly produced. The thermal stressor is heat and the mechani-cal stressor is pressure.

Polymerization involves the combining of small molecules to form large mole-cules. The type of substance involved in a polymerization reaction is known as a monomer. Molecules of monomers have double bonds that easily break and then form new single bonds with the open bonds of nearby monomer molecules. The result is a larger molecule composed of multiple monomer molecules bonded together. The poly-merization process is always exothermic and in most instances will occur very easily. As a result, most monomers are mixed with a polymerization-inhibiting chemical known as an inhibitor or stabilizer.

Polymerization Components

Monomer

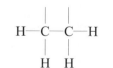

Monomer with Open Bonds

Polymer
(minus hydrogens)

Under normal production circumstances, the polymerization process occurs within the highly controlled environment of a chemical reactor vessel. This environment allows the reaction to proceed at a highly controlled rate. As polymerization continues, the heat of the reaction tends to increase the rate of reaction. When this occurs in a reactor vessel, it can be controlled.

However, should this reaction take place within a shipping container, there is no control. Because of this, the reaction speeds up and becomes known as a runaway polymerization reaction (*runaway* also gives you a good idea of an appropriate tactical option in such a case). This reaction releases a tremendous amount of heat and results in the physical expansion of the polymer. The heat and the expansion both contribute to a rapid increase in the pressure within the container. Often the heat of reaction is so great that it starts to pyrolyze the polymer. Pyrolysis produces decomposition gases that further increase the pressure. At this point, it is highly likely that the container will fail violently or even explosively.

Thermal stressors in containers take the form of low or high temperature conditions. Generally, low temperature conditions tend to embrittle the material of construction used in the container. Embrittlement weakens the container and leads to its failure. One major exception to this effect on the material of construction is aluminum. Aluminum actually strengthens when it is cooled. This is why cryogenic transports and cylinders increasingly use aluminum in their construction.

A low temperature thermal stressor can obviously be found when dealing with cryogenic containers. However, it can also be found when dealing with compressed gases or liquefied compressed gases. Low temperatures become a consideration when containers such as gas cylinders develop leaks. The breach in the container that forms the leak normally allows the product to escape. As the product escapes, it goes from a higher pressure to a lower pressure.

As the gas drops in pressure, a series of interactions occurs. These interactions are described in the Ideal Gas Law (a combination of Boyles' Law, Charles' Law, Gay-Lussac's Law, and Dalton's Law). As the pressure within a container decreases and the volume of the gas remains constant, its temperature must also decrease. Further, when the gas goes from the higher pressure within the constant volume found inside the container, to a lower pressure with an almost infinite potential volume, its temperature must decrease. In other words, the gas cools within and outside the container.

This cooling effect can be extreme and to such a degree that the gas, its container, and its immediate surroundings cool to the boiling point of the gas. This means that the following temperatures are possible:

1. Ammonia can cool to $-36°F$.
2. Chlorine can cool to $-37°F$.
3. Propane can cool to $-44°F$.
4. Hydrogen chloride can cool to $-121°F$.
5. Nitrogen can cool to $-320°F$.
6. Hydrogen can cool to $-422°F$.

Although some of these temperatures would have nominal impacts, others could severely embrittle a container. However, all will embrittle the elastomers used to make chemical protective clothing. Thus, they are a major potential contributor to PPE failure.

High temperature thermal stressors to a container can occur through two mechanisms: indirect and direct. Indirect mechanisms involve heat transmission by radiation, convection, or conduction, while direct mechanisms involve direct flame impingement.

Indirect mechanisms involve no direct flame contact. Rather, the heat produced by processing, chemical reactions, or a nonimpinging fire causes some type of effect on the contents of the container. The three primary ways to affect the contents include thermally induced chemical reactions (decomposition or polymerization), an increase in vapor pressure, or simple product expansion within the container. Remember, vapor pressure and volume increase directly with the temperature of the product.

Through these indirect mechanisms, a secondary stressor also develops. This secondary stressor is a mechanical stressor present in the form of pressure. This secondary mechanical stressor may then induce failure of the container. So it is not the indirect thermal stressor itself that leads to possible container failure, but rather the secondary mechanical stressor (pressure) that is responsible for a failure.

Depending on the type of container, the exact product involved and the circumstances surrounding the incident, this type of container failure can range from relatively minor pressure venting to a violent explosion. If the product is flammable, the release of potential energy in the form of a fireball simply compounds the danger of the situation. Unfortunately, the exact nature and violence of the failure are often impossible to predict.

Container failure as a result of this indirect mechanism is most commonly seen in atmospheric or possibly low pressure containers (Figure 4-69). For example, a steel 55-gallon drum exposed to radiant heat from a fire will tend to expand due to an increase in product vapor pressure. Typically, the heads of the drum expand and take on a rounded appearance. This is commonly known as rounding out.

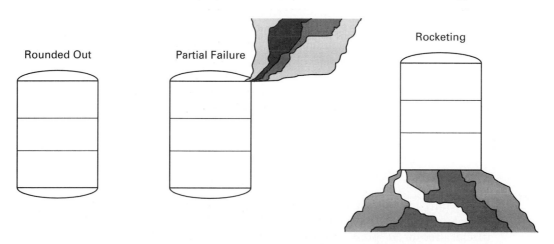

Figure 4-69   Container Failure

If the vapor pressure increases enough, a seam at one end commonly fails. This failure ranges from partial (a small opening between the wall and head of the drum) to complete head separations. Complete separation normally results in some degree of rocketing caused by the rapid escape of product and pressure. This type of failure has also been seen in older, small diameter, weak wall-to-bottom weld, and vertical storage tanks and many other types of ferrous atmospheric containers.

Partial or rocket type failures are not necessarily BLEVEs (Boiling Liquid Expanding Vapor Explosion, also known as Blast that Levels Everything Very Effectively), however. Rather, they may simply be a vapor pressure explosion resulting from the overpressurization of the container. Simply stated, the liquid does not have to be heated to its boiling point for the pressure to exceed the limits of the container. Obviously, atmospheric containers are the most susceptible to this type of failure. In some instances, this type of failure is accompanied by a large cloud of product. This cloud may simply be atomized liquid produced as it escapes while under pressure, and not by the instantaneous vaporization of a liquid heated above its boiling point.

Further, it should also be obvious that this type of container failure is much less common in high pressure containers due to their design features. The design features include an MAWP greater than 100 psi, thicker wall cross-sectional dimensions, and the typical rounded ends that more evenly distribute the pressure force.

Direct thermal stressors involve direct flame impingement on the container. In terms of thermal stressors, flame impingement is the most critical. Flame impingement can lead to rapid container failure by burning, melting, cracking, thinning, or tearing of the container's shell, depending on the specific type of container and its material of construction. Further, the potential high amount of heat conducted to the product can rapidly cause expansion, increase in vapor pressure, boiling, decomposition, or polymerization. Again, regardless of which occurs, a substantial amount of this heat energy is converted from a thermal stressor to the mechanical stressor, pressure.

Direct thermal stressors are a concern with any type of container, but they are especially troublesome when the container is a liquefied compressed gas container. There are multiple reasons for this high degree of concern. First, liquefied compressed gas containers are constantly exposed to a mechanical stressor in the form of pressure.

Second, the potentially high degree of heat energy transfer to the product will only act to exacerbate the existing mechanical stressor. Unless the container is protected by some type of fixed cooling system (e.g., deluge sprinklers or master stream devices) or is fitted with some type of thermal protection system, this type of thermal stressor can cause a BLEVE in as little as 10 minutes after impingement begins.

Boiling Point is Reached When:
Product Vapor Pressure = Atmospheric Pressure

Third, under most ambient temperature conditions, the liquefied gas is already heated above its boiling point. The only reason it does not boil normally is that the artificial atmospheric pressure within the container is quite high, thus preventing boiling. If the container's integrity is compromised, the artificial atmospheric pressure will no longer exist. Thus, the violent boiling or instantaneous vaporization of the product and extremely violent container failure will result.

Thermal Stressors and BLEVE

There are four primary mechanisms that cause BLEVEs: thermal, chemical, critical temperature, and mechanical. To fully understand the relationship between the stressors and actual damage, it is important to examine each mechanism in detail.

The traditional thermal BLEVE most commonly involves a liquefied compressed gas container that receives a direct thermal stressor. In this case, the container receives flame impingement to the vapor space located above the level of the liquid. Vapor cannot absorb nearly as much heat as liquid, so the container shell rapidly heats. As a result, the heat within the exposed metal increases dramatically until the metal starts to yield (Figure 4-70). The metal becomes softer, stretches, and often forms a bubble from the pressure within.

Eventually, the container's metal skin cannot withstand the pressure and begins to tear. As the container tears, the pressure within rapidly drops. Because the liquefied compressed gas is in the liquid state only from the pressure within the container, the liquid starts to vaporize almost instantly. This vaporization simply converts more potential energy into kinetic energy that can propel pieces of the container over tremendous distances.

Do not misunderstand this information. Liquids other than liquefied compressed gases can also create BLEVEs (e.g., steam boilers). However, substances that are liquids under ambient conditions generally are found in relatively weak containers. As previously stated, these lightly constructed containers most commonly fail due to overpressurization as a result of internal vapor pressure and/or thermal expansion well before the liquid reaches its boiling point. Flame impingement below the liquid level, or in some cases indirect thermal stressors, can also induce chemical or critical temperature BLEVEs.

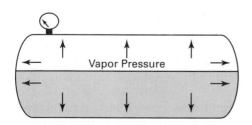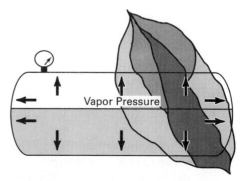

Figure 4-70 Thermal Stressors

In the case of a chemical BLEVE, the thermal stressor induces a chemical reaction in the lading such as degradation or runaway polymerization. Both reactions dramatically increase the pressure within the container and cause it to fail violently and produce a BLEVE.

In the case of a critical temperature BLEVE, the thermal stressor heats a liquefied gas to its critical temperature. When considering the liquefaction of a gas, critical temperature is defined as the minimum temperature to which a gas must be cooled before it can be liquefied through the application of pressure. When considering the heating of a liquid, critical temperature is defined as the maximum temperature to which a liquid can be heated before it will instantaneously convert to a gas.

Should the lading be exposed to sufficient thermal energy, the liquid can be heated to its critical temperature. If additional thermal energy is added to the liquid, it instantaneously converts to a gas and expands. The expansion produces significant pressure within the container causing a BLEVE.

It is important to note that the critical temperature BLEVE does not require heating the container above the liquid level. Rather, heating below the liquid level will produce maximum heating of the lading and the potential for the occurrence of this type of BLEVE.

It is also important to note the enhanced thermal protection found in today's tank car fleet and the positive impact it has had on thermal BLEVEs in this type of container. All tank cars that carry liquefied compressed gases are now required to have thermal protection in the form of four inches of mineral type wool, covered by a metal shell, or sprayed-on insulation. This thermal protection works very well in the prevention of thermal, chemical, or critical temperature BLEVEs in rail tank cars. The last rail tank car BLEVE of the types described thus far that this author has been able to identify occurred in 1986. This is not to say that this type of BLEVE is not going to occur, but it does speak highly of the protection afforded by these simple protective measures.

This incident data also provides insight into the protection of non-tank car containers. The simple fact is that if the container can be cooled so the metal shell does not heat excessively or the lading is kept reasonably cool, the likelihood of a BLEVE is substantially reduced. In the case of fixed containers, the use of fixed cooling systems such as deluge sprinklers or automatic master stream devices could substantially minimize the probability of BLEVEs involving this type of container.

The final BLEVE mechanism is the mechanical BLEVE. A mechanical BLEVE is a result of a mechanical stressor acting on the container. In this case, the mechanical stressor results in enough damage to the container that it loses its structural integrity. In other words, the container is so damaged that it can no longer withstand the pressure from within. The specific types of damage include dents, fractures, cracks, gouges, and cold working. All of these will be discussed later in this unit.

The classic example of a mechanical BLEVE occurred in Waverly, Tennessee, where a mechanical BLEVE resulted from a rail burn (gouge and cold working) on a pressure tank car containing propane. The day of the incident, the temperatures were in the low-to-mid thirties. Several days later during the recovery process, the temperature climbed to approximately 65°F when the tank car failed. The tank car's original test pressure was 300 psi. However, the car failed at an estimated 85 psi due to the loss of structural integrity caused by the 1/8-inch-deep gouge and cold working. Needless to say, mechanical stressors can have a profound effect on structural integrity.

GENERAL METAL PROPERTIES AND TREATMENTS

For the rest of this section, the discussion will focus on metal containers, their properties, and assessment of damage. Some of this information is a review from earlier in this chapter, but is provided in a different context.

A series of properties affects the behavior of metals when they are exposed to a stressor. Further, there is a series of metal treatments that also affect their behavior. By understanding the basic properties and how the various treatment processes interact, one is better able to identify the impact of stressors. The primary properties discussed include ductility, elongation, and tensile strength. The primary treatments discussed include annealing and tempering. These are not the only properties of metal that come into play. Rather, they are the predominant and most understandable.

Ductility

Ductility is the ability of a metal to form by pounding or drawing. Thus a ductile metal is also malleable. When a stressor acts on the metal, the metal will give. This means it will bend and generally thin when exposed to elongation forces. On the other hand, nonductile metals do not give and do not elongate very well. This property has a great impact on the total strength and the degree of brittleness of the container. High ductility means lower total strength. However, high ductility also means greater yield (mechanical change) and less immediate metal failure.

Elongation

Elongation is the ability of a metal to stretch and maintain its integrity. As stated earlier, when a substance is placed under tension, it tends to elongate. Ductile materials tend to elongate rather well, although they will thin at the same time. At some point however, the metal will eventually fail. When it starts to totally lose its strength, it has reached the yield point. Generally, a substance with good elongation properties will be ductile and will have a relatively low total strength. On the other hand, a substance with poor elongation often has a relatively high strength but is more brittle.

Tensile Strength

Tensile strength is the ability of a metal to withstand tension and elongation forces. Commonly, metals with high tensile strengths (meaning they resist elongation) also have high total strength. However, they tend to be more brittle and are susceptible to cracking as a result. Further, when high tensile strength metals yield, they rapidly fail.

Annealing

Annealing is one of the most common forms of metal treatment used today. It involves the heating of metal (or glass) to a specified temperature for a specified time. Following the heating, the metal is allowed to cool slowly.

The annealing process produces a stronger, more uniform metal when it is complete. The reason is that all metals have crystals within them. Before they are annealed, the metal crystals are in a haphazard arrangement similar to those found in cast iron. By heating the metal to the specified temperature and then slowly cooling, the crystalline structure acquires a more uniform nature. This uniform structure imparts additional strength to the metal.

Any subsequent heating and cooling cycle that occurs in an uncontrolled fashion will return the crystalline structure to a more haphazard arrangement. This is why the sections of metal on either side of a weld bead are the weakest points. This is also the reason that most rail tank cars are required to undergo post-welding heat treatment.

Additionally, the rapid cooling of mild or carbon steels to temperatures of $-40°F$ or below reverses the annealing process. Specifically, an extremely grainy, haphazard crystalline structure results. The structure is so grainy that it takes on an appearance similar to cast iron. As a result, these metals become extremely brittle and remain extremely brittle even after they have returned to normal temperatures. This situation is a potential concern should such metals be exposed to low temperature thermal stressors. These types of stressors are associated with releases of pressurized gas, liquefied compressed gas, or cryogenic liquid.

Most commonly, MS are annealed. In most instances HSLA and SS, as well as AL are not.

Tempering

Tempering is the process of hardening a metal through a different heat treatment or quenching process. The two processes are known as nonquench tempering (NQT) and quench tempering (QT). In some instances they are referred to as nonquenched and tempered or quenched and tempered, respectively. In both cases, the metal is heated to an extremely high temperature then cooled rapidly. The heating and rapid cooling of QT produce a metal that is quite hard and strong, but also can be somewhat or even extremely brittle.

NQT involves heating the metal to a specified temperature for a specified period of time. The metal is cooled relatively slowly through the use of some type of heated media such as a melted salt. This tempering process produces metal that is hardened, but not as hard as QT metal. Also, NQT steels are not as brittle as QTs.

The importance of these differences is extremely valuable when performing damage assessment. For example, many MC 331s are made of QT HSLA. Over the past few years there have been several incidents involving MC 331 propane cargo tanks on interstate highways. Rollover accidents occurred in both cases, and the head of each tank struck a bridge abutment. One occurred in Memphis, Tennessee, and another occurred outside White Plains, New York. In both of these cases, the head of the tank was cracked and ejected from the rest of the container (Figure 4-71). The remaining tank section

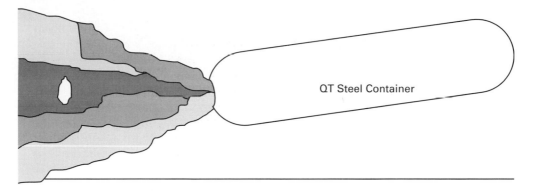

Figure 4-71   Ruptured Tank Made of QT Steel

rocketed a considerable distance, all the while spewing liquid propane. In both cases, the spilled liquid propane ignited and resulted in major fires and multiple fatalities.

When considering each incident from a damage assessment perspective, these results are predictable. Simply stated, QT HSLA is extremely hard and durable, but it is also rather brittle. The vehicles were traveling at highway speeds and were subjected to massive mechanical stressors when they hit the immovable bridge abutments. Since the QT HSLA is brittle, this is the perfect scenario to cause the metal to crack. Also, the containers were continually exposed to a pressure type of mechanical stressor. When all of these data are considered together, the outcomes of both incidents make perfect sense.

METALS AND THEIR PROPERTIES

High Strength Low Alloy (HSLA)

As mentioned earlier, HSLA is a high carbon steel that may contain some nickel, manganese, or chromium. Its most common applications involve high strength. As a result, it finds common use in high and ultrahigh pressure containers for compressed gases or liquefied compressed gases. HSLA is very hard and strong and has minimal ductility. Since it is so hard, it does not dent or abrade easily. When denting occurs, HSLA has a tendency to crack. It is found as QT or NQT types.

Another example of the differences between QT and NQT involves rail tank cars. Pressure rail tank cars produced after 1966 are most commonly made of NQT HSLA. Before 1966 they were made of QT. When assessing these tank cars, a tool known as a dent gauge helps to determine the severity of dents in the tank shell. NQT HSLA is able to withstand a more severe dent than is QT. This is why the dent gauge has two different concave (inward) and a convex (outward) radii (Figure 4-72). A pre-1966 QT is considered critically damaged if the dent radius is less than four inches, while a post-1966 car must have a dent radius of less than two inches.

Further, HSLA is not compatible with many corrosives unless it is protected by a lining or cladding. It is highly rigid due to its low ductility and hardness. As a result, it

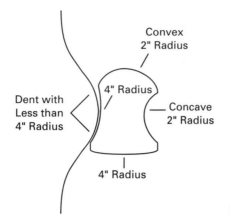

Figure 4-72 Dent Gauge

has a high tensile strength and does not elongate well. Further, it does not accept leak control plugs well because it does not give. When plugging is attempted, the plug merely shreds as it is driven into the breach.

Mild Steel

Mild steel is ductile relatively soft, and bends easily. As a result of its ductility, it has good elongation characteristics. However, during elongation, it will thin. The greater the thinning, the greater the likelihood of cracks or fractures forming on the convex side of the bend (Figure 4-73). Fractures are cracks that do not run the full thickness of the metal; cracks run completely through the metal.

Since MS is ductile, flexible, and elongates well, it is highly responsive to plugging operations for the control of leaks. Because the metal gives, a plug can easily enter the breach in the metal. It is also flexible and tends to spring back to its original shape, within reason. As a result, MS tends to "bite down" on the plug and hold it in place.

MS is not compatible with corrosives as it will rapidly corrode. It must be lined or clad if it is to contact these types of substances or environmental conditions. (As it reacts, it will release hydrogen gas.)

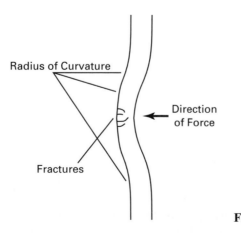

Figure 4-73 Fractures in MS

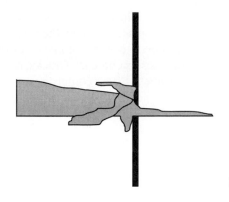

Figure 4-74   Plug Damage

Stainless Steel (SS)

SS has high strength and is very hard; it is similar to HSLA but not as brittle. It is also quite expensive, so its most common use is when high resistance to corrosion is required. There are a variety of alloys with differing corrosion resistance properties. As such, not all corrosives are compatible with any specific alloy.

Due to its hardness, it does not dent, crack or abrade easily. Also due to its hardness and subsequent lack of give, it does not accept leak control plugs well (Figure 4-74).

Aluminum (AL)

Aluminum is lightweight and has good impact strength. However, it is not as strong as ferrous metals when used in most containers. Routinely, specifications require from 25 to 50% greater cross-sectional thickness when it is used instead of ferrous metals. It also has relatively low reactivity with many substances. Since AL is relatively soft, it dents and abrades, especially when involved in sliding impact, such as high speed rollovers.

AL behaves differently than many of the other metals when exposed to thermal stressors. When exposed to low temperature stressors, it is unusual because it becomes harder. Generally, when exposed to high temperature stressors, it has a marked tendency to melt rather than explode. This is not to imply that an aluminum container will not explode, only that the likelihood is lower.

When it comes to leak control, AL also does not take a plug well. However, this is not due to hardness but rather to its amount of give. The give results from the softness of the metal and occurs when a plug is inserted. As a result, AL tends not to "bite down" on the plug. Consequently, the plug is often "spit out" of the breach.

DAMAGE TYPES

There are many types of damage found on a real container involved in a real incident. However, this damage can be identified as one of ten primary types. These types are scores, dents, gouges, cold working, fractures, cracks, punctures, tears, corrosion, and complete failure. Scores, dents, gouges, cold working, and fractures do not go completely through the container by themselves, although they can weaken a container enough for some type of complete failure to occur. Complete compromise is when the container is breached entirely and releases its contents. Cracks, punctures, tears, corrosion, and complete failure are responsible for such compromises.

Scores

Scores (sometimes called scours) are superficial damage done to the exposed surface of a container. They result from some type of sliding action between the container and some other object, similar to sand paper rubbing on wood. They result when a container slides over a surface or an object slides over the container's surface. Scores are analogous to abrasions. Commonly, they are called street burns when they occur on highway cargo tanks.

When dealing with hard containers (e.g., HSLA and SS), it is not uncommon to find scores that simply remove the paint from the surface of the container. This type of superficial scoring normally produce no actual damage to the container. However, high speed slides can produce gouges and cold working along with scoring.

When soft containers (e.g., MS and AL) sustain damage from sliding, it is not uncommon to find scores accompanied by dents, gouges, punctures, and tears.

Dents

Dents involve a blunt impact that causes deformation of a container's surface. In pliable containers (including ductile metals and plastics), denting is generally not a major problem unless it is deep, parallel to the long access of the container, or very large. In brittle containers, the deformation can be sufficient to induce complete failure. As shown before, dents produce curves that can thin the container's skin and lead to fractures or cracks.

Further, when the denting process is accompanied by sufficient force, it is not uncommon to find tears or large punctures. For example, a common occurrence in rail incidents involves the impact of a coupler with the head of an adjacent car. This scenario still occurs even with top and bottom shelf couplers. In pressure cars, there is normally only nominal, if any, damage to the product tank, due to the required presence of head shields. However, very few nonpressure cars are so equipped. As a result, when the couplers disengage, they can strike the bare head of the tank. This author has witnessed several such impacts that have resulted in major breaches in the tank car and subsequently a large loss of product.

Gouges

Gouges are tight, deep grooves cut into the skin of the container (Figure 4-75). These grooves are a result of a sliding motion over or by a tight radius object. The sliding action literally scrapes material from the shell of the container in the same fashion the tip

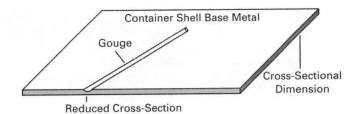

Figure 4-75   Gouge in Container Shell

of a pocket knife can scrape wax from a candle. The tight radius object can include a curb, the corner of a concrete median barrier, a rock, the edge of a rail, or even the inner edge of a rail wheel and it can result in specific types of gouges.

Curbs and concrete median barriers produce long gouges often accompanied by tight radius dents. Rails also produce long gouges that may be accompanied by tight dents and are commonly called rail burns. The inner edge of a rail wheel can produce a semicircular gouge commonly called a wheel burn. Any of these scenarios can involve gouging and may also be accompanied by cold working.

Obviously, softer materials most commonly used in atmospheric or low pressure applications will be more easily and rapidly gouged when involved in this type of situation. However, hard materials are also susceptible to this type of damage. Since harder materials are used in high and ultrahigh pressure applications, gouges can be extremely dangerous.

For example, it has been found that a 1/8-inch-deep gouge can compromise the structural integrity of a pressure rail tank car; likewise a 1/16-inch-deep gouge can compromise the structural integrity of an MC 331. The reason is simple. The process of forming a gouge removes material from the cross-sectional thickness of the container. The cross-sectional thickness is necessary for the container to withstand the pressure from within. When the gouge removes material, it weakens the container. The more material removed, the weaker the container. The higher the pressure within the container, the more likely the container is to fail.

Cold Working

Cold working (Figure 4-76) is another type of damage that reduces the shell's cross-sectional dimension but is more complex than a simple gouge. Unlike a gouge, cold working does not evulse shell material; rather it redistributes it.

A wheel burn provides a good example. As the inner edge of the wheel rubs against the container, it can abrade a given amount of material. However, it will also push and move material into new locations in a fashion similar to cold forging. When this type of cold working is examined closely, redistributed material often is located next to the thinned area.

Further, the redistribution and cold working of the shell material produce additional negative effects involving annealing and heat treatment. Cold working realigns

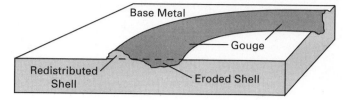

Figure 4-76   Cold Working

the crystalline structure of the metal, thus reproducing a haphazard arrangement. This results in more brittle and less strong metal that is subject to possible failure. As a result, this type of damage produces a double whammy of reduced cross-section and embrittled, weakened metal.

Fractures

As mentioned before, fractures are small fissure that most commonly occur on the convex sides of a dent. They are a result of tensile forces that thin or overstress the material. Although they do not run the full thickness of the container's shell, they nonetheless reduce its cross-sectional dimension. Depending on the specific material of construction involved, fractures may indicate a complete compromise of the container's structural integrity.

High and ultrahigh pressure containers are subject to failure when fractures occur. This is due to the ever present mechanical stressor, pressure. Unfortunately, fractures commonly occur on the inside of the container and cannot be identified through external, visual inspection.

Cracks

Cracks are the same as fractures with the exception that they extend through the entire cross-section of the container. They are most common in hard, brittle materials because softer ones tend to puncture or tear rather than stop at the crack stage. Unfortunately, many cracks are not visible to the naked eye until product escapes through the shell. As such, they are critically important when found on high and ultrahigh pressure containers because they indicate that the container has totally lost its structural integrity and may fail at any moment.

Punctures

Punctures are breaches caused by the penetration of some relatively pointed object. The force of the object causes an extremely deep, tight radius dent. The dent is so severe that it causes the material to rapidly thin, fracture, then crack. When the crack forms, the object now penetrates the breach. Such objects can include fork lift tines, guard rail, bolts, nails, and pipes; they may remain impaled within the container or may pop out.

Punctures are most frequently found in MS or AL because they are soft enough to allow the needed deformation and thinning. However, punctures can also occur in HSLA and SS but normally require extremely high energy levels to occur. The hardness of these metals and their high tensile strengths tend to limit this type of damage.

Tears

Tears are breaches that result when a crack, dent, or puncture is accompanied by a sliding action. This sliding action causes the material to experience tensile forces known as shear forces. These forces cause the container to separate, resulting in a linear breach that can also form a V-shaped breach.

Again, tears are most common with softer materials of construction and are uncommon in the harder materials. One reason for this situation is that the harder materials tend to crack as opposed to tear. Most commonly, when tearing occurs in high or ultrahigh pressure applications, it is a result of a massive internal pressure spike (jump in pressure) caused by filling gaseous systems with the liquid form of the product or rapid heating caused by a water hammer effect (adiabatic shock). This situation often results in a pressure induced explosion of the container or some similar type of violent occurrence.

When such tears occur, there is so much energy involved that the container will commonly deform substantially. Further, the edges of the tear are extremely sharp and can be as thin as one molecule. Obviously, extreme caution is warranted when working near the torn material.

Corrosion

Corrosion is a chemical attack on, or incompatibility with, the material used to construct the container. The chemical attack involves a corrosive and/or oxidizing chemical. As previously mentioned, such chemicals can range from strong acids or bases, such as hydrochloric acid or sodium hydroxide, to strong oxidizers like fluorine, chlorine, or nitric acid, to routine substances such as road salt and water.

Corrosion can damage the container from the outside or inside. It decreases the cross-sectional dimension by eating away at the shell. Corrosion can sometimes be detected by surface discoloration, shell or paint flaking, pinhole leaks, and in some instances accumulation of scale. However, it is possible that there may be no visual signs. As such, corrosion damage may not be readily identifiable even though it is present.

Complete Failure

With the possible exception of scores, all of these types of damage can compromise a container's ability to hold its lading. Compromises can run the gambit from minor pinholes to total disintegration. As such, the exact nature of the compromise depends on

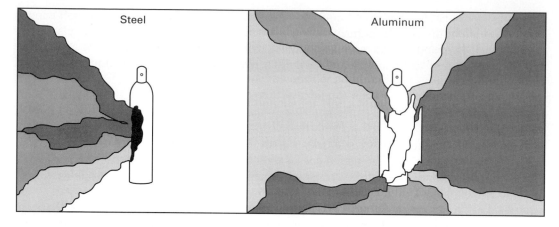

Figure 4-77   Complete Failure in Steel and Aluminum Tanks

the type or types of damage to the container, the type and magnitude of the stressor or stressors acting on the container, and the internal pressure of the container.

Complete failure is when a container literally comes apart because it lost its structural integrity. It can range from the shredding effect seen when a loaded paper bag is dropped, to the shattering of a glass jar, to the rocketing and shrapnel produced when a compressed gas cylinder fails.

When considering containers, (other than atmospherics) complete failure is routinely a violent event. This is especially true of high and ultrahigh pressure containers or metallic atmospheric containers exposed to fire. At this point, it should be obvious that the pressure within is quite capable of inducing such complete failures. Should the container suffer a sufficient impact force or flame impingement, it can easily develop a crack. Then the pressure within the container can cause the crack to rapidly expand and extend to the point where the container physically comes apart.

The exact nature of this failure not only depends on energy, stressors, and forces, but it also depends on the material of construction used in the container. For example, standard high pressure (2,000 psi or more) compressed gas cylinders can be constructed of steel (no more than 0.5% carbon) or aluminum alloy. When exposed to a fire, both types of cylinders will eventually fail quite violently. However, the steel cylinders tend to simply fail and rocket away from the fire. On the other hand, testing has shown that aluminum cylinders tend to come apart into at least five pieces of shrapnel (Figure 4-77).

ADDITIONAL DAMAGE CONSIDERATIONS

Finally, there are several additional considerations that require attention as part of a thorough damage assessment. Further, these data assist in the next step of the GEDAPER process, Estimating Potential Course and Harm.

Container Complexity

The complexity of the container is yet another consideration. There are two types of container complexity: simple and complex. A simple container has a shell that is the only integral structural component of the container. Some examples include, a 5-gallon bucket, 55-gallon drum, DOT 406, and a majority of other containers. Simple containers may have multiple walls such as those found in insulated cargo tanks or rail tank cars. However, the outer wall simply acts to hold insulation in place and protect it from damage.

Complex containers, on the other hand, have a minimum of two walls, both of which are integral to the designed performance of the container. The compromise of the outer shell can cause complete failure of the container. Cryogenic containers, with their thermos-like construction, are primary examples of complex containers. Their outer shell is integral to the entire containment system because it maintains the vacuum most commonly found within the insulating space between walls. Should the outer shell be compromised, the vacuum will dissipate along with its insulating properties.

Such a compromise can produce dire consequences since the liquids inside cryogenic containers are so cold. Heat will rapidly transmit to the shell of the product tank and raise the temperature of the laden. The temperature increase will cause the liquid to warm and boil more rapidly. This results in a pressure increase. The input of so much heat could result in violent container failure. Should flame impingement on the inner tank accompany such a compromise, BLEVE is highly probable.

Location of Failure

The location of container failure or anticipated failure needs to be identified. There are five locations where a container may fail: the wall, openings, valving, piping, and relief devices. The exact location of the failure can be critical in estimating the probable behavior of the container and potential for control of the leak and will be discussed more fully in Chapter 5.

Container Stability

The overall stability of the container is also quite important. It should be rather obvious that the location of the failure and the degree of failure are considerations in determining container stability. The physical attitude and location of the container can also be critical. Just as the stability of a vehicle is important when preparing to perform vehicle extrication, the physical stability of an involved container, structure, or exposure is also important.

For example, a highway cargo tank teetering over an embankment is not stable. The tank could slide down the embankment or move when entry team approaches. Either occurrence could result in injuries or complicate the incident situation. It is also important to remember that the stability of the container may change dramatically as the incident proceeds. The center of gravity will shift as product drains from the container. Obviously, this could affect its stability. Further, incident occurrences and stressors such as flame, radiant heat exposure, or the continued release of product may act to rapidly destabilize the container.

Although damage assessment is a complicated and possibly dangerous process, it is nonetheless important. The information gathered through this process becomes an integral piece of the puzzle when estimating the potential course and harm of the incident. As such, it is also integral to the determination of operational strategies and tactics.

INFORMATION MANAGEMENT

As the potential volumes of information about the product, container, and environment are collected, it must also be recorded. This recording process provides a means to identify, review, and document the data. For the data to be of most benefit, it must be recorded in some type of standardized format.

A standardized format provides many benefits. Although the benefits are many, some of the more important ones follow. First, a standardized format provides for rapid reference and recall. Second, the format identifies needed information. As such, it acts as a memory jogger and helps assure thorough data collection. Third, when data for a given section is not available it identifies an area requiring research, interviews with technical experts, assignment to a reconnaissance team, and so on. Finally, it provides a documentation of data used in making operational decisions. This type of data can be vital to answering questions about the incident approach used or actions taken.

As such, a set of product, container, and environmental data sheets is provided following the conclusion of this chapter. These data sheets, as well as others provided with other chapters and in Appendix D, are in use by active hazmat teams and are provided as a model for the development of your own sheets. Remember, forms should be user-friendly so they will be accepted and used as a tool. Further, actual incident experience has proven the value of these forms in documenting incident data.

SUMMARY

This chapter addressed the first step in the GEDAPER process, the Gathering of Information. It has discussed:

1. The need for information in any problem solving process
2. The three types of data that are needed including physical, cognitive, and technical
3. The three incident components about which information must be gathered including product, container, and environment
4. Primary product data including physical properties, chemical properties, quantities, and presence of release
5. Primary types of containers including fixed and transportation
6. The pressure ranges of containers and their implications

7. The five modes of transportation including highway, rail, air, water, and pipeline and their associated DOT Regulations found in 49 CFR Parts 100 to 185

8. The differences between bulk, nonbulk, and intermodal containers

9. The various types of fixed containers

10. Environmental considerations including occupancy/location, topography, time, conduits, geology, exposures, confinement, meteorology, and environmental media

11. The role of damage assessment regarding information gathering

12. The relationship between damage, energy, matter, and forces

13. The primary stressors, including mechanical, chemical, and thermal

14. Metals, their properties, and their treatments

15. Potential types of damage

16. Additional damage assessment considerations

At this point, we turn our attention to how this information is used to help estimate the potential course and harm of the incident, the next step in the GEDAPER process. Chapter 5 will address the use of information that has been gathered.

SAMPLE DATA SHEETS

INCIDENT NUMBER:_____ PREPARED BY:_____
        YR/MN/DAY/NO.

SCIENCE OFFICER:_____

ADDITIONAL SCIENCE PERSONNEL:_____

_____

Separate forms must be completed for each product involved.

PRODUCT NAME:_____

SYNONYMS:_____

FORMULA:

    [ ] Structural-_____

    [ ] Empirical-_____

Identification Numbers: U.N. Class/Division-_____, U.N. Ident. Number-_____

                        CAS-_____, STCC-_____

                        EPA Registration-_____

                        EPA Establishment-_____

NFPA 704 DESIGNATION:
      Health [ ],   Flammability [ ],   Reactive [ ],   Special Hazards [ ]

HAZARD COMMUNICATION/HMIS DESIGNATION:
      Health [ ],   Flammability [ ],   Reactive [ ],   Special Hazards [ ]

RELEASE STATUS: [ ]No Release, [ ]On-going Release, [ ]Complete Release, [ ]Unknown

QUANTITY: Reportable quantity (RQ)-_____

      Released-_____, Available for release-_____

FLAMMABILITY PROPERTIES			
Reference Sources	1	2	3
	pg.	pg.	pg.
LEL			
UEL			
flash point			
ignition temp.			
decomposition (y/n)			
explosion potential			

PHYSICAL PROPERTIES

Reference Sources	1	2	3
	pg	pg	pg
odor			
odor threshold			
color			
physical state			
physical form [ ] particulate [ ] granule [ ] slurry/gel [ ] cryogenic [ ] liquefied comp. gas			
boiling/condense pt.			
freezing/melting pt.			
sublimation (y/n)			
specific gravity			
vapor density			
vapor pressure			
Reid vapor pressure			
water solubility			

REACTIVITY PROPERTIES

Reference Sources	1	2	3
	pg	pg	pg
oxidizer (y/n)			
pyrophoric (y/n)			
corrosive (y/n)			
pH anticipated			
MSST			
SADT			
ability to explode (y/n)			
ability to polymerize (y/n)			
radioactivity [ ] alpha [ ] beta [ ] gamma			

TOXICITY PROPERTIES

Reference Sources	1	2	3
	pg	pg	pg
TLV			
PEL			
IDLH			
STEL			
CEILING			
LD_{50}			
LC_{50}			
exposure routes (I) inhalation (D) ingestion (S) skin abs./cont.			
carcinogen (y/n)			
mutagen (y/n)			
teratogen (y/n)			
target organ/s			
symptoms of exposure			
first aid			

COMPATABILITIES	INCOMPATIBILITIES
PPE:	PPE:
substances:	substances:

PROTECTION DISTANCES	
Isolation [ ]	Evacuation [ ]
small qty.	small qty.
large qty.	large qty.

MONITORING DATA

Anticipated Atmosphere Hazards:

flammable [ ], oxidizer [ ], oxygen deficient [ ], oxygen enriched [ ], corrosive [ ], radiation [ ], other: _____

Monitoring Factors:

Relative response	R.R. factor:	source:
Ionization potential	i.p.:	source:
Action levels (based on relative response)	10% LEL with RR Factor:	20% LEL with RR Factor
Minimum O_2 function level		

INSTRUMENTATION USED AND READINGS							
Time	hr.	hr.	hr.	hr.	hr.	hr.	hr.
Instrument	Reading	Reading	Reading	Reading	Reading	Reading	Reading
CGI							
% O_2							
pH paper							
Colorimetric Tubes (specify)							
Tube 1							
Tube 2							
Tube 3							
Dip Stick (specify)							
Radiation (specify)							
PID							
FID							

PRODUCT DATA SHEET

CONTAINER DATA SHEET—SCIENCE GROUP

INCIDENT NUMBER: _____ **PREPARED BY:** _____

Separate forms must be completed by Science for each container involved as needed.

TYPE OF CONTAINER:

Portable Container: applies [ ], does not apply [ ]
    Non-bulk [ ]—(less than: 119 gal.—liquid; 882 lb.—solid; 1,000 lbs. water capacity—gas)
        bag [ ], bottle/jar [ ], box [ ],
        drum [ ], type:
            fiber [ ], steel [ ], stainless steel [ ], plastic [ ]
            35 gallon [ ], 55 gallon [ ]
        cylinder [ ], type:
            liquefied compressed gas [ ], compressed gas [ ], cryogenic [ ]
            other/specify: _____
    Bulk [ ]
        portable tank [ ]
            Spec. 56 [ ], Spec. 57 [ ], Spec. 60 [ ]
    Intermodal [ ]
        container/trailer [ ]: COFC [ ],/TOFC [ ],
        Tank [ ]: IM 101 [ ], IM 102 [ ], SPEC 51 [ ]
        Capacity: gallons _____, pounds _____, cubic ft _____
Fixed Container [ ]:
    Atmospheric [ ],
        fixed/cone roof [ ], floating roof [ ], internal floater [ ], retrofit floater [ ], dome [ ]
    Low Pressure [ ]
        dome roof [ ]
    High Pressure [ ]
        horizontal pressure [ ], pressure sphere [ ], reactor/process vessel [ ],
        other/specify: _____
    Ultra-High Pressure [ ],
        tube bank, reactor/process vessel,
        other/specify:
    Capacity: gallons _____, barrels _____, cubic ft. _____

TRANSPORTATION: applies [ ], does not apply [ ]
    Mode: highway [ ], rail [ ], air [ ], water [ ], pipeline [ ]
    Highway: apply [ ], not apply [ ]
        box [ ], van [ ], refrigerated [ ], flatbed [ ], dry bulk [ ], MC 306/DOT 406 [ ],
        MC 307/DOT 407 [ ], MC 312/DOT 412 [ ], MC 331 [ ], MC 338 [ ], tube trailer [ ]
        other/specify _____

Rail: apply [ ], not apply [ ]
    flat [ ], box [ ], hopper/gondola [ ], dry bulk [ ], tube [ ],
Tank car [ ],
    Non-Pressure [ ] (low pressure)
        DOT 103 [ ], DOT 104 [ ], DOT 111 [ ]
    Pressure [ ]
        DOT 105 [ ], DOT 109 [ ], DOT 112 [ ], DOT 114 [ ]
    Misc. [ ]
        DOT 113 [ ], DOT 115 [ ], DOT 106 [ ], DOT 107 [ ], DOT 110 [ ]
    other/specify _____
Air: apply [ ], not apply [ ]
    passenger craft [ ], cargo craft [ ], work craft [ ], military [ ]
Pipeline: apply [ ], not apply [ ]
    liquid [ ], gas [ ], slurry [ ], other/specify _____

This Section Applies to all container types.

CONTAINER MAWP (design pressure):
    atmospheric (0-5 PSI) [ ], low pressure (5-100 PSI) [ ], high pressure (100-3000 PSI) [ ],
    ultra-high pressure (3000 UP) [ ]

RELIEF DEVICES:
    none [ ], spring loaded (valve) [ ], rupture disk (vent) [ ], fusible plug/link [ ]

MATERIAL OF CONSTRUCTION:
    Non-metallic: [ ],
        paper [ ], cardboard [ ], wood [ ], glass [ ], plastic [ ],
        other/specify _____
    Metallic: [ ]
        aluminum (AL) [ ], mild steel (MS) [ ], austinitic stainless steel (SS) [ ]
        high strength low alloy (HSLA) [ ], quench tempered (QT) [ ], NQT [ ],
        (pre 1966/515-B & 212B—use 4 in minimum radius for rail dent gauge, post 1966/
        TC-128—use 2 in minimum radius for rail dent gauge)

COMPARTMENTS: yes [ ], number _____, no [ ],
    Capacity and arrangement of each compartment:
        # 1 Compartment: capacity _____ lading: _____
        # 2 Compartment: capacity _____ lading: _____
        # 3 Compartment: capacity _____ lading: _____
        # 4 Compartment: capacity _____ lading: _____

CODES OF CONSTRUCTION: 49 CFR [ ], NFPA [ ],
    page and section number (if available): _____

SPECIFICATION MATERIAL THICKNESS:
    wall/shell/barrel _____, head _____

Gross Weight _____, Net/Tare Weight _____

DAMAGE ASSESSMENT

STRESSORS:

Thermal: radiant [ ], impingement [ ], chemical reaction [ ]
Chemical: corrosive [ ], acid [ ], base [ ], oxidation [ ], specify _____
Mechanical: impact [ ], friction [ ], pressure [ ], substance expansion [ ],
    source/s of pressure _____

TYPE AND DEGREE OF DAMAGE:

Damage present: [ ]
    thermal [ ], deformation [ ], expansion [ ], dents [ ], burns (rail/street) [ ],
    scores [ ], gouges [ ]
    additional information: _____

Dents in rail and pressure containers:
    dent radius _____, dent depth _____

Potentially critical dents present [ ]:
    large & parallel to long axis [ ], greater than 1/4 container diameter [ ]
Scores and gouges in rail and pressure containers:
    depth—1/16″ [ ] (little damage), 1/8″ [ ] (product transfer), 1/4″ [ ] (critical)
    weld involved (y/n) [ ],
        gouge or other removal of bead material only [ ]:
        gouge to weld bead removing base metal (potential critical container) [ ]
        additional information: _____

Breach: [ ]
    location:
    opening/s [ ], shell/wall [ ], piping [ ], valving/attachments [ ], relief devices [ ]
    additional information: _____
    _____

    type and degree:
    corrosion [ ], thermal burn-through [ ], pin-hole [ ], split/tear [ ], crack [ ],
    complete failure [ ]
    additional information: _____
    _____

CONTAINER COMPROMISE:

Is structural integrity presently compromised? **(Y/N),** By what stressor? thermal [ ],
chemical [ ], mechanical [ ]
May structural integrity become compromised? **(Y/N),** By what stressor? thermal [ ],
chemical [ ], mechanical [ ]

Net Thickness: Container thickness minus depth of damage = net thickness
        Specification thickness _____
Is net thickness less than specification thickness? **(Y/N)**

Rail and pressure:
Container is critical [ ], non-critical [ ]

**If container is critical, immediately consider tactical options including withdrawal,
evacuation, etc.**

ENVIRONMENTAL DATA SHEETS—SCIENCE GROUP

INCIDENT NUMBER: _____ **PREPARED BY:** _____
                    YR/MN/DAY/NO.

BASIC INCIDENT INFORMATION

Location: _____
_____

Occupancy or Transportation Type: _____

Initial Entry Time (all times must use military time): _____
Updated Entry Times: _____, _____, _____, _____, _____, _____, _____, _____, _____

Situation Status (upon arrival):
Spill (release) (y/n)
    type of containment, solid [ ], liquid [ ], gas [ ], size contaminated area _____
Leak (y/n)
Fire (y/n)
    fuel: product [ ], container [ ], exposures [ ]
Explosion (y/n)
    possible [ ], occurred [ ], on-going [ ]
Additional information: _____

CONFINEMENT

within a structure [ ], outside [ ],
confinement devices: dikes [ ], retention pond, detention pond [ ], retention tanks [ ]
other/specify _____

CONDUITS

drainage ditch/swale [ ], storm sewers [ ], gullies [ ]
other/specify _____

EXPOSURES

People/Populations:
    Victims: involved [ ], contaminated [ ], injured [ ], entrapped, number _____
    Population/Occupancies Endangered:
        residential [ ], commercial [ ], mercantile [ ], industrial [ ], mixed [ ], hospital [ ],
        nursing home [ ], school [ ], prison [ ], transportation corridor [ ],
        other/specify _____
        _____

Structures/Property
    types: structures [ ], processes [ ], containers [ ], vehicles [ ], water wells [ ],
    closed water storage/treatment [ ], sewage treatment [ ], food production/handling
    facilities [ ]
    other/specify _____

EXPOSURES (CONT.)
Natural
bodies of water: stream [ ], river [ ], pond [ ], lake [ ], open reservoir [ ], wetlands [ ], estuary [ ], ground water [ ]

soils: sand [ ], gravel [ ], clay [ ], compacted ground [ ], asphalt [ ], concrete [ ]

living organisms: dead animals/plants present [ ],

    animal: mammal [ ], fish [ ], bird [ ], endangered species [ ], farm animals [ ]

    plant: agricultural [ ], aquatic [ ],

other/specify _____

WEATHER

Meteorologic readings should be taken every 15 minutes. In critical situations, readings may be needed at intervals of less than 15 minutes. In non-critical situations, intervals may be longer.

On-Scene Weather Station

TIME	hr.	hr.	hr.	hr.	hr.	hr.	hr.
temperature							
humidity							
dew point							
wind: direction speed							
bar. pressure							

NOAA

TIME	hr.	hr.	hr.	hr.	hr.	hr.	hr.
temperature							
humidity							
dew point							
wind: direction speed							
bar. pressure							

Other Source (specify):

TIME	hr.	hr.	hr.	hr.	hr.	hr.	hr.
temperature							
humidity							
dew point							
wind: direction speed							
bar. pressure							

5 Estimating Potential Course and Harm

CHAPTER OBJECTIVES

Upon completion of this chapter, the student will be able to:

1. Explain the role of estimating in the operational decision-making process
2. Identify and describe the three steps involved in hazard analysis
3. Relate the three hazard analysis steps to the estimating process
4. Describe the three incident elements
5. Identify the incident components associated with each incident element
6. Estimate the potential course and harm of the three incident elements if provided with a scenario of actual incident
7. Identify and describe the four primary types of releases
8. Explain release mimicry and transformation
9. Explain the role of pyrolysis and combustion products in the estimating process
10. Describe the role of the incident estimate and mode of operation determination
11. Describe how the estimating process helps identify the need for additional incident information

ESTIMATING THE INCIDENT'S COURSE AND HARM

The Estimating Process

Once information about the product, container, and environment has been gathered, and a thorough damage assessment is complete the question is, "What do we do with all of this information?" Most concisely speaking, the information must be examined (analyzed), put into a usable format (assessed), and used to identify appropriate response actions (strategies and tactics).

Estimating involves a series of individual predictions that are brought together (synthesized) to identify an overall picture. To comprehend this process more fully, as well as the linkage between the gathering of information and the estimating of course and harm, an understanding of analysis and assessment is crucial. Further, this understanding helps explain how information gathering and estimating fit within the overall decision-making process (command sequence).

Analysis is a process of examining something by identifying its fundamental features (incident components) and determining each feature's basic characteristics. Once identified, these features and characteristics are examined to identify their relationships to each other and to the system as a whole. The step of gathering information incorporates the first two phases of analysis. Namely, it breaks the incident into components (ie., the fundamental features or pieces) and seeks to identify each component's basic characteristics. The final phase of analysis, identifying the interrelationships between the components, is the first phase of estimating.

Assessment is the estimation or evaluation of the importance, character, or amount of something. In this case, assessment involves estimating the importance, character, and amount of damage, hazard, vulnerability, and risk associated with the incident.* Specifically, assessment's intent is to extrapolate the potential course and harm of the incident. Here again, it is necessary to define terms including hazard, vulnerability, and risk.

It is important to note that the planning process encompasses a process known as *hazard analysis.* Hazard analysis is subdivided into three primary steps. These steps are known as hazard identification, vulnerability analysis, and risk analysis. The approach set forth herein reflects this systematic methodology for emergency-response evaluation as opposed to the planning function.

Hazard Assessment

Hazard assessment (i.e., hazard identification) is the process of identifying the following:

1. What substance(s), container(s), and environment(s) are involved
2. Where they are located
3. What has happened or may possibly happen
4. How much and how many are present (the quantity)
5. What the situation can do (chemical properties, physical properties, stability, etc.)

Much of this type of data is derived while gathering information. As such, hazard assessment is a rather rote process (with the exception of identifying what it can do). For the most effective performance of this aspect of hazard assessment, the individual must be able to synthesize based on the data collected. Synthesis is the opposite of

---

*NOTE: The term *damage* and the process of damage assessment were discussed in detail at the end of Chapter 4. As such, no further discussion should be needed. However, considering the definitions of *analysis* and *assessment,* it should be evident why damage assessment bridges both gathering information and estimating potential course and harm.

analysis because it involves taking the information derived from the analysis process and putting it back together to garner a true and realistic view of the entire situation. To most effectively synthesize, an individual must be able to extrapolate.

Extrapolation is the estimation of the value or importance of a variable outside its observed range. Further, extrapolation involves inferring or predicting something that is not known from data that is known. In other words, the individual uses available information to predict potential outcomes, occurrences, impact, and so on, although they have not occurred or displayed themselves at this point.

Such mental manipulation of data requires extensive, indepth knowledge, training, and experience. It must have its basis in sound knowledge, understanding, and thinking. Regarding levels of learning and education, it is at the highest possible level. As such, the ability to perform synthesis through the incorporation of extrapolation is the true difference between a hazardous materials technician and a hazardous materials specialist. This ability is often the difference between an excellent strategist and tactician, and one that barely makes the grade.

It is extremely important to consider that not all people are capable of or are good at synthesis or extrapolation. Some people do not have the background knowledge, level of experience, or simply a mind that works in this fashion. As such, not everyone is suited to be a hazardous materials specialist.

Vulnerability Assessment

Vulnerability assessment or analysis is the process of identifying a geographic area potentially affected by the incident's occurrences. Once this geographic area is identified, everyone and everything within the area are vulnerable to impact. Depending on the specific incident and situation, the vulnerable zone (geographic area) may range from a few feet to many miles. Within the zone, one may find a plethora of potential vulnerable parties. Some of the potential vulnerabilities include responders, victims, containers, structures, other substances, sensitive environments (e.g., bodies of water and wild life), and so on.

Contrary to the belief of many outside the emergency-response community, the identification of the vulnerable or potentially vulnerable zone of a hazmat incident normally involves a large dose of judgment and experience. Although there is a multitude of technical support systems that range from the *Isolation Table* in DOT's North American Emergency Response Guide (ERG) to computer plume models, the data that they generate are only guidelines. There is a multitude of variables that dramatically alter the actual vulnerable areas from the "text book answers" as provided by these technical supports. Field tests and actual incident experience have shown that even the most sophisticated models have limitations and cannot incorporate all the potential variables.

Finally, most of these projections or guidelines are based on a supposed *worst case scenario*. However, it is necessary to define *worst case*. Is this the worst case for responders, victims, the public within 500 feet, the public beyond 500 feet, or the environment? Worst case depends on your perspective and is very difficult to define.

For example, the worst case in SARA Title III planning is a release that will take ten minutes to empty the entire container. Is that worst case a greater problem than a catastrophic failure that releases all the product right now? The answer to that question is that it all depends on the exact incident and situation. This is why vulnerability assessment can be very tricky and imprecise, to say the least.

However, this is not to indicate that the information provided by these systems is not worth considering. Rather, this data must be factored into the synthesis and extrapolation processes while considering the true magnitude of the hazard. Specific considerations include the incident conditions and situation as well as a myriad of other variables encountered during any emergency response.

Risk Assessment

Risk assessment (sometimes known as risk analysis) is probably the most misrepresented and misunderstood of these three types of assessment. Specifically, risk is the chance of injury or loss. This means that risk assessment is the process of identifying the *probability* that a certain course or harm will occur. This brings us to the crux of another very important aspect of the estimating process: during every incident there are *multiple possible occurrences and outcomes* (course and harm). The specific possible outcomes and their probability of occurring revolve around the incident situation and specific control options (strategies and tactics) chosen by the responders.

Putting Hazard, Vulnerability, and Risk Assessment Together

Consider the incident situation where a 55-gallon drum has fallen from the back of a pickup truck. The hazard and vulnerability assessment indicated the following:

1. The incident occurred at an intersection in the middle of town. (hazard and vulnerability)
2. There is a 55-gallon drum containing a flammable liquid involved. (hazard)
3. The liquid consists of a mixture of mid-level molecular weight liquid hydrocarbons, alcohols, and ethers as well as some benzene, toluene, and xylene. (hazard)
4. The drum has a small puncture on the bottom that is releasing a trickle of product. (hazard)
5. The product is forming a small pool (5 feet in diameter) in the middle of the intersection. (hazard and vulnerability)
6. The product:
   - Is flammable (hazard and vulnerability)
   - Is toxic by inhalation, ingestion, and skin absorption (hazard and vulnerability)
   - Is a central nervous system depressant, a pulmonary toxicant, a defatting agent, and a suspected carcinogen (hazard)
   - Has a TLV-TWA of 300 ppm and an IDLH of 3,000 ppm (hazard)

- Has vapors that can cause dizziness, altered levels of consciousness, and suffocation (hazard)
- Is in a container that may explode if exposed to the heat of a fire (all containers may explode *if* exposed to the heat of a fire) (hazard)
- Has vapors that present an explosion hazard anywhere since the vapors may travel to a source of ignition and flash back (vulnerability)
- Can cause structural turnouts and self-contained breathing apparatus (SCBA) to provide limited protection for responders (vulnerability)

7. If there is fire involving tank cars or trucks, isolate an area of 1/2 mile in all directions. (vulnerability)

Considering the hazard and vulnerability data, the worst case scenario would seem to indicate that this is an extremely hazardous and quite possibly life threatening situation (which granted, it could be, if it were *completely* mishandled). Based on the data provided and considering a worst case scenario, first responders and some hazmat personnel may estimate the following course and harm:

1. The product is highly toxic and dangerous.
2. Level A or Level B chemical protective clothing is needed to handle this incident.
3. There may be a large fire and explosion that can affect a substantial area.
4. This is too dangerous for operations level personnel to handle.

Consequently, responders could identify a plan of action that includes evacuating one block in all directions, calling for a hazmat team for their chemical protective clothing and expertise, establishing a perimeter of at least 500 feet from the incident, and waiting. The question now becomes, "Is this plan of action appropriate?"

The answer is no. Why? The incident estimate was not based on sound synthesis and extrapolation, considering the true magnitude of the hazard for this incident situation. In other words, it did not accurately identify the true nature of the incident, nor the probable course of events based upon the incident situation.

When considered in light of the following information, this scenario takes on an entirely different complexion. The common name for this flammable liquid is gasoline. The estimate, although possible in a large quantity release, is highly improbable given the situation provided. Additionally, if appropriate defensive actions are taken, the probability of the worst case prediction taking place is nominal to non-existent. Such defensive actions may include the use of turnouts and SCBA while diking to prevent further spread, foaming the spill, or placing absorbent material to handle the puddle.

Yet another consideration is that should the original Plan of Action (POA) be chosen, the original estimate of course and harm may well become a self-fulfilling prophecy. Specifically, the inaction identified in the POA would allow the product (both the liquid and its vapors) to spread over a greater area. In turn, the spread would increase the probability of vapor finding an ignition source. Should an ignition source be found, it would initiate combustion, possibly an explosion or any of the other identified negative impacts.

This brings us to the crux of the matter regarding risk and probability. Specifically, not only must the worst case scenario be identified, but it must be identified consistent with the incident situation as it is found. It also must be assessed consistent with a reasonable expectation of its course and harm, and it must be assessed consistent with potential strategic and tactical options for its control. Further, the risk assessment process must seek to identify the most probable course(s) and harm(s) of the incident.

Some may still ask, "How can an incident have multiple possible courses?" The answer involves the interaction between the incident's course and the strategic and tactical options used for its control. Everyone has heard the saying, "For every action, there is an equal and opposite reaction." This is true for strategies and tactics used to control any given incident regardless of its type (e.g., fire, hazmat, and EMS).

Most commonly, emergency operations require a series of strategies and tactics be performed in a coordinated and controlled sequence. This is the same approach that is applied to the advancement of hose lines, ventilation, rescue, application of water, and so on during a structural fire attack. Furthermore, different tactical operations produce different outcomes. Consider a direct versus an indirect application of water, or horizontal versus vertical ventilation, for example.

As a result, all control options must be assessed regarding both their positive and negative impacts on the efficacy and efficiency of the operation. Also, these options must be assessed regarding their impact on the incident course and harm as well as incident occurrences before and during their implementation.

For example, the decision to apply a foam blanket to the gasoline drum spill will produce both positive and negative reactions. On the positive side, the foam blanket should suppress the flammable vapors produced by the liquid (if an appropriate type and quantity of foam are used). On the negative side, the additional liquid from the foam will spread the gasoline over a larger area. Depending on the exact situation and conditions, the additional spread may be a minor negative compared to the possibility of ignition. In yet another situation, the additional spread may create an extremely negative condition possibly by contaminating some highly sensitive exposure.

Nevertheless, it may be determined that another option such as adsorption provides the same speed and degree of vapor suppression without the potential for further spread and without requiring reapplication. As such, the potential harm is minimized. The obvious implication is that the use of the adsorbent is both more effective and efficient, so it is the *best* choice. All of these steps are part of the estimating process.

The Ongoing Nature of Estimating

The previous discussion provides an insight into the phases, steps, and initial scope of the estimating process. However, it does not fully encompass its entire scope. As has been indicated, the information gathering process is continually ongoing during the incident, and thus so is the estimating process.

Before identifying any strategy or tactic as appropriate, or before implementing it as part of the plan of action, its potential efficacy, efficiency, and implications must be evaluated. As new or updated information becomes available, it must be evaluated. The evaluation is to determine any impact on the course and harm of the plan of action that are in place. The completion of any strategies and tactics requires consideration of its implications. The estimating process is never ending. Unfortunately, all too commonly an initial single estimate of course and harm is identified and there is no additional consideration throughout the remainder of the incident.

The Role of Estimating

To this point, the process of estimating course and harm has been described, but the reason for using the process has not. Most precisely stated, estimating enables the identification of general and then specific response options appropriate for the handling of the existing and anticipated incident situation.

In essence, estimating allows the incident commander and the subcommands to anticipate how the incident will most likely unfold, and it identifies additional probable occurrences. By identifying such occurrences, personnel can identify potential operations (strategies and tactics) that will help to control the incident and minimize its impact. In other words, estimating provides the means for responders to adopt proactive management of the incident.

Consider the example of firefighters responding to a reported single family dwelling fire at 2:30 on Sunday morning. During the response to the scene, personnel normally start to identify potential course and harm by considering something like this:

> It's 0230 hours, Sunday morning, and the response is to a dwelling fire. As such, there is a high probability of occupants trapped inside.

This is an initial estimate of one potential harm of the incident. Upon arrival, personnel find that the dwelling is of balloon frame construction and there is a moderate to heavy fire condition in the basement. The following is thus a well-founded estimate or extrapolation:

> A basement fire in a balloon frame dwelling has a high probability of extending rapidly through vertical and horizontal concealed void spaces.
>
> Based on the degree of involvement, extension may already exist.
>
> Heat and smoke from the basement fire will rapidly extend to the attic, creating the potential for flashover in the attic.
>
> Due to the type of construction and its behavior during fire conditions, there is a higher probability of trapped occupants.

When considered together, these individual predictions form the basis for a relatively high probability estimate. As a result, responders identify that strategies of rescue, ventilation, and confinement as the initial strategies that must be implemented rapidly. The specific tactics and the priority of their implementation can vary, depending on the exact incident situation and the availability of resources.

Rescue addresses:

1. The potential for trapped victims
2. Confinement by providing information on interior conditions and identifying additional needs as a search is conducted

Ventilation (vertical and horizontal) addresses:

1. Heat and smoke conditions in the attic and potentially occupied areas
2. Aiding rescue operations
3. Horizontal concealed space spread by providing a vertical path for the heat and smoke

Confinement addresses:

1. Rapid spread and extension
2. Trapped victims by providing time for rescue operations
3. Minimizing further extension, if it is combined with opening concealed spaces

Additionally, this example shows how estimating also helps identify anticipated outcomes of the strategies and tactics outlined in the plan of action (e.g., ventilation should reduce heat and smoke conditions in the attic and occupied areas). As such, estimating helps to identify anticipated operational milestones. By identifying milestones, estimating further assists in evaluating the efficacy and efficiency of the plan of action. If these milestones are not reached, or if they require an inordinate amount of time to be reached, it generally indicates a flaw in the plan of action.

When an estimate of course and harm is not developed, responders are forced to simply react to ongoing incident occurrences. Instead of assuming a proactive stance and managing the incident, they are relegated to a reactive stance. In essence, responders are not managing the incident, rather the incident is managing the responders. Such a situation presents a much higher probability of inefficient, ineffective, and unsafe practices because full planning and review of operations have not occurred before implementation.

INCIDENT ELEMENTS

The final phase of analysis involves identifying the interrelationships between the incident components. Again, the incident components for a hazmat incidents are product, container, and environment. These components are defined as things about which information must be gathered. Incident elements (Figure 5-1) are incident occurrences that involve two or more of the incident components. The interrelationships between the incident components are found in the incident elements. The elements of a hazmat incident include spills, leaks, and fires.

A spill is when any product escapes its intended location (container) and enters a surrounding environment. In other words, a spill is released product. It requires the

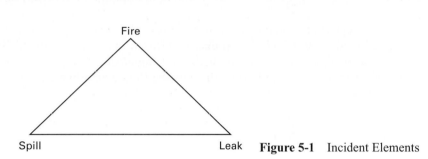

Figure 5-1   Incident Elements

presence of some type of leak (container failure, breach, opening, etc.) or there would be no means for the product to escape. The spill involves the incident components of product and environment.

A leak is the failure, breach or opening through which product is able to escape. A leak may or may not be accompanied by a spill (if no product escapes, there is no spill). The leak involves the incident components of the container and product.

A fire is a rapid oxidation chain reaction that liberates heat, gases, and in most cases, light. It may be the result of, cause of, or concurrent event to a leak and spill. The fire may involve any of the incident components (i.e., product, container, or environment).

An important point to constantly remember is that leaks and spills are not the same thing. Webster's defines a leak as, "An unintentional hole, crack, or the like, through which water, air, etc., enters or escapes." Webster's further defines a spill as, "To run or escape from a container, especially by accident or careless handling . . . , to scatter." Although this distinction seems technical, for the purposes of estimating as well as strategic and tactical option, it is very important.

As with any emergency, there will often be an interrelationship between all three of these elements. Yet each element is distinct and to some degree independent. Although each can affect the behavior of the others, each element behaves in a relatively predictable fashion. Each element will require different equipment, materials, and activities to be controlled.

Additionally, the potential degree of danger facing responders who attempt stabilization activities can vary significantly from element to element. Once the elements have been considered individually, it is much easier to make an accurate estimate as to their probable course and harm, as well as identify potential control options.

SPILL

The spill element of an incident involves the product and the environment into which the product has escaped, is escaping, or may potentially escape. As a result, the interrelationship between product and environment are the primary considerations. To effectively estimate the potential course and harm of the spill, the specific type or types of spill must be identified.

To determine the type of spill, the physical state of the product and the environmental media into which the product is released must be considered. There are three possible physical states of the product: solid, liquid, or gas (vapors from liquids are considered gases). There are four possible media for the environment: air, water, surfaces, or subsurface.

Environmental Media

The air media encompasses any atmosphere. Air is obviously present in any incident that occurs outside. It is also found within structures. Further, it is found within storm sewers, chemical reactor vessels, sumps, drainage trenches, among other places.

The water media encompasses streams, rivers, lakes, and ponds. However, it may also be present within a storm or sanitary sewer. It may also be present within retention ponds, holding tanks, and sumps.

The surface media encompasses any solid surface, regardless of porosity, location, or size. That means a surface can be a table top, a floor, a street, the ground, the inside of a dry storm sewer, and so on. As discussed in environmental considerations, surfaces exhibit a wide range of characteristics. Some are relatively impervious while others act like sponges. Some are smooth and even while others are rough and undulating. Yet all are considered surfaces.

The subsurface media encompasses the soil found beneath a surface. As a result, product that escapes directly into the soil from its container (e.g., pipe, underground storage tank, storm sewer, or sanitary sewer) results in a subsurface spill. A subsurface spill can also result from the process of product percolation into the soil. As mentioned in Chapter 4, this process is highly dependent on the geologic makeup found within a given area.

It is helpful to correlate the physical states of the product with the potential environmental media into which they may be released. When these variables are put into the form of a matrix, twelve spill (release) types are possible (Table 5-1).

Although the matrix indicates that there are twelve different types of spills, some extenuating circumstances make the actual number far less. The first circumstance encompasses all three types of subsurface releases. Although they do occur, it is almost impossible for a standard emergency responder to estimate subsurface course and harm. Further, the options for handling these types of spills require equipment and approaches normally handled only by environmental agencies or cleanup contractors. As

Table 5-1   Possible Spill Types

Physical state	Air	Water	Surface	Subsurface
Gas	Gas/air	Gas/water	Gas/surface	Gas/subsurface
Liquid	Liquid/air	Liquid/water	Liquid/surface	Liquid/subsurface
Solid	Solid/air	Solid/water	Solid/surface	Solid/subsurface

such, when a subsurface spill is found or suspected, it is vital to inform the appropriate agency in that area. Should the spill become apparent in one of the other environmental media, responders should handle this form of the spill in the best practical way.

The second circumstance involves mimicry, which happens when one spill type behaves like another. Solids that enter a body of water will behave like a liquid. If the solid has a specific gravity less than one and is not water soluble, it will float on the surface and move downstream. If the solid has a specific gravity greater than one and is not water soluble, it will sink to the bottom. Depending on the body of water's movement, the solid may flow downstream, on the bottom (if the particles are of small enough size). Finally, if the solid is water soluble, it will dissolve and behave like a water soluble liquid. As a result, all solid/water releases are considered to mimic liquid/water releases.

The third circumstance involves transformation, which occurs when a spill starts as one type and then transforms into another. Liquids released directly into the air would form a liquid/air release. However, the liquid will very rapidly fall downward until it contacted a surface or water. As such, the liquid/air release would transform into a liquid/surface or liquid/water release. Conversely, a gas/surface release would immediately transform into a gas/air release.

Similarly, consider a solid/air release when the solid is in a fine particulate such as powder or dust. Initially, when the fine particles enter the air, they will behave like a gas (i.e., they move with the air currents). As such, this type of spill mimics a gas/air release. However, after a time, the particles will settle to the ground or some other surface. At that point, it becomes a solid/surface release.

As a result of mimicry and transformation, we find that there are really only four spill types. These four types include gas/air, liquid/surface, liquid/water, and solid/surface (Table 5-2).

Another point to remember, is that both liquid/surface and liquid/water releases can each take two forms. The first form is known as a membrane. A membrane spill results when a minimal quantity of liquid forms a thin coating, normally under a quarter of an inch deep, and covers the surface or the water. The second form is known as a pool. A pool result when a larger quantity of liquid forms a pool, normally greater than a quarter of an inch deep, and covers the surface or water.

The final spill circumstance involves multiples. Multiples are spills where two or more different types of release occur simultaneously. For example, *all* liquid/surface releases will also produce gas/air releases. This is because all liquids produce vapors. Granted, for some liquids, due to their low volatility, the vapors may present little or no

Table 5-2   Four Types of Spills

Physical state	Air	Water	Surface	Subsurface
Gas	**Gas/air**	Gas/water	Gas/surface	Gas/subsurface
Liquid	Liquid/air	**Liquid/water**	**Liquid/surface**	Liquid/subsurface
Solid	Solid/air	Solid/water	**Solid/surface**	Solid/subsurface

problem. However, they must still be considered and factored into the estimate. Additionally, solids that sublime, such as iodine and camphor, also produce vapor. As such, they will not only produce a solid/surface release, but also a gas/air release.

Yet another example of a multiple is when a liquid such as diesel fuel is released from a highway cargo tank onto the shoulder of a roadway. This produces a liquid/surface release. The liquid then flows into a storm sewer and enters a stream. This produces a liquid/water release. Since the product is a liquid, it produces vapor, and thus also produces a gas/air release.

The reason it is so important to identify all the release types is that each spreads in different ways and directions, and thus potentially produces different vulnerable zones, potential exposures, possible impact locations, etc. All of these possibilities must be considered as part of the estimate because they will affect the strategies and tactics employed to manage the incident.

For example, consider a situation where a volatile liquid is released from a rail tank car. The liquid flows downhill. The downhill direction happens to spread the liquid upwind. The vapor produced by the liquid will follow the wind uphill. As such, responders must consider where the liquid will flow and where the vapors will flow. These considerations will help to identify the size, shape, and extent of the incident zones and perimeter.

Additionally, the specific tactical options available to control the release apply only to a specific type of release. For example, diking is a tactical option for a liquid/surface release but not for a gas/air release. Booming is a tactical option for a liquid/water release (assuming the liquid that is not water soluble and has a specific gravity of less than 1.0) but not for a liquid/surface release.

As such, it should now be evident that the identification of specific spill types is an essential aspect of estimating the course and harm. Also it is an essential aspect of determining appropriate strategic and tactical options.

Additional Product Considerations

Although the initial product consideration for estimating the spill is its physical state, a whole spectrum of additional product properties and behaviors must also be factored into the equation. Some of the most critical product considerations involve its health, flammability, and reactivity hazards. The degree of toxicity is based on the TLV-TWA, and IDLH must be assessed in conjunction with the routes of exposure. The flammability of the substance considering volatility, vapor pressure, flash point, flammable range, and ignition temperature requires identification. The reactivity with other substances, incident stressors, and environmental conditions must also be assessed. Critical types of reactivity such as instability, corrosivity, radioactivity, and water reactivity require indepth consideration.

Physical properties must also be considered. Possible physical forms such as compressed liquefied gas, cryogenic liquid, fine particulate, and molten solids often have significant impact. The quantity of product released, the rate of release (e.g., gallons per minute) and the amount of product available for release (i.e., the amount remaining in the container) are paramount considerations.

The rate of vaporization as indicated by vapor pressure will have an impact on the extent and area of the gas/air release. Further, vaporization rates may provide insight into the spread of product as a clue to its viscosity. Table 5-3 lists vapor pressures for common substances to help with comparisons. These vapor pressure values are determined at standard temperature and pressure (approximately 68°F and one atmosphere).

Density and solubility properties must also be considered. Density, in the form of specific gravity or vapor density, can dramatically affect the spread and potential tactical options appropriate for controlling product spread.

Specific gravity of a liquid or solid will determine whether it floats or sinks when released in water. A specific gravity of less than 1 indicates the substance will float on the surface (assuming it is not water soluble). If it floats, it will spread very rapidly over the surface of the water. Whereas, a specific gravity of greater than 1 indicates the substance will sink (again, assuming it is not water soluble). If the substance sinks, the downstream spread of the product can be difficult to identify and assess. Obviously, the water solubility of the substance will also be critical to determining the exact nature and rate of spread.

The vapor density of the substance will also dramatically affect how it spreads as well as the effectiveness of potential control options. A vapor density less than 1 indicates that the substance is lighter than air and will rise. Most commonly, this is advantageous because the substance will tend to dissipate itself (assuming the release occurs outside). A vapor density greater than 1 indicates that the substance will tend to follow the contour of the surface and potentially collect in low areas. Additionally, such substances are very poor candidates for natural ventilation when they are released within a structure. As such, some type of mechanical ventilation, either positive pressure (PPV) or negative pressure (NPV), will be needed to dissipate such releases.

Several final considerations include whether the release is ongoing or complete. In other words, has all the product already escaped the container? Finally, is fire already involved in the release? If so, specific considerations must be examined. These considerations are covered in the discussion of estimating the course and harm of the fire.

Table 5-3   Vapor Pressures of
Common Substances

Substance	Vapor pressure
Liquid (often viscous)	1 mm Hg
Fuel oil #4	2 mm Hg
Water	25 mm Hg
Isopropyl alcohol	33 mm Hg
Ethyl alcohol	40 mm Hg
Gasoline	180 mm Hg
Acetone	180 mm Hg
Ethyl ether	440 mm Hg

Additional Environmental Considerations

As was true with the product, there are several additional environmental considerations beside the media into which the release has occurred. One of the first consideration is the degree of confinement. Basically, is the release within a structure or outside? This simple point can produce major variables involving potential spread, degree of health and safety hazards, available control options, and others associated with the release. Further, are there existing confinements or conduits, and how will they affect spread?

Natural environmental considerations such as topography, meteorology, and geology, will directly impact the spill estimate. So will exposures and their degree of sensitivity to the release. For example, 50 gallons of diesel fuel spilled on an open stretch of highway has different potentials than 50 gallons of diesel fuel spilled directly into a wetland. The release of a large volume of propane will have a different impact on a desolate stretch of interstate highway than on a downtown urban street.

By using the spill matrix and in conjunction with the additional considerations for both the product and the environment, an extensive and thorough estimate of the spill is possible. The findings of this overall estimate must then be related to their potential impact on responders, victims, the public, and the environment.

LEAK

The leak element of an incident involves the product and the breach, failure, or opening in the container through which the product has, is, or may escape. As a result, the interrelationship between product and container are the primary considerations. The process of estimating the potential course and harm of the leak is two-fold.

First, the leak estimating process strives to identify the nature and extent of the leak (e.g., breach or opening), as well as the probability that the leak will result in container failure. In this circumstance, container failure is the loss of the container's structural integrity. The loss of structural integrity results in total or almost total release of the container's contents. As such, a major effort is made to identify the potential for the leak (breach) to become larger, and to identify the potential for violent or explosive container failure.

Violent container failure is a result of some type of pressure-related failure that causes a rapid release of product. This rapid release of energy and product presents a potential for injury to personnel located in the immediate vicinity of the container. This is especially true of personnel who are performing leak control tactics.

Explosive container failure is a result of a more significant release of energy and product that not only threatens personnel in the immediate area but also extends a significant distance from the container. Furthermore, explosive failure will cause the container to react violently. Such violent reactions include significant container movement, deformation, and damage that can include rocketing of major sections of the container or the formation of shrapnel. Most commonly explosive failure is a result of one or more of the following situations:

1. Severe structural damage to a high pressure or ultrahigh pressure container, causing it to become critical
2. High radiant heat exposure or direct flame impingement on a container that causes or induces:
   a. Structural weakening through heating of the metal
   b. A liquefied compressed gas or cryogenic liquid to reach its critical temperature
   c. Self polymerization of the product
   d. Self accelerating or runaway decomposition *(runaway also identifies a sound tactical option)*
3. Product contamination or environmental conditions (temperature, humidity, etc.) that can initiate:
   a. Self polymerization
   b. Self accelerated or runaway decomposition
   c. Hypergolic ignition (almost instantaneous ignition of the materials upon contact)
4. Product to container incompatibility (true chemical stressor)

The types of product presenting the greatest potential for explosive container failure include liquefied compressed gases, cryogenic liquids, monomers, organic peroxides, high concentration oxidizers, unstable substances, and so on. The types of containers presenting the greatest potential for explosive container failure include high pressure and ultrahigh pressure. However, it is important to note that in a fire situation, almost any type of product found in almost any type of closed container possesses the potential for violent to explosive failure.

Second, the leak estimating process strives to identify the potential for successful leak control operations. For example, high and ultrahigh pressure leaks are notorious for being difficult and often impossible to control unless or until the pressure has been relieved. Further, high and ultrahigh pressure leaks are also quite dangerous because they have the potential to inject the substance into the skin of personnel, even through gloves and other PPE.

One option that is helpful and effective in the leak estimating process is the identification of the specific type of leak involved. Again, a matrix can help categorize the leak. In this case, the two variables are the physical state of the product and the maximum allowable working pressure of the container.

The pressure rating system is the same as the one discussed in Chapter 4 and is as follows:

Classification	Pressure range
Atmospheric	0 to 5 psi
Low pressure	5 to 100 psi
High pressure	100 to 3,000 psi
Ultrahigh pressure	>3,000 psi

Based on these criteria, the matrix looks like Table 5-4.

Table 5-4   Pressure Rating Matrix

Physical state	Atmospheric	Low pressure	High pressure	Ultrahigh pressure
Gas	Gas/atmos.	Gas/L.P.	Gas/H.P.	Gas/UH.P.
Liquid	Liquid/atmos.	Liquid/L.P.	Liquid/H.P.	Liquid/UH.P.
Solid	Solid/atmos.	Solid/L.P.	Solid/H.P	Solid/UH.P.

Unfortunately, this leak matrix is not as helpful as the spill matrix. Primarily, the matrix provides some insight into the potential for the leak to expand, the potential rate of release of product, and the potential violence involved in container failure.

Normally, atmospheric and low pressure breaches do not tend to expand (get larger) unless one of two conditions is present. The first is product and container incompatibility. Such an incompatibility is a true chemical stressor, in which case the product literally eats away at the container. The second condition is the presence of an outside stressor. Such outside stressors are either thermal (i.e., from heat, fire, or cold) or mechanical (e.g., the weight of another container or vehicle, pressure induced by decomposition, or the polymerization of the product).

As discussed in the section on damage assessment and earlier in this section, breaches in high pressure and ultrahigh pressure containers can expand themselves simply from the internal pressure within the container. If this pressure is accompanied by any type of external stressor (i.e., chemical, mechanical, or thermal), an expanding leak must be considered and potentially anticipated.

The potential rate of product release through the leak is rather obvious. The higher the internal pressure of the container, the greater the rate of product release. Additionally, the larger the leak, the greater the rate of product release.

Finally, the higher the container's MAWP, the greater the potential for violent failure of the container. Also, the higher the MAWP, the greater the potential spread of the product in a short period of time. Further, the higher the MAWP, the less likelihood that leak control tactics will succeed.

Additional Product Considerations

The primary additional product considerations include its chemical and physical properties. The most important physical properties include the product's ambient physical state and any form variation present while it is in the container, specifically, if the substance in the form of a liquefied compressed gas, cryogenic liquid, molten solid, etc. If the product is in the form of a liquefied compressed gas or a cryogenic liquid, expansion ratio is a definite consideration.

Expansion will affect the pressure within the container as well as the force of the product escaping through the breach. Additionally, any attempt to transfer product from the vapor space will have little or no effect on the total quantity of product remaining. Further, it will act to decrease the pressure within the container only while the transfer is in progress.

It is also critical to identify the reactivity and flammability characteristic of the product. As indicated earlier, instability and potential reactivity with the container are critical pieces of the puzzle. If fire is involved, the flammable range of the product is vital in determining the potential for the fire to propagate back into the container. For example, ethylene oxide has a flammable range of 3 to 100%. Because of this property, it is conceivable that a flame front of burning product outside the container could draw back into the container as the pressure of the release decreases. The result could be the ignition of the product within the container and a subsequent explosive container failure.

Additional Container Considerations

Additional container considerations include the following:

1. The specific type and configuration of the container
2. The material(s) of construction:
   a. Paper, cardboard, fiberboard, plastic sheet, woven plastic, etc.
   b. Glass, porcelain, ceramic, etc.
   c. Rigid plastic, wood, etc.
   d. Steel (mild or HSLA)
   e. Stainless steel
   f. Aluminum
   g. Other metals or alloys
3. The location of the leak
   a. Openings or closures including caps, gaskets, site gauges, hatches, manways, etc.
   b. The shell, wall, or heads
   c. Valves, devices that control loading or discharge of product
   d. Relief devices [spring loaded, frangible (rupture disk), or fusible (heat activated)]
   e. Piping
4. The capacity of the container
5. The stability of the container both physically (e.g., position, location, center of gravity) and structurally
6. Stressors that are acting upon or have acted upon the container

FIRE

The fire element of an incident involves a rapid oxidation reaction (chemical chain reaction) that produces heat, gases, and often light. The fuel for the oxidation reaction may include one or more of the incident components (i.e., product, container, or environment). The following are five primary considerations in estimating the potential course and harm of the fire:

1. What is burning?
2. What are the potential impacts of the fire itself?
3. What are the potential pyrolysis or combustion products?
4. What are the potential impacts of the pyrolysis or combustion products?
5. What are the potential impacts of firefighting activities versus allowing a controlled burn?

What is Burning?

It is critical to identify which incident components are actually on fire. Is it the product, the container, the environment, or some combination? This determination is so important because it may well determine how the fire is best handled. For example, if the fire involves the environment where the container and the product it holds are located, the product and container may simply be exposures. As such, an aggressive fire attack and exposure protection may prevent or eliminate the hazmat involvement. On the other hand, if the fire involves the product, it is a whole different story.

A good example is a tractor trailer hauling explosives. A fire limited to the engine compartment of the tractor is an environmental fire. As such, the explosives and their containers are an exposure to the fire. Depending on the extent of the fire, the separation of the tractor from the trailer or an aggressive attack may prevent any involvement of the explosives. However, if the fire has penetrated or threatens to penetrate the trailer, the situation is a loser. All personnel and members of the public must evacuate the area. A perimeter must be established, and entry must be denied.

Some may ask how a container can burn. The answer is straight-forward: Some containers are constructed of ordinary combustibles such as plastic, wood, and paper. Obviously, these containers burn easily. However, suppose the container is constructed of metal. Metal does not burn—or does it? Truly, it is difficult to cause ferrous metals to burn, but what about magnesium? It will burn. Further, so will ferrous metals if a strong oxidizer is involved. This author has witnessed metal pumps, housings, and piping burn during an incident involving liquid oxygen where there was an explosion and fire.

Potential Impacts of the Fire

Once the involved incident components are identified, it is necessary to identify the potential and probable impacts of the combustion. Specifically, one must identify how the fire will affect the product, the container, and the environment. The following are some of the primary concerns:

1. Will the fire lead to involvement of the product, container, or environment?
2. If such involvement occurs, how will that affect each component?
3. Can the fire cause damage to the product, container, or environment?
4. Is there the potential for violent or explosive container failure?

5. Can or will the product explode (e.g., explosives, organic peroxides, accelerated decomposition)?

6. Will the fire affect responders or the public? If so, how and over how great an area?

7. Is there the potential for mass fire, explosion, or BLEVE?

8. Is the container fitted with some type of thermal protection (e.g., fixed sprinklers, a deluge system, thermal protection, or insulation) that will minimize thermal effects to the container and product?

9. What are the chemical and physical properties of the product?

10. Is the product pyrophoric?

11. If the product is a gas, can its source be controlled?

12. How toxic is the product, and what are the routes of exposure?

13. What are potential pyrolysis or combustion products?

14. Will the fire produce excessive amounts of highly toxic combustion product or contamination?

15. How does the toxicity of the product compare to that of the pyrolysis or combustion products?

Consider the following examples:

Example 1.   A DOT 111J100W4 carrying ethylene oxide (C_2H_4O) has a leak from a transfer line located on the top of the car. The temperature is 55°F. The released product has ignited and a relatively lazy flame is burning parallel to the top of the car. A DOT 111J100W4 tank car has a test pressure of 100 psi and also has thermal protection, top and bottom shelf couplers, and a head shield. Ethylene oxide is highly toxic through inhalation, ingestion, and skin contact. It has a TLV-TWA of 1 ppm, is carcinogenic, and is a suspected mutagen and teratogen. It is also highly flammable with a flash point less than 0°F, a boiling point of 51°F and a flammable range of 3 to 100%. It is miscible with water and has a vapor pressure of 1,095 mm Hg.

Obviously, this is a potential BLEVE situation. However, there are several bits of information that are critical to an effective estimate of the fire's potential course and harm, as well as to the identification of potential response options. First, the temperature is 55°F, and the boiling point of the liquid is 51°F. This is only slightly above the boiling point. As such the pressure from within the car is relatively low (in the range of 16 to 20 psi) in comparison to the test pressure of the car.

Additionally, the car has thermal protection that will minimize thermal conduction to the tank car shell and the product within. Thermal protection must keep the tank shell (not the external metal covering) from reaching 800°F after the protection system is exposed to a pool fire of 1,600°F, plus or minus 100°F, for a minimum 100 minutes. Further, this protection must keep the tank shell from reaching 800°F when the protection system is exposed to a torch fire of 2,200°F, plus or minus 100°F, for a minimum of 30 minutes.

The estimate of potential fire course and harm indicates that a BLEVE is possible. However, if the car's thermal protection is intact, it should withstand the fire for a relatively long period of time as opposed to a nonprotected container. The estimate would indicate that although there is the potential for a BLEVE, the likelihood of one occurring rapidly is low. If there is a sustainable water supply available within a reasonable distance, the estimate would indicate that the potential BLEVE situation is all but eliminated. This will allow time to consider additional fire control options and to gather more information from the rail line, shipper, and manufacturer.

Example 2.   An open, 55-gallon drum of 2,2-dichloroethyl ether is on fire. Dichloroethyl ether is a flammable liquid with a flash point of 131°F (DOT system lists flammable liquids with flash point to 140°F). It is highly toxic by ingestion, inhalation, and skin absorption, and it has a TLV-TWA of 5 ppm and an IDLH of 250 ppm.

When involved in fire, this substance will produce highly toxic pyrolytic or combustion products including hydrogen chloride (HCl) and phosgene ($COCl_2$). Hydrogen chloride is highly toxic through inhalation and ingestion, and it is a powerful irritant (i.e., tissue destructive) to skin and mucous membranes. It has a TLV-TWA and CEILING of 5 ppm and an IDLH of 100 ppm. Phosgene is extremely toxic through inhalation and ingestion, and it is a strong irritant to skin and mucous membranes. It has a TLV-TWA of 0.1 ppm and an IDLH of 2 ppm.

When estimating the incident's potential course and harm, a comparison of the toxicity of the product itself and the combustion products finds that the combustion products are even more toxic that the product itself. This is especially true for inhalation and skin contact regarding the public and possibly responders. As such, a fire involving this product is an excellent candidate for extinguishment to control the course of the fire and to minimize its potential harm to the public and responders. That is assuming there is some agent such as an appropriate foam or an adsorbent that can be used to suppress the product's vapors once the fire is extinguished.

Example 3.   There is a fire coming from a failed spring-loaded relief device on a cylinder of silane (SiH_4). Silane is highly toxic through inhalation and is a strong irritant to skin and mucous membranes. It is pyrophoric and highly reactive with halogens, halogenated agents, oxygen, hydrogen, and nitrogen. Pyrophoric ignition has been known to occur after substantial leakage occurred, resulting in explosive ignition. The combustion product is almost exclusively silicon dioxide (the equivalent of high purity sand).

Due to these properties, the estimate of potential course and harm indicates that it is essential to control the flow of gas before any extinguishment is attempted. Otherwise, reignition and possibly an explosion will be the probable course of the incident. As such, if the flow of gas cannot be controlled, a controlled burn may be the only appropriate option to minimize the potential harm of the incident and also to control its course.

Example 4.   A 1,000-gallon spill of acrylonitrile (CH_2:$CHCN$) is on fire. The spill is created when a tank is overfilled. Acrylonitrile (also called vinyl cyanide) is highly toxic by inhalation, ingestion, skin contact, and absorption. It has a TLV-TWA of 2 ppm, an IDLH of 4 ppm, is a suspected carcinogen, and has a vapor pressure of 83 mm Hg. When acrylonitrile burns, it creates highly toxic products that include nitrogen dioxide (NO_2) and possibly some hydrogen cyanide (HCN). Nitrogen dioxide is toxic by inhalation and ingestion, and it is a strong irritant to mucous membranes and skin. It has a TLV-TWA of 5 ppm and an IDLH of 50 ppm and is nonflammable. Hydrogen cyanide is also highly toxic through inhalation, ingestion, and skin absorption. It has a TLV-TWA of 5 ppm and an IDLH of 50 ppm. However, hydrogen cyanide has a flash point of 0°F and a flammable range of approximately 6% to 40%.

This situation presents a perplexing set of data. The product is extremely toxic all by itself, yet the potential combustion products are also highly toxic. Depending on the availability of oxygen for the combustion process, the hydrogen cyanide will be consumed by the fire and then also produce nitrogen dioxide.

Considerations such as the proximity of exposures including people and occupancies will be a vital concern. Additionally, the porosity of the soil and the proximity of any bodies of water will also be vital concerns. In this case, it is very difficult to estimate whether intervention in the form of extinguishment or nonintervention in the form of allowing a controlled burn will produce less harm. The final answer will depend on the specific incident site factors, availability of resources, and other factors specific to the situation.

Hopefully, these examples provide some insight into the kinds of data that must be examined when estimating the course and harm of the fire and the implications of various fire control options. Further, these examples identify the need to ascertain detailed information regarding the incident components.

Pyrolysis and Combustion

It is important to note that there is a difference between pyrolysis and combustion before continuing. Pyrolysis (pyro: heat; lysis: cutting or breaking) is the process of breaking a substance apart through the application of heat. In other words, pyrolysis does not require the substance to burn. Rather, the substance need only be *exposed* to heat. Combustion is the process of rapid oxidation. In the case of most liquids and solids, combustion results from the pyrolysis of the substance involved. The molecular fragments that result from pyrolysis routinely form the fuel for the combustion process. One exception to this combustion model is the combustion of flammable solids such as the metals. In this case, the oxidation reaction occurs directly on the surface of the metal without any pyrolysis.

Organic peroxides are but one example of an extreme case of low-temperature pyrolysis. Various organic peroxides start to pyrolyze (decompose) at very low temperatures. The most unstable members start to pyrolyze at temperatures as low as approximately 20°F. At that temperature, the molecules of the peroxide start to decompose. The decomposition produces heat, and oxygen, and it fragments the peroxide's organic radicals. Since the decomposition produces heat, it is exothermic and is self-sustaining. The temperature, above which this form of pyrolysis begins, is known as the

Maximum Safe Storage Temperature (MSST). If the peroxide is not cooled rapidly, it will continue to heat.

At some point it will reach its Self-Accelerating Decomposition Temperature (SADT), which is where the peroxide rapidly decomposes, releasing a maximum amount of heat, oxygen, and radical fragments that can ignite. Routinely, the ignition results in explosive combustion (due in part to the presence of oxygen).

However, this example is not intended to imply that all pyrolysis processes result in any form of combustion, let alone explosive combustion. Nor does it imply that the substance must be flammable or even unstable. Rather, this example is used to reinforce the fact that the substance does not have to burn for pyrolysis to occur.

Another example of noncombustion pyrolysis is that of polyvinyl chloride (PVC). When PVC is heated to around 350°F, it will pyrolyze regardless of whether or not it is exposed to flame. As the PVC pyrolyzes, it produces hydrogen chloride (HC1). This means that even though PVC is self-extinguishing and may not be burned in a fire, it can easily produce large quantities of HC1 when exposed to heat.

Hopefully, these examples, as well as the earlier examples, provide some insight into the potential complexity of the analysis, assessment, synthesis, and extrapolation. There is a federal agency that can provide some assistance in this respect. The Agency for Toxic Substance and Disease Registry (ATSDR) is a division of the Center for Disease Control in Atlanta, Georgia. ATSDR has a host of technical information available to emergency responders concerning toxic substance exposures and treatments.

Additionally, they employ a group of combustion chemists who can provide data on potential pyrolysis and combustion products, and their potential impacts and implications. Their 24-hour emergency number is (404) 639-0615. All of their emergency response personnel have computers at home. These computers are tied into the agency's database and provide high speed, on-line data to response and health personnel.

Pyrolysis and Combustion Products

There are many potential pyrolysis and combustion products possible when chemicals are involved in fire. Further, they vary greatly from those anticipated when ordinary Class A combustibles burn. This section will identify and discuss some of the more common products, the types of substances from which they arise, and some of their potential impacts. The pyrolysis and combustion products to be discussed include:

1. Phosgene
2. Hydrogen chloride and hydrogen fluoride
3. Hydrogen cyanide
4. Dibenzofurans
5. Metal oxides
6. Nonmetal oxides

Phosgene.   Phosgene ($COCl_2$) is also called carbon oxychloride or carbonyl chloride. It is extremely toxic through inhalation and ingestion, and it is an extreme irritant to mucous membranes and skin. It has a TLV-TWA of 0.1 ppm and an IDLH of

2 ppm. In low concentrations, it has a sweet odor but is stifling in higher concentrations. It was used as a chemical weapon during World War I and is a pulmonary agent.

Phosgene is a combination of a carbonyl group bonded to two chlorine atoms. A carbonyl group is a carbon atom that is double bonded to an oxygen atom (as is carbon monoxide).

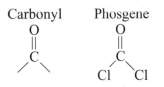

Carbonyl   Phosgene

Phosgene is produced through the combustion of and, in some instances, simple pyrolysis of chlorinated hydrocarbons. It is also produced when free chlorine or chlorine oxides contact moisture, high heat, carbon monoxide, and carbon particles as are found in a structural fire situation. As such, fires involving chlorine, chlorinated compounds, chlorinated salts such as chlorates or chlorites, and trichlorocyanurates present perfect situations for the production of phosgene. Halogenated hydrocarbons that contain chlorine are also susceptible to its production when they are pyrolyzed or contact heated metal surfaces. Such compounds include freons, halons, and similar substances.

An interesting sidelight is the use of carbon tetrachloride (also called carbon tet or tetrachloromethane) as a fire suppressant. During World War II, carbon tet was used as a very effective extinguishing agent onboard military aircraft. Unfortunately, it not only effectively extinguished the fire, but it also had the nasty habit of extinguishing the crew as well. This was due to the production of phosgene as the carbon tet pyrolyzed and interrupted the chemical chain reaction as well as when it contacted hot metal surfaces within the aircraft.

Firefighters have encountered phosgene when air conditioning or refrigeration lines have failed during overhaul operations following structural or vehicular fires. The most common symptoms of such exposures include severe respiratory irritation as well as severe burning and irritation to any exposed skin. After such exposures, firefighters have expressed an understanding of why phosgene is sometimes likened to mustard gas.

Finally, it is important to understand that halogenated substances containing fluorine, bromine, or one of these in combination with chlorine produce a series of related compounds. These compounds include carbonyl fluoride (COF_2), bromophosgene ($COBr_2$), and chlorofluorocarbonyl ($COFCl$).

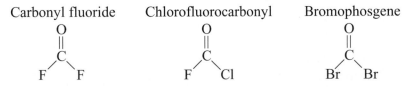

Carbonyl fluoride   Chlorofluorocarbonyl   Bromophosgene

Although these substances are not quite as toxic as phosgene, all have high toxicity and are strong irritants to mucous membranes and skin. These substances are created by mechanisms similar to those that produce phosgene.

Hydrogen Chloride and Hydrogen Fluoride.   Hydrogen chloride (HCl) is simply one hydrogen atom covalently bonded to one chlorine atom. Hydrogen chloride is highly toxic through inhalation and ingestion, and it is a powerful irritant (meaning it is tissue destructive or necrotic) to skin and all mucous membranes. It has a TLV-TWA and CEILING of 5 ppm and an IDLH of 100 ppm. It is the anhydrous form of hydrochloric acid. As such, when it contacts mucous membranes or moist tissues, it ionizes and hydrolyzes with the water to produce hydrochloric acid. This reaction accounts for its high toxicity.

Hydrogen chloride is most commonly produced when non-chlorine saturated, chlorinated hydrocarbons are pyrolysed. These chlorinated hydrocarbons have had only some of the hydrogen atoms on the hydrocarbon replaced with chlorine. When the formulas of these substances are examined, they are found to contain both hydrogen and chlorine along with the carbon. Because of the chlorine and carbon content within their molecules, many of these same chlorinated hydrocarbons will also produce phosgene.

Probably the most recognized example of an HCl-producing hydrocarbon is the polymer polyvinyl chloride (PVC). PVC is notorious for the production of massive volumes of HCl when it is pyrolyzed or exposed to flame. It also reinforces the point that the substance does not have to burn, but only must pyrolyze to produce HCl. Even relatively short-term exposures to temperatures above 450°F will cause PVC to pyrolyze and evolve HCl.

It is important to note that not only can hydrogen chloride be produced but also hydrogen fluoride (HF) if the polymer or the hydrocarbon contains fluorine or a combination of fluorine and chlorine. The fluorocarbons or chlorofluorocarbons are the primary hydrocarbons that contain such arrangements.

Hydrogen fluoride is also highly toxic by inhalation, ingestion, skin contact, and absorption. It has a TLV-TWA of 3 ppm and an IDLH of 30 ppm. HF is also quite water soluble. When it contacts moist tissues or mucous membranes, it ionizes and produces hydrofluoric acid. Unlike HCl, HF is not only tissue destructive (also called necrotic) when it forms hydrofluoric acid, it is also a systemic toxin due to the physiologic activity of the fluoride ion. There is rapid absorption and uptake of the fluoride ion through the skin and mucous membranes, including the gastrointestinal tract and the respiratory system.

HF is suspected of interfering with energy production at the cellular level. It can rapidly induce electrolyte swings within tissues by creating insoluble calcium and magnesium fluorides.

The resulting hypocalcemia can result in muscle rigidity (tetany) and cardiac arrhythmia that can be fatal. Some relatively common substances that contain chlorine, and thus can produce HCl, include:

1. Vinyl chloride monomer
2. Allyl chloride
3. Di-, tri-, and tetrachloroethane and ethylenes
4. Chlorinated freons and halons

5. Chlorinated benzenes, toluenes, and xylenes
6. Acid chlorides
7. Chlorosulfonic acid
8. Hydrochloric acid

Some relatively common substances that contain fluorine, and thus can produce HF, include:

1. Teflons
2. Fluoroform
3. Fluorosulfonic acid
4. Vinylidene fluoride
5. Difluorophosphoric acid
6. Trifluorostyrene
7. Fluoroethane
8. Tetrafluoromethane (fluorocarbon 14)

Hydrogen Cyanide.   Hydrogen cyanide (HCN) is one hydrogen atom covalently bonded to a carbon atom that is triple bonded to a nitrogen atom. Hydrogen cyanide is flammable with a flash point of 0°F and a flammable range of approximately 6% to 40%. HCN is highly toxic through inhalation, ingestion, and skin absorption. It has a TLV-TWA of 5 ppm and an IDLH of 50 ppm.

HCN is a systemic asphyxiant that rapidly attacks at the cellular level. Specifically, HCN shuts down the oxidative phosphorylation process and its associated adenosine triphosphate (ATP) production within the mitochondria of the cells. This process provides the energy needed for each cell to function. Due to their continuous high level of activity and need for a consistent energy supply, the cardiovascular and central nervous systems (CNS) are the first systems affected. According to a U.S. Army Research and Development report,* the inhalation of 270–300 ppm causes immediate symptoms and death "in less than one minute."

Hydrogen cyanide is produced when organic substances containing nitrogen are pyrolyzed or burned. Such substances include the following:

1. Nitriles
2. Cyanates
3. Isocyanates
4. Ureas
5. Urethanes

*NOTE: See Nicello (1993), "ALS Hazmat," NFA Course Material.

6. Wool

7. Leather

Since HCN is flammable, it may be consumed by the fire if there is sufficient oxygen present. It will most likely be produced when the substance pyrolyses or when there is a large fire that is oxygen limited. In the case of the polymers such as the urethanes and natural proteins (e.g., wool, and leather), the highest concentrations of HCN are often produced after the fire has been controlled and before all smoldering is extinguished or latent heat dissipates.

Dibenzofurans.   Dibenzofurans are a wide ranging group of compounds composed of two benzene rings connected by a cyclic ether bond arrangement (the same type of arrangement found in tetrahydrofuran).

Dibenzofuran

Until recently, dibenzofurans were little known compounds outside the world of chemists. However, in recent years furans in general and dibenzofurans in particular have been identified as physiologically active compounds. An association between the unsaturated, cyclic ethers and various cellular problems including carcinogenesis, teratogenesis, and mutagenesis has been identified. As such, many are suspected carcinogens as well as possible reproductive toxins.

In the case of combustion chemistry, the dibenzofurans have been identified in only the past decade or so. They form when a wide variety of the larger hydrocarbons burn in oxygen-controlled and oxygen-deficient situations. Interestingly, the dibenzofurans have been identified as major components of exhaust, especially from diesel engines. Dibenzofurans can be produced when large, complex aromatic compounds (e.g., dibenzos, and napthalenes) or aromatic polymers (e.g., styrenes, and urethanes) are involved in fire.

Metal Oxides.   As the name implies, metal oxides are composed of a metal and one or more oxygen atoms that have ionically bonded. Again, this is a wide ranging group of compounds whose toxicity depends on the specific metal involved. The toxic members of this group are toxic through inhalation and ingestion, and they are often severe irritants to skin and mucous membranes.

There are two general hazards of these compounds. The first involves those that are water soluble and will react with water to produce a caustic, metal hydroxide solution. For example, sodium oxide that is produced when sodium metal burns will react with water to produce a sodium hydroxide (NaOH or lye) solution. The reaction that produces the metal hydroxide is generally quite exothermic. As such, these soluble members are necrotic by both caustic (corrosive) attack and thermal damage.

The second hazard involves metal toxicity. Such toxicity is not dependent on water solubility. Both inhalation and ingestion routes of exposure are most common. Metals such as cadmium, lead, tin, titanium, mercury, beryllium, magnesium, and chromium present potential toxicity including metal fume fever and various negative impacts to other body systems. The central nervous, respiratory, digestive, hepatic (liver), and nephritic (kidneys) systems are the most common target tissues.

Metal oxides form whenever metal or metal-containing compounds, (mainly salts) burn. Under the right conditions, especially when they are in the dust or powder form, almost all metals will burn. The most common salts that will burn or become involved in fires are the oxysalts (e.g., nitrates, chlorates, phosphates, sulfates, and chromates), hydrides, phosphides, and carbides.

Nonmetal Oxides.   Nonmetal oxides are composed of a nonmetal plus one or more covalently bonded oxygen atoms. They form anytime a nonmetal or combustible substance, other than metals, burn. Some of the most common nonmetal oxides include the following:

1. Carbon oxides (CO and CO_2)
2. Nitrogen oxides (nitrogen dioxide, NO_2 being the most common)
3. Sulfur oxides (sulfur dioxide, SO_2, being the most common)
4. Phosphorous oxides (phosphorous pentoxide, P_2O_5, being the most common)

It is important to note that although this list includes the most common nonmetal oxides, it is far from complete. Further, as is seen in the case of carbon, the nonmetal oxides can have varying amounts of oxygen within their molecules. Because of this, the nonmetal oxides are often seen with a subscript "x" following the oxygen (e.g., NO_x and SO_x). This notation indicates that there are multiple oxides possible, all with differing oxygen contents.

The nonmetal oxides are generally toxic through ingestion and inhalation, and they are strong irritants to skin and mucous membranes. They primarily attack by ionizing when they contact water or moisture and form oxyacids. Oxyacids include nitric acid (HNO_3), sulfuric acid (H_2SO_4), phosphoric acid (H_3PO_4). As was noted with the oxides themselves, the oxyacids can have varying oxygen contents as well. For example sulfurous acid has a formula of H_2SO_3, and nitrous acid has a formula of HNO_2.

Additionally, some of the nonmetal oxides are systemic toxins as well. One example is carbon monoxide (CO), a systemic asphyxiant. When CO is inhaled and absorbed through the lining of the lungs, it binds to the hemoglobin at the sites normally occupied by oxygen. CO has about 200 times the affinity for hemoglobin as does oxygen. As such, when enough CO occupies the space of oxygen on the hemoglobin, the blood becomes hypoxic.

A second example of a systemic nonmetal oxide toxin is arsenic. When arsenic-containing compounds burn, they produce arsenic trioxide (As_2O_3) or arsenic pentoxide (As_2O_5). These arsenic oxides also combine with water or moisture and produce an arsenic-type acid, and thus are skin and mucous membrane irritants. Such exposure can cause dermal sensitization and dermatitis. Inhalation and ingestion can cause severe inflammation of mucous membranes and possible exfoliation (sloughing off) of the tissue. Another major concern is that arsenic is also a human carcinogen and mutagen.

In fires involving halogenated oxysalts, the halogenated oxyradicals (i.e., FO_x, ClO_x, and BrO_x) will produce their corresponding oxides. These oxides will also mix with water to form the various halogenated oxyacids (e.g., fluoric, fluorous, chloric, and chlorous). Of special note are the fluorine oxyacids. They will display similar hazards as discussed with hydrogen fluoride.

Again, although this listing of pyrolysis and combustion products identifies the more common, resultant substances, it is by no means complete. However, it provides an initial starting point. Further, through the identification of the appropriate products, key information is provided for potential air monitoring plans.

For example, consider a fire involving tons of 30 agricultural chemicals ranging from pesticides to fertilizers. It is extremely difficult, if not impossible, to identify all the potential pyrolysis and combustion products. However, based on the constituents of these chemicals, the production of nonmetal oxides of nitrogen, sulfur, and phosphorous, as well as organic fragments are significant and high probability products.

As such, the use of colorimetric tubes for the potential nonmetal oxides and Photo Ionization Detectors (PID), and Flame Ionization Detectors (FID) for organic fragments provides a broad spectrum sensing capability. The reason is simple; these substances are used as tracers to identify general levels of all contaminants produced by the pyrolysis and combustion processes. As a result, significant hard data is provided as to the degree of contamination spread and its potential concentrations when the readings are taken at various distances downwind from the scene. This data can also be used to identify the types of downwind public protection needs.

Finally, data on the toxicity, routes of exposure, target tissues, and symptomology can help determine the nature of any potential public exposures. For example, if nonmetal oxides of nitrogen and sulfur are probable products, possible victims should display symptoms including eye, upper respiratory, and possibly skin irritation. If victims complain of lightheadedness, nausea, or other symptoms that do not include these types of irritation, the likelihood is that they received some other type of exposure, or the symptoms are psychosomatic.

Anticipated Impacts of Pyrolysis and Combustion Products

Once the probable pyrolysis and combustion products are identified, their potential impacts must be evaluated. Specifically, the quantity produced, their toxicity, routes of exposure, symptomology, and potential for residual contamination must be examined. Factors such as dead or dying plants and animals must also be considered

From these data, the potential impacts on the public, response personnel, property, and the environment must be estimated. This estimate will help to clarify the need and probable success of firefighting options. Although this is an easy statement to make, it is sometimes a difficult determination to make. Finally, this estimate cannot be the sole basis for an intervention or nonintervention decision. This estimate must be considered in the context of the previous estimates about the potential fire impact on the product, container, and environment. Additionally, it must be examined considering the potential impacts of the firefighting activities themselves.

Anticipated Impacts of Firefighting

The firefighting activities undertaken to control the fire may produce substantial impacts in and of themselves. The primary potential impacts include the following:

1. Run-off and environmental contamination
2. Personnel contamination
3. Mixing and involvement of other chemicals
4. Inducement of additional chemical reactions
5. The cooling or heating of the combustion process

Run-off and Environmental Contamination

Run-off is always a potential problem when water-based agents are used to extinguish a fire. The extent of the problem depends on the nature and magnitude of the threat presented by the fire. In some situations, the run-off and the environmental contamination that it will cause is a greater hazard and more costly than the destruction caused by the fire. In the case of agricultural chemicals, there have been innumerable cases where the environmental cleanup that results from firefighting run-off cost millions of dollars— a cost that far exceeded a total fire loss.

One such incident involved an open quarry within 500 feet of the fire building that housed 254,000 tons of agricultural chemicals. The quarry had an inflow and outflow of ground water. Had a major firefighting effort produced run-off, it would have entered the quarry, the groundwater, and the aquifer. Such contamination would have required a cleanup that would have been astronomically expensive, time consuming (potentially years), and of questionable effectiveness.

Obviously, not all firefighting activities will result in contaminated run-off. However, the estimate of the fire control activities must always address the possibility and determine its credibility.

If the water-based firefighting activities will produce contaminated run-off, the estimate of course and harm must include the effects of steps taken to assure minimum impact of the run-off.

As such the influence of dikes, retentions, or diversions to control the spread of run-off and its subsequent environmental contamination must be factored into the estimate.

Personnel Contamination

The degree, type, and implications of contamination to firefighting personnel must be considered as a primary safety issue when estimating the potential impacts of firefighting activities. In some instances, the contamination to personnel will be nominal in form and quantity and thus of no major concern.

However, in other situations, personnel contamination may be the primary harm agent for the entire incident. Data on toxicity, routes of exposure, and the ability to decontaminate personal protective equipment (PPE) is a vital factor in estimating the potential course and harm of firefighting activities. In some instances, turnouts will be completely and irrevocably contaminated. This is obvious harm.

The possibility of covering the turnouts in an outer envelope composed of disposable chemical clothing is an option that's impact must be considered. Further, the use of positive pressure ventilation, may clear the area so firefighters receive minimal contamination. In other instances, aggressive fire attack and extinguishment in combination with the double envelope may minimize or prevent excessive contamination. All of these options must be factored into the estimate.

Mixing and Involvement of Other Chemicals

When hose lines are placed in service, there is always a possibility that the streams will agitate and move small containers of various materials. Such movement can easily result in the mixing and the potential involvement of previously uninvolved substances.

When there are multiple products or containers involved, consideration of potential mixing or possible involvement is imperative. In some instances, it is possible that mixing or involvement is of no great concern. In other instances it may be critical. Should such potentials be identified, the estimate must include consideration of the potential impacts. Further, the considerations of run-off, environmental, and personnel contamination must be examined as to their potential impact and harm.

Inducement of Additional Chemical Reactions

A primary consideration is the effect of extinguishment operations and potential reactions with the extinguishing agents. Will firefighting cause additional reactions to occur? Will mixing result in chemical reactions or explosions? Are available extinguishing agents appropriate for the substance involved?

When considering extinguishing agents, it is vital to determine the potential for reactions that range from accelerated combustion, to toxic or flammable gas or vapor production, to explosions. Most personnel are well aware of the explosive reaction that results from the application of water to a burning metal fire. However, they are not aware that carbon dioxide acts as an accelerant (an oxidizer) and may also produce explosions.

The carbon dioxide does not have to be the primary agent to cause this reaction. It may simply be the propellant for a dry chemical agent. Further, it may be produced by decomposition of a primary agent such as a carbonate salt like soda ash (Na_2CO_3 or sodium carbonate) or a bicarbonate salt like sodium bicarbonate ($NaHCO_3$).

Further, most personnel are not familiar with the water gas reaction. When temperatures exceed approximately 1,800°F, the oxygen is torn from the water molecules.

The oxygen performs as an oxidizer while the remaining hydrogen becomes another fuel for the combustion reaction.

Consideration must be given to the quantity and form of the combustion product as well. For example, consider burning magnesium (or any flammable metal for that matter). There is a big difference between one pound and five hundred pounds of burning magnesium. The amount of extinguishing agent is different; the amount of toxic magnesium oxide (MgO) produced is different; the magnitude of an explosion is different; and so on.

Further, a fire involving filings and shavings is different from a fire involving an ingot that is 2 feet $\times$ 2 feet $\times$ 8 feet. The rate of fire spread and total volumes are different. A small quantity burning on a pile is different from a foot-deep pool of molten, burning magnesium. The greater the quantity and its depth, the greater the potential for metal extinguishing agents to liquefy, decompose, sink in the molten mass, and so on.

The exact location and incident circumstances are also important to consider. There is a difference between phosphorus burning within a tank car and the same quantity burning on the ground. Water application into the tank car can and has led to massive steam explosions, while application on the ground can spread pieces over a large area.

A caution regarding reactions and explosions is in order. Firefighting activities are questionable to the extreme when it is known or is possible that they will result in explosions. This statement may seem obvious. However, periodically, fire departments choose to perform just such tactics. Recently, the most frequent examples have involved magnesium or other flammable metal fires. In the last eight years, there have been at least three major magnesium fires where the departments employed master stream applications. In each case, these firefighting activities resulted in multiple explosions that either injured or at least placed personnel in physical jeopardy. None of the fires were extinguished before the magnesium and the building or scrap yard were consumed.

Such actions are indefensible, especially should injury or additional damage result. Further, such actions indicate a lack of understanding and willful disregard (part of the definition of gross negligence) for the potential harm they can produce. Maybe it was best stated in the movie, *Forest Gump:* "Stupid is as stupid does."

Further, these actions disregard the priorities of the fire service, namely, Life safety, Incident stabilization, and Protection of property and the environment (LIP). Based on the fundamental tenant, "If little is to be gained, little should be risked," why risk personnel and the surroundings when little is to be gained?

Cooling or Heating of the Combustion Process

Fire involving pesticides are the prime example of cooling the combustion process and its potential implications. It is common knowledge that the application of water to a pesticide fire reduces the temperature and in turn alters the combustion products. These new products have higher toxicity than those produced from the higher temperature burn. Further, they have a great tendency to deposit themselves on objects, people, and almost anything else in the surrounding area. As such, the cooling effect is of little or no value.

Another example is the application of cooling streams to a cryogenic container when there is no fire. First, cryogenic containers are normally the equivalent of large thermoses. Application of water to the outer skin does not produce cooling for the inner tank. Likewise, should the outer skin be compromised, the application of water will act as a heating agent. However, should fire be impinging on the outer skin, the application of water is called for and effective.*

This discussion of estimating the potential course and harm of the fire element of a hazmat incident has focused on the many considerations involved. In most cases, when a chemical incident involves fire, the fire is a complicating factor. Unfortunately, all too often, initial companies follow standard fireground strategic and tactical approaches and commit to fire control operations without fully assessing or even understanding the potential impacts and harm of their actions. Further, all to often, this type of unplanned or nonevaluated action puts hazmat personnel in the unenviable position of having to identify and implement operations that will at best only minimize the impact of previous actions as opposed to successfully controlling the incident.

Do not misunderstand the intent of this section. It is not to imply that all hazmat incidents involving fire are mishandled. Nor is it meant to imply that all fires involving chemicals should be left to burn. Rather, this entire discussion indicates that careful analysis, assessment, and synthesis are needed to estimate the potential course and harm of the incident as well as the impacts of strategic and tactical fireground operations.

The Role of Preplanning

Preincident planning is an integral part of the estimation process, especially for fixed facilities. It is also an integral part of the estimating process for transportation corridors (although such planning is more generic). Further, its importance is the foundation for all SARA Title III notification and planning requirements. At its most fundamental levels, preplanning for both fixed and transportation incidents, should provide at least broad general estimates as to the probable areas that are vulnerable to a chemical release. In many instances, potential strategic and tactical options are also identified and dramatically assist in the estimating process.

Specific estimates regarding spill flow patterns, direction of spread, possible dispersion zones, identification of evacuation routes, key control points, access points, unseen conduits, and so on, are invaluable. Based on the preincident estimate, key decisions can be made before any incident occurs. For example, through the joint planning process identified in the SARA, CAER, and TRANSCAER programs, both the public and private sectors have identified facility and transportation corridor specific response options based on indepth estimates of potential course and harm. As a result, specific action criteria have been developed.

*NOTE: When the skin of a cryogenic container starts to lose its strength, it will not deform outward. Rather, it will normally deflect inward due to the vacuum commonly used in the annular space.

Intervention Versus Nonintervention

Now that an initial estimate of the potential course and harm is completed, a determination regarding the mode of operation is appropriate. Specifically, this determination involves identifying whether intervention activities should be undertaken. Most incidents are suitable for intervention to some degree or another. However, a few incidents will be determined to be beyond the capabilities or just too dangerous to intervene. In either case, there are certain strategies that lend themselves to all incidents, while others only lend themselves to those involving intervention.

As such, the mode of operation (intervention or not) helps to identify potentially appropriate strategies that in turn help to identify appropriate tactics. Again, the estimating process, by its nature, helps to identify at least strategic level approaches and often tactical levels as well.

THE ONGOING ESTIMATING PROCESS

Once the initial incident estimate is completed, the estimating process is far from finished. The reason for this is simple: An incident is not a static situation; rather, it is dynamic. The dynamic nature of an incident is what requires the estimating process to be ongoing. As the incident evolves, conditions, situations, and tactical operations cause changes in the incident status, components, and elements. These sometimes subtle or not so subtle, changes result in new or additional information that must be factored into the estimate.

Routinely, basic bits and pieces of information about the incident components (product, container, and environment) are missing from the picture. The retrieval of this missing data is routinely accomplished by a reconnaissance team (first entry team), interviews with technical experts, review of technical data sources, and so on. As this data becomes available, it must be factored into the incident estimate to identify any potential implications that may result.

However, too often, crucial bits of information are missed or never factored into the overall picture. This kind of oversight can lead to potentially disastrous results. To help prevent such shortcomings, a systematic approach is required. Specifically, the use of systematic, sequential, and coordinated approach to the information gathering and estimating processes are of great assistance to the identification of missing data.

Thorough and systematic documentation results when a protocol is established using written check lists and data sheets. These check lists and data sheets are an integral part of information management function of the overall incident command system. Such data sheets readily help identify missing information through their data entry requests and questions. This last aspect of the sheets is of vital interest. By identifying that data is missing, these forms help identify possible strategic considerations and often tactical assignments for the recon or other entry teams.

Further, not only do these forms help document acquired data and identify needed data, they also help to identify and explain why actions were taken during the incident. For example, following an incident, some pointed questions arise, similar to the following:

1. Why were specific operations undertaken?
2. Why did it take so long?
3. Was it necessary to evacuate the building?
4. Did the hazmat team overreact?

At some point, every team has faced questions about their actions and operations. These questions may be from within the department, from politicians, or in the form of a lawsuit. Good documentation of the raw data along with an indepth estimate can be the most powerful tools to counter such questions.

Unfortunately, all too often, responders have not fully documented the raw data on which the estimate is based. Even fewer follow a sequential and systematic approach to the estimating process. As a result, when hazmat responders are accused of overreacting or underreacting they may not have sound information and documentation to refute such accusations. Good documentation of the raw data and an effective estimating process are both fundamental to safe, sound, and logical scene operations.

As in Chapter 4, a model estimating sheet is provided following the conclusion of this chapter. This sheet, as well others provided with other units and the full set by ICS position found in Appendix D, are in use by active hazmat teams and are provided as a model for the development of your own sheets. Remember, make any forms user friendly so they will be accepted and used as the tool they are intended to be. Further, actual incident experience has proven the value of these forms.

SUMMARY

This chapter addresses the second step in the GEDAPER process, namely estimating the potential course and harm of the incident. It has discussed:

1. The role of the estimating process as it relates to operational decision making
2. The involvement of hazard identification, vulnerability analysis, and risk analysis as part of hazard analysis
3. How hazard identification, vulnerability analysis, and risk analysis relate to the estimating process
4. The three incident elements of spill, leak, and fire
5. How the incident components of product, container, and environment relate to the incident elements spill, leak, and fire
6. The four primary types of releases: gas/air, liquid/surface, liquid/water, and solid/surface
7. Release mimicry and transformation

8. The role of pyrolysis and combustion, as well as their products, in the estimating process.

9. How the estimating process helps identify the need for additional incident information

After estimating the potential course and harm, the next step in the process is determining appropriate strategies. Chapter 6 will address this topic.

INCIDENT COURSE/HARM ESTIMATE
COMPLETED BY HAZMAT COMMANDER OR SCIENCE GROUP

INCIDENT NUMBER: _____ PREPARED BY:_____
                            yr/mn/day/no

SPILL

    Present [   ]

    Type: gas/air [   ], liquid/surface [   ], liquid/water [   ], solid/surface [   ]

    Anticipated spread:_____

_____

    Anticipated impact:

        responders:_____

_____

        victims:_____

_____

        public:_____

_____

        exposures- structures [   ], other containers [   ], other substances [   ],
            production/processes [   ], animals [   ], vegetation [   ].

        additional projections:_____

_____

_____

_____

LEAK

    Present [   ], Anticipated [   ], Possible [   ].

    Type:_____

    Anticipated Course: remain static [   ], expand [   ], container failure [   ]

    Anticipated Failure Type: explosive [   ], violent [   ], non-violent [   ].

    Anticipated Harm of Failure to:

        responders:_____

_____

        public:_____

_____

        other containers:_____

_____

        other substance:_____

_____

        other exposures:_____

_____

FIRE

Present [   ], Possible [   ], Anticipated [   ]

Anticipated Course: remain static [   ], spread to exposures [   ], intensify [   ], result in explosion/s [   ]

Anticipated Harm of Controlled Burn: highly contaminated smoke [   ], possible explosion/s [   ], threaten exposures [   ]

Anticipated Harm of Controlled Burn on:

responders:_____

_____

public:_____

_____

other containers:_____

_____

other substances:_____

_____

other exposures: _____

_____

Anticipated Harm of Suppression:

highly contaminated smoke [   ], contaminated run-off [   ], mixing of substances [   ], water reactions [   ], explosions [   ],

Anticipated contamination spread to:

responders [   ], public [   ], structures [   ], production process [   ], surface water [   ], animals [   ], plants[   ], ground water [   ]

Anticipated Harm of Suppression on:

responders:_____

_____

public:_____

_____

other containers:_____

_____

other substances:_____

_____

other exposures: _____

_____

2 OF 2   COURSE/HARM ESTIMATE

6 Determining Strategic Goals

CHAPTER OBJECTIVES

Upon completion of this chapter, the student will be able to:

1. Define the term *strategic goals*
2. Identify and describe the eight hazardous materials strategic goals
3. Describe the relationship between the estimating process and the determination of appropriate strategic goals
4. Explain the process of strategic goal prioritizing
5. Explain the relationship between strategic goals, tactical objective, and tactical methods

DETERMINING APPROPRIATE STRATEGIC GOALS

By the time this step of the GEDAPER process is reached within the command sequence, information has been gathered, analyzed, and assessed. An estimate of the potential course and harm of the incident has been identified through synthesis and extrapolation of data. The estimate not only identifies the potential course and harm of the incident left unchecked but also identifies potential intervention options and their impacts. As a result, the mode of operation with respect to intervention or nonintervention has also been identified.

All of these steps in the traditional size-up process use the data gathered about the product, container, and environment as their foundation. This raw data provides the basis for determining the incident type, magnitude, and situation, as well as helping to identify the prominence of each incident priority. Finally, these steps help identify the general goals for successful intervention in the incident.

Strategic Goals

According to Webster, a goal is the result or achievement toward which effort is directed. The goals for an incident (or any endeavor for that matter) are broad and general milestones or accomplishments that are both desired and needed to manage the situation successfully. In the case of emergency response, these milestones are called *strategies* or, more accurately, *strategic goals.*

Further, strategic goals form the first level of the Plan of Action (POA) or action plan. Initially, strategic goals identify, in broad and general terms, milestones necessary to at least start bringing the incident under control. As the incident evolves and various identified strategic goals are met, additional strategic goals often must be identified and accomplished to stabilize and potentially gain complete control of the incident.

As this implies, strategic goals normally must be prioritized and routinely implemented in a sequence dictated by the incident. In other words, command personnel must identify the strategies that are needed immediately and then identify additional strategies to finally resolve the problem.

Consider a structural fire in an old, unprotected, multistory apartment building. The main body of fire is on the first floor in the primary stairwell. There are occupants trapped on the upper stories of the building. To accomplish the strategic goal of *Rescue,* it may be essential to accomplish the strategic goals of *Extinguishment* and *Ventilation* first. This is because, once the fire is controlled in the stairwell (*Extinguishment*), and the heat, smoke, and other combustion products are removed (*Ventilation*), *Rescue* is accomplished quickly and effectively.

Under other circumstances, the strategic goal of *Confinement* is deemed to be appropriate due to limited personnel and the need to accomplish *Rescue* and *Ventilation*. Once *Ventilation* is accomplished, *Rescue* may become the priority, or *Confinement* may be replaced with *Extinguishment* because sufficient personnel are now available.

To reiterate, as initial strategic goals are accomplished, others may now become necessary. For example, during an aggressive interior extinguishment effort, overhaul is not an issue. However, once the fire is knocked down, overhaul replaces extinguishment as a primary goal.

The Role of Tactics

At this point it is essential to understand the difference between strategies and tactics. *Tactics* are "the procedures of deploying operational forces in a fashion necessary to gain the advantage" (Webster's). Each strategic goal has a finite number of tactical options available to accomplish that goal. To fully identify the number of options available to accomplish a goal, it is essential to understand that tactical operations are subdivided into two levels based on their specificity (Figure 6-1).

The first level is the *tactical objective*. Tactical objectives are actions (operations) that must be performed to accomplish a strategic goal. They are moderately specific and demonstrable (i.e., one can identify when they are completed).

The second level is the *tactical method*. Tactical methods are very specific and directed. As the name implies, they identify the specific methodology used to perform the tactical objective(s) needed to accomplish the strategic goal.

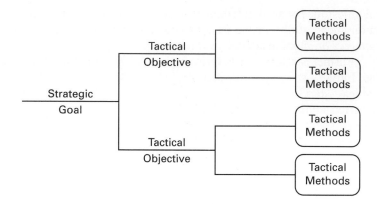

Figure 6-1   Tactical Operations

Consider this example. The strategic goal of ventilation has four primary tactical objective options (Figure 6-2). These options include the following:

1. Horizontal natural ventilation—ventilation in which the natural air currents move the contaminated air from the upwind side and out the down wind side
2. Horizontal mechanical ventilation—ventilation in which some mechanical device (e.g., Heating Air Condition Ventilation [HVAC], smoke ejectors, and Positive Pressure Ventilation [PPV] fans) forces movement of the contaminant
3. Vertical natural ventilation—ventilation in which the natural buoyancy of contaminant (due to heat, vapor density, etc.) moves it vertically to the outside
4. Vertical mechanical ventilation—ventilation in which some mechanical device forces movement of the contaminant

This is not to indicate that there are only four ways to perform ventilation. This is obviously not true because there are many different methods to accomplish each of the tactical objectives.

For example, mechanical ventilation methods include options such as the use of the buildings HVAC system, smoke ejectors, or PPV fans. Either positive pressure or negative pressure methods are options. Additionally, the exact number of devices, their

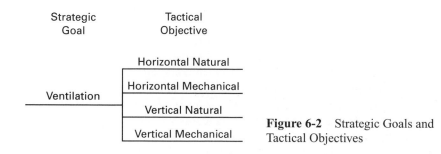

Figure 6-2   Strategic Goals and Tactical Objectives

placement, and the personnel (units) assigned to perform the task will vary. This means at the tactical method level there are often many potential options available to meet the tactical objective.

Strategic Goal Systems

For strategies and tactics to function effectively and efficiently during an emergency operation, all the players must understand the options available for their use. They must understand where and how they fit into the operation. Further, they must use common communications procedures and terminology, as well as common personnel roles, responsibilities, and functions.

To accomplish this end requires the use of a system, or more accurately stated, a series of systems. One of the components of many of today's incident management systems is the Incident Command System. As stated earlier, ICS is an extremely functional and effective resource management system. Also, ICS is only part of overall incident management.

Specifically, each type of emergency operation requires its own strategic goal system. For example, we have already identified that fire-ground operations require and have their own goal systems (e.g., RECEO VS or REVAS) (Table 6-1). EMS operations have their own system as well (e.g., Assess, Stabilize, Triage, Extricate, Package, and Transport). Likewise, hazardous materials response has its own system.

The value of using a uniform set of strategic systems by an organization or region should be rather obvious. Such systems fill the same role as the play book for a football team. It identifies the plays or options available for their use. The specific way the play is implemented depends on the specific defensive or offensive alignment the opposition uses. Additionally, the way and speed with which the plays are implemented varies depending on the score, the remaining time, and so on. Further, the exact play book used depends on the game being played. For example, a team would not use a baseball play book if they were playing basketball or football.

This analogy is beneficial when considering emergency operations. Specifically, the play book for a fire is different from the play book for EMS, and both are different from the play book for hazmat. When a jurisdiction has a set of strategic goal systems in place, the incident commander merely has to identify the type or types of incidents

Table 6-1 Goal Systems

Fire RECEO VS	Fire REVAS	EMS ASTEPT
Rescue	Rescue	Assess
Exposures	Exposures	Stabilize
Confinement	Ventilation	Triage
Extinguishment	Attack	Extricate
Overhaul	Salvage	Package
Ventilation		Transport
Salvage		

involved and then use the appropriate system(s) to identify the strategic options. Hopefully, the benefit of such a systems approach is evident at this point.

Additionally, many incidents involve two or more strategic goal systems. Consider the following examples. A structural fire involving trapped occupants will require the use of both fire and EMS goal systems. A hazmat incident involving a victim who is injured when a hose breaks while a chemical tank truck is being unloaded, will require the use of both hazmat and EMS goal systems. Finally, a structural fire at a chemical production facility involving an explosion and injured employees will require the use of fire, hazmat, and EMS strategic goal systems.

Characteristics of Strategic Goal Systems

All of these systems identify a series of optional strategic goals. The term *optional* indicates that not all the goals are necessarily required for each incident. As such, it is important to keep several points in mind.

First, the order in which the goals are listed is by no means the order in which they must be addressed during an incident. The order of implementation depends upon the specific incident situation. As such, there is no hierarchy implied by the numeric order in which they are presented. Very commonly, they are placed in an order that will provide a mnemonic memory tool.

Second, not all goals need be addressed during any individual response. Again, only the goals needed for the specific incident require coverage. For example, a minor food-on-the-stove fire could easily not require *Rescue, Confinement,* or *Salvage,* because *Extinguishment* and *Ventilation* are all that are needed.

Third, certain goals are defensive, others are offensive, while the rest can be either offensive or defensive (depending upon the exact tactics chosen to achieve the goal). For example, *Confinement* operations are designed to contain a fire to the area already affected. As such, *Confinement* is a defensive strategy. On the other hand, *Extinguishment* operations are designed to seek and put out the fire. As such, *Extinguishment* is an offensive strategy.

Finally, certain goals tend to be needed for every response. The reason for this is simple: These goals are basic and fundamental to the management of the specific type of incident. For example, *Extinguishment* is needed any time there is an actual fire. As such, it is needed on almost every fire response (other than when the fire went out by itself).

It is now time to examine the eight strategic goals used during hazmat response. In the following section, each goal will be identified and described, and the optional tactical objectives will be identified. A full discussion of the primary tactical objectives is provided in Chapter 7.

HAZMAT STRATEGIC GOALS

There are eight strategic goals used with the GEDAPER process. This is not to imply that other goal systems cannot be used. Rather, these goals are used as a functional model because they have been found to address thoroughly and effectively operational needs. The eight strategic goals are as follows:

1. Isolation
2. Notification
3. Identification
4. Protection
5. Spill control
6. Leak control
7. Fire control
8. Recovery and termination

Isolation

Isolation is the process of securing and maintaining physical control of the incident scene. The tactical options within this goal revolve around identifying physical areas that have differing and varying degrees of hazards. It is an integral aspect of scene management and safety because it establishes a physical buffer between the hazard and potentially affected exposures (e.g., people, property, and the environment). The vulnerability analysis plays a major role in this goal.

Isolation is a strategic goal that is required in all hazmat incidents, to some degree or another. In almost all instances, it is considered to be defensive and thus may be performed by first responders. The primary tactical objectives are as follows:

1. Perimeter establishment
2. Zoning (Zone Establishment)
3. Initial public protection
4. Denial of entry
5. Withdrawal

Notification

Notification is the process of alerting and communicating with mandated or needed resources. One of the most basic aspects of this goal is to alert appropriate governmental agencies and private sector groups and representatives that an incident has occurred, and to request appropriate assistance. The assistance includes personnel, resources, expertise, and information, depending on the group or agency. Further, notification encompasses alerting the public of an incident and providing necessary emergency public information. The primary tactical objectives are as follows:

1. Establish communication links
2. Request assistance

3. Incident level identification

4. Emergency public information

Certain tactical options of notification are required for every hazmat incident (establishment of incident level at the least). Others may be required, depending on the jurisdiction and the magnitude of the incident. Notification is a defensive strategic goal and thus can be performed by first responders. Further, under the OSHA regulations, this is one strategy *all* emergency responders, including first responder awareness level, must be able to perform.

Identification

Identification is the process of identifying, confirming, and obtaining information about the product involved in the incident. This process routinely includes the same steps regarding the container and possibly the environment. Further, as the incident evolves, the process involves identifying the location, extent, and spread of contamination. The primary tactical objectives are as follows:

1. Recognition and Identification (R&I)

2. Data retrieval

3. Interview

4. Review of plans and surveys

5. Reconnaissance

6. Monitoring

7. Sampling

The R&I tactic must be performed in all incidents to determine the type of incident involved. Some of the tactics for this goal are defensive (e.g., R&I and interview) while others are offensive (e.g., sampling), others can be offensive or defensive, depending on the tactical methods (e.g., reconnaissance and monitoring). As such, first responders may perform some of these operations while technicians or specialists are required for others.

Protection

Protection is the process of assuring the safety of the public and response personnel. Although a small portion of this goal addresses the public, a majority targets on-scene response personnel. Primary tactical options for achieving the strategic goal of protection are as follows:

1. Secondary public protection

2. Personal Protective Equipment (PPE)

3. Decontamination
4. Pre-entry briefing
5. EMS and first aid
6. Safety assessment
7. Pre-entry medical monitoring
8. Reassess zones

Protection can be either offensive or defensive depending on the specific tactical method chosen for the tactical objectives. As such, first responders can perform some of these tactics while others require technicians or specialists.

Spill Control

Spill control is the process of stopping, limiting, or controlling the spread of the product through the environment. The specific spill control options depend on the specific type of spill involved. As such, there are specific tactical options for gas/air, liquid/surface, liquid/water and solid/surface releases. It is important to note that even though several types of releases have the same tactical objective, the tactical methods used to accomplish the objectives are different. These methodology differences will be discussed in more detail in Chapter 7. The primary tactical objectives for each release type are as follows:

1. Gas/air release
   a. Ventilation
   b. Dissolution
   c. Dispersion
   d. Diversion
2. Liquid/surface release
   a. Diking
   b. Diversion
   c. Retention
   d. Adsorption
   e. Neutralization
   f. Absorption
3. Liquid/water releases
   a. Damming
   b. Diverting
   c. Booming
   d. Absorption
4. Solid/surface releases
   a. Blanketing

Spill control is generally considered defensive in nature because most of the tactics do not require direct contact with the product. However, spill control operations that occur within a confined area (e.g., within a structure, vault, or tank) or require the use of special thermal or chemical protective clothing are considered offensive. As such, spill control can be performed by first responders or technicians depending on the specific circumstances and hazards associated with the operation.

Leak Control

Leak control is the process of slowing or preventing the release of product through the breach or opening in the container. There are two types of leak control tactical options: direct and indirect.

Direct objectives hold the potential of being one of the most dangerous operations to perform. This is because the performance of direct control options requires entry personnel to be in close proximity to the compromised container. As a result, personnel are almost assured of becoming contaminated with the product, and they have no physical safety buffer because they are so close to the container. On the other hand, indirect objectives hold much less potential for danger because, as the name implies, they are performed remotely from the container. Primary tactical objectives for achieving the strategic goal of direct and indirect leak control are as follows:

1. Direct leak control objectives
   a. Plugging
   b. Patching
   c. Overpacking
   d. Crimping
   e. Product transfer
   f. Valve actuation
2. Indirect leak control objectives
   a. Remote shut-offs
   b. Emergency shut-offs
   c. Product transfer
   d. Product displacement

Leak control is almost exclusively offensive by definition. The only exceptions are the use of emergency or remote shut-offs that do not require any product contact to accomplish actuation. As was discussed in Chapter 2, in some instances the manual shut-off of control valves is considered defensive and within the realm of operations level personnel who have been trained to perform the task. However, leak control is almost exclusively considered offensive.

Fire Control

Fire control is the process of minimizing the total impact of a fire involving or threatening to involve chemicals. The primary considerations concern the safety of responders, the public, and the environment. The primary tactical objectives are as follows:

1. Extinguishment
2. Controlled burn
3. Exposure protection
4. Withdrawal

When fire control involves substances such as fuel type liquids and gases, it is considered a defensive operation. Should the fire involve substances such as flammable solids, fire control may or may not be considered defensive. As such, this is one goal that is rather nebulous in its interpretation and classification.

Recovery/Termination

Recovery/Termination is the process of incident wrap-up both operationally and administratively. Recovery involves operation returning to the preincident conditions (as best as possible) regarding the scene and resources. Recovery has two phases: operational and administrative. The primary tactical objectives of each phase are as follows:

1. Operational recovery
   a. Oversight of cleanup
   b. Oversight of product transfer
   c. Oversight of container righting and handling
   d. Demobilization
2. Administrative recovery
   a. Inventory control
   b. Restocking
   c. Financial restitution

Termination involves the documentation, evaluation, error correction and investigation of incident operations. The primary tactical objectives are as follows:

1. Debriefing
2. Hazard communication
3. Critique
4. Investigation
5. After action analysis
6. After action report
7. Follow-up

It is important to note that, although several of the tactics identified under recovery/termination do not occur on the incident scene or even during the emergency phase of the incident, they are still addressed under this goal. The reason for this is straightforward. Specifically, most of these nonscene operations are mandated by federal regulations or are essential to the continuation of service. As such, they are included under this goal.

Once the determination of appropriate strategic goal is made, the focus of decision making turns to tactics and resources. Chapter 7 examines the primary tactical objectives for each strategic goal and some of the technical considerations regarding the exact tactical methods and resources to employ.

Following the summary is a sample plan of action that identifies each of the strategic goals and their primary tactical objectives. Again, this check sheet is provided as a model for your use. It has been found to work very effectively in actual incident situations.

SUMMARY

This chapter addressed the third step in the GEDAPER process, determining strategic goals. It has:

1. Defined the term *strategic goals*
2. Identified and described hazardous materials strategic goals of Isolation, Notification, Identification, Protection, Spill Control, Leak Control, Fire Control, and Recovery/Termination
3. Described how the estimating process flows directly into the determination of appropriate strategic goals
4. Explained the process of strategic goal prioritizing
5. Explained the relationship between strategic goals, tactical objectives, and tactical methods

PLAN OF ACTION

INCIDENT#:_____ PREPARED BY:_____

ISOLATION:
[   ] Establish Perimeter_____
[   ] Establish Zones_____
[   ] Deny Entry_____
[   ] Initial Public Protection_____
[   ] Withdrawal_____

NOTIFICATION:
[   ] Notify Appropriate Authorities_____
[   ] Notify Hazmat_____
[   ] Request Mutual Aid_____
[   ] Contact CHEMTREC_____
[   ] Contact NRC_____
[   ] Provide Status Report_____
[   ] Establish Staging_____

IDENTIFICATION:
[   ] Use Documentation_____
[   ] Placards and Labels_____
[   ] Reconnaissance_____
[   ] Interview_____
[   ] Review Plans_____
[   ] Monitoring_____

PROTECTION:
[   ] Decontamination_____
[   ] PPE_____
[   ] Secondary Evacuation/In-Place_____
[   ] EMS and First Aid_____
[   ] Safety Assessment_____
[   ] Pre-entry Briefing_____
[   ] Pre-entry Medical Monitoring_____

SPILL CONTROL:
RELEASE TYPE—[   ] GAS/AIR, [   ] LIQUID/SURFACE, [   ] LIQUID/WATER, [   ] SOLID/SURFACE

Gas/Air:
[   ] Ventilation_____
[   ] Dispersion_____
[   ] Dissolution_____
[   ] Blanketing_____

PLAN OF ACTION

LIQUID/SURFACE:
[　] Diking_____
[　] Absorption_____
[　] Adsorption_____
[　] Retention_____

LIQUID/WATER:
[　] Damming_____
[　] Diversion_____
[　] Booming_____
[　] Absorption_____

Solid/Surface:
[　] Blanketing_____

LEAK CONTROL:
[　] Remote Shut-Offs_____
[　] Emergency Shut-Off_____
[　] Plugging_____
[　] Patching_____
[　] Product Transfer_____
[　] Overpack_____
[　] Crimping_____
[　] Other_____
_____

FIRE CONTROL:
[　] Extinguishment_____
[　] Controlled Burn_____
[　] Exposure Protection_____
_____
[　] Withdrawal_____

RECOVERY/TERMINATION:
[　] Clean-Up Oversight_____
[　] Product Transfer Oversight_____
[　] Container Righting/Removal_____
[　] Release of Callbacks/Mutual Aid_____
[　] Debriefing_____
[　] Hazcom_____
[　] Critique_____
[　] After-Action Analysis_____
[　] After-Action Report_____
[　] After-Action Follow-Up_____

PREPARED BY:_____ DATE:_____

PLAN OF ACTION

7 Assessing Tactical Options and Resources

CHAPTER OBJECTIVES

Upon completion of this chapter, the student will be able to:

1. Explain the relationship of strategic goals, tactical objective, and tactical methods
2. Identify the primary tactical objectives for each of the eight strategic goals
3. Explain the role of resources in the determination of appropriate tactical options
4. Describe the intent of each tactical objective listed in this chapter
5. Explain the interconnections between the isolation tactical objectives of perimeter and zone establishment, denial of entry, initial public protection, and withdrawal
6. Describe the advantages of incident leveling and notification needs
7. Explain the role of communications links as part of notification
8. Explain each of the six primary tactical objectives available to achieve the goal of identification
9. Describe the functioning, strengths, and weaknesses of the five primary types of monitoring equipment
10. Identify and explain the nine primary tactical objective available to achieve the goal of protection
11. Identify and describe the seven personnel decontamination mechanisms
12. Describe the limitation of thermal and chemical protective clothing
13. Describe the primary components of a site safety plan
14. Explain the role that spill typing plays in identifying appropriate tactical options for the goal of spill control
15. Identify the primary tactical objectives for each spill type
16. Explain the difference between direct and indirect leak control tactical objectives and their implications on the safety of entry team personnel
17. Identify and describe the five primary tactical objectives available to perform direct leak control

18. Identify and describe three of the primary tactical objectives available to perform indirect leak control

19. Identify three alternative extinguishing agents available to accomplish fire control

20. Describe the strengths and weaknesses of the four primary types of firefighting foams

21. Explain the expansion ratio options for firefighting foams, their strengths, and weaknesses

22. Describe the difference between recovery and termination

23. Identify and describe the two phases of recovery

24. Describe the role of the three phases of termination

ASSESSING TACTICAL OPTIONS AND RESOURCES

Once appropriate strategic goals have been determined for an incident, an assessment of possible tactical objectives and methods is the next step in developing the POA. Along with the assessment of tactical options, there must be an assessment of resource needs. Hopefully the reason is obvious. After all, it is nonsensical to identify a tactical option as appropriate if the resources needed to carry it out are not available.

Resources are one of the imponderable and most fluid aspects of operational strategies and tactics. The reason is simply that no two response agencies are the same. Resources that are considered routine and run-of-the-mill in one jurisdiction may be unheard of luxuries in another. As such, when discussing specific resources in this chapter, a very general approach is employed. If quantities are identified, they are the bare minimums required.

Appendix E contains a minimum inventory developed by the Commonwealth of Pennsylvania. The inventory is part of Pennsylvania's Hazmat Team certification program. It is not provided as the resource answer for all jurisdictions. Rather, it is provided to answer general questions and provide a starting point or benchmark for determining resource needs.

This chapter addresses each of the eight strategic goals and then the primary tactical objectives and methods available to meet the particular goal. However, it is impossible to identify all potential methods, and in some instances, objectives. As such, there is room for the reader to identify additional methods and possibly objectives and to incorporate them into his or her own unique system.

Further, the depth and degree of coverage provided any individual objective or method varies depending upon its complexity and technical nature. Some of the tactics are simple, straightforward activities that require little technical understanding or complex implementation. Others require substantial technical knowledge and evaluation before appropriate implementation is tried. These are the criteria used to determine depth of coverage.

Further, the objectives and methods identified in this unit are all within the existing standard of care. However, at various times, the "accepted" methodologies and ways of identifying those methodologies will be questioned. Some of the accepted methodologies stem from rules of thumb that in some instances have no validity in truth. In other instances, these accepted methodologies are the simple way of addressing the problem.

However, as the discipline of hazardous materials emergency response has grown and evolved, more understanding and factually-based considerations have emerged. Often accepted methodologies are based on hazardous waste operations. Unfortunately, although there are conceptual similarities, there are many significant and fundamental operational differences.

The discussions in this chapter are based on sound science, practice, and street experience. They are intended to provide realistic approaches to the multitude of intangibles that responders routinely encounter during emergency-response operations. It is hoped that this approach will raise questions and stimulate discussions that will further our understanding.

ISOLATION

The process of securing and maintaining physical control of the incident scene	**Primary Tactical Objectives:** Perimeter Establishment Zoning Initial Public Protection Denial of Entry Withdrawal

Isolation involves tactics that accomplish securing and maintaining physical control of the incident scene. The purpose for this goal and its tactics is three fold. First these operations prevent the spread of contaminant of a greater physical area than is necessary. Second, they delineate work areas by their relative hazard. Third, they minimize accidental crossing between hazard area by establishing physical barriers and operational control measures.

ISOLATION—DENIAL OF ENTRY

Denial of entry involves preventing the movement of the public and response personnel into areas to which they are not trained, protected, or needed. As this definition implies, this tactic involves the establishment of prescribed procedures that provide access control as well. This is necessary because not everyone is prevented from entering the various zones once information has been gathered to determine the nature and extent of the hazards (site characterization).

Initially, denial of entry starts with the first individual or responder to arrive on the scene. Most commonly this will be police, EMS, or fire personnel. At a fixed facility, denial of entry may be the responsibility of security, maintenance, or other workers. It involves identifying there is a chemical problem (or a potential problem) and identifying that only properly trained, equipped, and protected personnel should get close to the incident. Once this is identified, the process of simply telling others not to enter the area is a rudimentary form of this tactic.

As the response swings into gear, denial of entry becomes more structured and methodical. First, a perimeter is established, often through the use of physical barriers such as tapes, highway cones, placement of vehicles, and barricades.

At this stage, denial of entry takes the form of basic crowd and traffic control. It is designed to indicate that outside this area things can proceed as usual. However, to enter the incident scene, people must have training and be part of an organized response function. Further, people who were within this area cannot simply leave without some type of evaluation or assessment to assure that they were not injured or contaminated.

As the response grows, so does the scope and complexity of denial of entry. As zones are established, movement between them must be restricted. Although the outermost zone is safe, movement inward and toward the incident is restricted. In this case, not only must personnel be properly trained, equipped, and protected, they must have a functional part in activities occurring within the other operational areas.

Moreover, not only is movement into and through the operational area closer to the center of incident controlled, movement back outward is also controlled. Before personnel leave the innermost operational areas, they must follow prescribed procedures to protect their safety and the safety of others.

To assure that all of these various control procedures are implemented, operational procedures, physical barriers, and oversight personnel must be designated. Operational procedures must be found in the form of Standard Operating Guidelines (SOGs) that designate how physical barriers are employed and who has oversight responsibility. Two fundamental pieces to this process are perimeter establishment and zoning.

ISOLATION—PERIMETER ESTABLISHMENT

The establishment of a perimeter, sometimes called *outer perimeter,* involves identifying a distance from an incident beyond which no one without proper training and personal protective equipment (PPE) should go. In essence, the perimeter separates the incident's operational area from its uninvolved surroundings. Commonly, the perimeter is established at the edge of the isolation distance and delineated by using physical barriers. Furthermore, first responders who arrive on the scene early in the incident routinely are responsible to establish the initial perimeter.

To a great extent, the establishment of the outer perimeter is an indicator of how the incident will evolve (Figure 7-1). Specifically, when first responders fail to identify the need for a perimeter, they usually have failed to identify the incident is a potential or actual hazmat incident. As a result, they employ some other type of strategy-tactic system that involves operations inappropriate for a hazmat incident.

The initial perimeter established by first responders, often becomes either the hot line (boundary between the hot and warm zones) or the cold line or inner perimeter, (boundary between the warm and cold zones). However, if first responders have identified an appropriate location, their perimeter becomes the actual incident perimeter.

On the basis of this description, the establishment of the perimeter sounds simple and straightforward, which it is. However, determining its location is often more complex. Specifically, the question of how far from the incident should the perimeter be established is difficult to answer.

Normally, the minimum distance from the incident to the perimeter should be 150 feet. This is assuming there are no incident conditions or information indicating the need for a greater distance. The 150-foot recommendation is based on the minimum distance identified in DOTs *North American Emergency Response Guide* (NAERG) guides and isolation table. As such, it provides at least well founded, initial baseline distance.

However, this does not imply that the distances identified in the NAERG, LEPC Facility plan, computer plume projections, and so on are etched in stone or are the only acceptable option. As stated earlier, contrary to the belief of many outside and, in some cases, inside the emergency response community, the identification of the incident perimeter and zones is one area of hazmat response that is somewhat an art. This is because this determination involves a large dose of judgment, operational knowledge, and experience.

Further, there is a multitude of technical support systems that range from the *Isolation Table* in NAERG to computer plume models, but the data that they generate are only guidelines. There is a multitude of variables that dramatically alter the actual vulnerable areas from the "text book answers" as provided by these technical supports. Field tests and actual incident experience have shown that even the most sophisticated models have limitations and cannot incorporate all the potential variables.

These variables include environmental conditions such as temperature inversions, ground and valley fog, topographic variables, and so on. Some of the field tests have

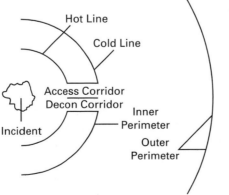

Figure 7-1   Perimeter Locations

raised definite concerns that require additional consideration. For example, one set of tests found concentrations of contaminant close to the release location were lower than the models projected. However, the same tests found concentrations farther from the release location that were significantly higher than projected.

Moreover, various modeling programs and systems use data derived from plain gases as opposed to hydrolyzed droplets or particulate fumes. Some are based on lighter-than-air contaminants while others are based on heavier-than-air contaminants. Most consider ambient condition releases and not fire conditions.

Additionally, most of these projections or guidelines are based on a supposed worst case scenario. As stated before, worst case depends on one's perspective and is very illusive to define. As a result, it makes perfect sense that none of the technical data sources identify exact distances for a situation that is less than the worst case scenario, since even that is difficult to define.

Fortunately, the answer to this quandary involves the acquisition of real time, on-scene data. These data, including instrumentation readings, changing scene conditions, tactical operations, and so on, are factored into repetitive examinations. As this statement implies, the locating and assessment of the perimeter and zones are routinely dynamic situations within the overall operation. Because of the dynamic nature of incident conditions and the relationship of the perimeter and zones, this discussion will continue under the tactical objective of zoning.

ISOLATION—ZONING

Zoning is the process of subdividing the operational area within the perimeter of the incident. These operational subdivisions are called zones and indicate the varying degree and extent of health and safety hazards found within each of them. Zones also identify the areas where specific types of operational activities and functions occur as well as identify areas that require the use of specified PPE.

Further, zones provide physical distance between incident hazards and responders. They provide a safety buffer between personnel and the hazards, much in the same way a collapse zone provides a safety buffer when structural collapse is a potential hazard during firefighting. Should something unexpected occur, the zones allow reaction time and distance necessary for personnel safety.

Traditionally, responders have been taught that there are three zones. Traditionally, responders have been taught that zones are similar to a bull's eye or a target by their graphic representation. Traditionally, responders have been taught zoning in the context of wide open areas or city streets. However, these traditional approaches are quite superficial and far from adequate when considering the complexities of zoning on an actual incident scene. There are many complicating factors that arise.

Consider the following examples as representing some possible situations that can be encountered and how they will affect zoning.

1. Approach from uphill and upwind is not possible
2. To be upwind, personnel must approach and set-up from downhill

3. The incident is on the second floor of a building
4. There are a basement, sumps, and a vault in the building
5. The release is between two three story buildings that are 150 feet apart and 1,000 feet long
6. The incident occurs on an interstate that runs through the heart of downtown

These are just a few of the myriad of situations and conditions that affect the establishment of zones. As a result, it is essential to examine zoning in greater depth and detail.

The Zones

Traditional training indicates that there are three zones used during a hazmat response. This is only partly correct because although there are three zones established within the outer perimeter, there is actually a fourth zone outside the perimeter. Even though this zone is important, there is no recognized name for it. This zone outside the perimeter is an important consideration during actual operations because personnel, resources, and oversight must be dedicated to establish, coordinate, and control movement through the perimeter.

Notice, there was no mention of the fourth zone being hazard free because it may very well be quite hazardous. Consider an incident on a six-lane interstate highway. The incident is small and requires a perimeter that encompassed the shoulder and first traffic lane. As a result, traffic is allowed to keep moving. However, is it safe to enter these traffic lanes? In an industrial setting, production equipment and processing is ongoing outside the perimeter. Again, is it safe to enter that area? The answer is no, even though both locations are outside the perimeter. Granted, although such situations are not the routine, they are not infrequent and consideration of their potential is needed.

Another issue is the potential shape of the perimeter and zones (Figure 7-2). Unlike the commonly held perspective, zones and the perimeter can take almost any shape as long as there is layering based on the hazards, among other factors (e.g., the hot zone, within the warm zone, within the cold zone, and within the perimeter).

There are four primary shapes. These shapes are the keyhole, teardrop, block, and irregular. Regardless of the specific shape, the hot zone is the innermost zone, and the cold zone is the outermost zone next to the perimeter.

Within the Perimeter: Three Zones and Subzones

As previously stated, there are three zones located within the perimeter and identified on the basis of their health and safety hazards. Unfortunately, there are no universally accepted names for these zones. The original names—exclusionary, contamination reduction, and safe zones—come from EPA terminology. However, this terminology is derived from hazardous waste site operations and does not fully address emergency operation needs. Consequently, this text employs the more commonly used and flexible names: hot, warm, and cold zones (Figure 7-3).

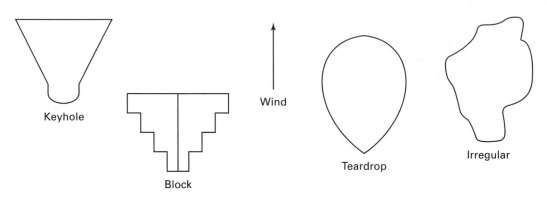

Figure 7-2   Primary Zone and Perimeter Shapes

Further, although the EPA terminology identifies principle functions and concepts of operation, it does not lend itself to clear and concise explanations of routinely used and accepted emergency operations. The primary operational need is subzoning located within each zone.

Subzoning is the process of breaking each operational zone into smaller areas with similar hazards, PPE requirements, operational functions, and so on. Although this term may not be familiar to many, we will examine the principle zone by zone.

Hot Zone.   The hot zone is the innermost zone within the perimeter. It is the location of the actual incident itself. This zone contains the highest degree and magnitude of both health and safety hazards on the entire scene. Personnel operating within this zone may require the highest level of PPE used on the incident scene depending on their specific function and tasks. Anything or anyone who enters the hot zone may become contaminated depending on their exact location. Thus, it is assumed that conta-

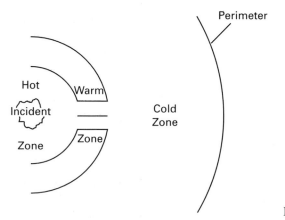

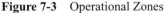

Figure 7-3   Operational Zones

mination has occurred. As a result, before people or objects leave the hot zone they must go through some process of decontamination.

The reason for this is simple. If a contaminated person or object is allowed to leave the hot zone, they will spread contaminant. Not only is the spread of contamination potentially dangerous, it expands the hot zone, which in turn complicates the operation.

For example, a worker is contaminated when the incident initially occurs. When responders arrive on the scene, the worker greets them outside the door. It is noted that the worker is exhibiting some slurred speech and other neurologic signs. EMS personnel rush to evaluate the victim. They load the victim and transport to the hospital. At the point when the victim arrives at the hospital and enters the emergency room (ER), the hot zone has been extended all the way from the incident scene to the ER.

Some may say that this type of situation does not occur anymore. Unfortunately, however, this type of inappropriate action occurs on almost a daily basis around this country.

The next question is whether this means that all people who were within what is now the hot zone require decontamination (decon). The answer is no. Even if workers or others who were in the area when the incident started require decontamination, the extent of the procedures usually vary depending on the person's proximity to the initial release. This type of information and a physical inspection are needed to fully evaluate their decontamination needs.

So how are such people handled? Obviously, they must be maintained within the hot zone because if they leave it before they are decontaminated, they simply take the hot zone with them. Here is where the first subzone within the hot zone comes into play. This first subzone is a place of refuge and is commonly known as the *area of safe refuge* or *safe haven* (Figure 7-4). It is located within the hot zone, near the boundary with the warm zone. If the refuge area were located deeper within the hot zone, it would have of high probability, if not a certainty, of being contaminated. As such, these people would be exposed to potentially severe health and safety hazards.

The safe refuge also provides an oasis for response personnel operating within the hot zone. Should there be some type of unusual occurrence, the entry team can retreat to the safe refuge. This does not mean they do not require decontamination or other routine exit actions. Rather, the safe refuge provides a safe location for them to receive assistance or instructions.

This logic sequence introduces the second subzone within the hot zone, known as the exclusion area or exclusionary subzone (Figure 7-5). This subzone is contaminated or will be contaminated should a release occur. The exclusion area starts where the actual contaminant starts based on visual indications or instrument readings. As such, the safe refuge area must be located well outside this subzone.

Additionally, it is highly advantageous to identify the existence and location of two other subareas within the exclusion area. These two subzones are the TLV-TWA and IDLH (or LEL) areas. They are simply locations within the exclusion area where the concentration of the contaminant has first reached the TLV and then the IDLH. The proximity of both of these subareas to each other, the exclusion line, the safe refuge area, and the hot line provides excellent tools for the assessment of the zones.

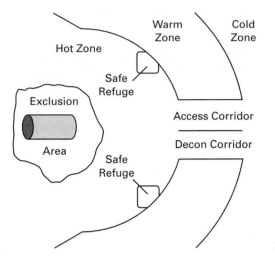

Figure 7-4   Areas of Safe Refuge

Further, changes in their location or size indicate changes in the incident situation. Such changes may indicate wind shifts, changes in humidity, an expanding leak, a diminishing release, an increase in pressure, and so on. In any event, such changes may indicate the need for immediate action or the need to gather additional information to determine *why* the changes occurred.

Since the exclusionary subzone is contaminated and contains the highest degree of health and safety hazards, operations within this zone require the maximum level of PPE required for the incident. Does this apply to the entire hot zone? No. If it did, the people in the safe refuge would be in serious trouble.

At this point, it is time to address several fundamental misconceptions held by many responders. Consider the following questions in the context of an actual emergency response.

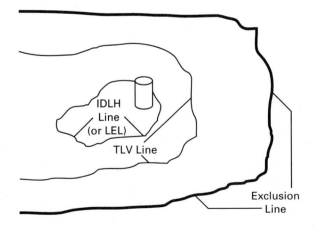

Figure 7-5   Exclusion Area and Lines

What type of PPE must be worn by responders who set up the physical barriers at the hot line (i.e., the line separating the hot zone from the warm zone)? Routinely the answer is "a minimum of full structural turnouts and positive pressure self contained breathing apparatus (SCBA)." This answer is wrong! The correct answer is none. The reason for this answer is the important point to the question. If personnel setting up the physical barriers need to use PPE, the hot line is way too close to the incident, and the hot zone is thus way too small.

Based on instrument readings, where should the hot line be established? Routinely, the answer provided goes something like this: "Where the instrumentation first provides a reading that indicates the presence of contaminant." Again, this answer is wrong. The correct answer is that the hot line should be established a reasonable distance before any readings are encountered. The reason for this is basically the same as in the first question. Additionally, based on subzoning, the need for an area of safe refuge and the exclusion area, it should be obvious that the hot line needs to be a considerable distance away from the first readings. This is especially true for instruments that have low sensitivity, such as combustible gas indicators (CGIs) or oxygen meters.

Would it be allowable to have personnel wearing full structural turnouts and positive pressure SCBA work in the hot zone? Routinely the answer provided is no. Again, this answer is wrong.* The correct answer is yes, with a caveat: The turnouts and SCBA must provide appropriate protection for the specific hazards to which the wearer will be exposed while performing his or her designated function. This is not to indicate that turnouts are appropriate in all situations or for all substances. Further, this does not mean that turnouts are even appropriate for all areas within the hot zone. This leads to the last question.

Is it possible to have personnel in turnouts and SCBA as well as personnel in chemical protective clothing working in the hot zone at the same time? The answer commonly provided is no. This answer is again wrong. The correct answer is yes, again with a caveat: The personnel wearing turnouts and SCBA are operating within an area where this type of PPE provides appropriate protection. Most commonly, this would be in the area outside the exclusion area. The reason for this goes back to the concepts of subzoning. The exclusion line is identified as the location where the contaminant is first detectable by instrumentation. Obviously, the instrumentation used must have high sensitivity and a continuous monitoring plan must be in effect.

Found within the exclusion line is the TLV or PEL area. This is where the concentration of the contaminant reaches the TLV-TWA. By definition, the TLV is the average concentration of a contaminant to which an individual may be exposed 8 hours a day, 40 hours a week. As such, personnel outside this area will not receive injurious chemical exposures. Further, even people entering the exclusion area and passing the

---

*NOTE: Whenever the term *turnouts* is used, it refers to OSHA or NFPA compliant structural fire fighting turnouts (whichever is applicable to the jurisdiction). Further, whenever the term *SCBA* is used, it refers to positive pressure self contained breathing apparatus.

TLV or IDLH lines may not require any higher level of protection, depending on the substance's hazards and properties.

Others may ask, "Under what type of incident situation would such duel PPE level entries occur?" One of the most probable situations would involve a liquid that has chemical hazards such as being skin absorptive or tissue destructive and also the thermal hazard of being flammable. Unfortunately, the PPE for chemical hazards provides no protection for thermal hazards, while PPE for thermal hazards provides limited protection for chemical hazards. Basically, this is a mutually exclusive situation.

To properly protect personnel, one of the hazards (either chemical or thermal) must be overcome. Most often, the thermal hazard is the one that can be controlled. In the case of liquids, the application of a foam blanket suppresses the vapors and thus normally controls the flammability hazard.

Consider the situation where a MC 307 highway cargo tank containing a liquid such as methylmorpholine is involved in an incident. The cargo tank has rolled over in a ditch. Liquid product is escaping from a belly valve and a hatch. The liquid is flammable, toxic through skin absorption, and tissue destructive.

Such materials include morpholine or allyl chloride. Morpholine is flammable (flash point of 95°F with a flammable range of 1.8 to 11%) and is toxic through inhalation, ingestion, and skin absorption (TLV-TWA 20 ppm). It is also an irritant to skin. Allyl chloride is flammable (flash point of −20°F with a flammable range of 3.3 to 11%) and toxic by inhalation, ingestion, and skin absorption (TLV-TWA 1 ppm, IDLH 300 ppm). It is also a strong irritant to skin.

Flammability is a hazard of both substances (an extreme hazard in the case of allyl chloride). To eliminate the flammability hazard as well as to minimize the skin destructive and absorptive hazard (especially during spill control operations), a foam blanket is ideal.

Foam application by responders wearing turnouts and SCBA is appropriate in this case. Also the use of turnouts and SCBA for spill control operations (where there is no contact with the liquid, especially after the vapor is controlled) is also appropriate. Such operations could be performed by properly trained and protected operations-level personnel since both are defensive, and the use of turnout and SCBA provides appropriate protection.

However, leak control operations require the use of chemical protective clothing (double enveloped over turnouts would be best). Since leak control requires personnel to work close to the container, they will most likely break the foam blanket. Even if they do not, foam drain necessitates periodic reapplication, thus back-up foam lines must be in place to provide protection (as has been the standard for decades). To accomplish these tactics, it is necessary to have hand line crews and a leak control team working in the hot zone at the same time (Figure 7-6).

What is more important is that all of these tactical actions are based on sound strategic, tactical, and personnel safety approaches. Further, they comply with all aspects of the standard of care as well as regulatory requirements, PPE selection and usage, industrial hygiene protocols, and overall health and safety approaches.

Why does this approach to this incident work? First, because it is based on health and safety principles as well as operational realities as opposed to "rules of thumb" or

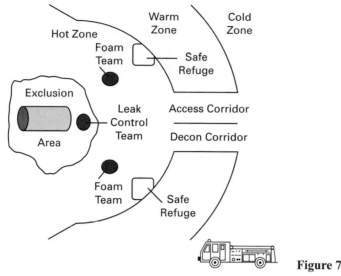

Figure 7-6   Subzoning

because "they say so." Second, because the process of subzoning enables the effective implementation of strategies and tactics without shortcutting safety or operational efficiency. They are based on existing hazards, vulnerability, and risk.

However, keep in mind that this is not meant to imply that the use of multiple levels and types of PPE is always justifiable simply on the basis of subzoning. Rather, one must consider the facts about hazards, personnel training, PPE selection criteria, PPE limitations, the incident situation, and so on before using these approaches.

Warm Zone.   At the most fundamental level, the warm zone is an intermediate area between the hot zone and the cold zone. Its dimensions vary. The warm zone should exhibit no health or safety hazards before the movement of personnel, victims, or equipment into this zone from the hot zone.

In most instances it extends only as far as necessary on either side of this zone's most upwind location. Beyond this area, the usefulness of the warm zone ceases. Extending this zone farther presents logistical problems in the form of a second control line that must be established and maintained. Normally, the hot line and the cold line simple merge to form the zone separation.

The reasoning behind this explanation becomes more clear when the function is described rather than the definition of this zone given. Specifically, the warm zone identifies an area where the access and decontamination corridors are located (Figure 7-7).

The access corridor is a subzone of the warm zone and is designated by physical barriers such as tapes, cones, and barricades. It is the only location on the incident scene through which access is granted into the hot zone. In essence, the access corridor is an extension of the cold zone, through the warm zone, to the hot line. Its length

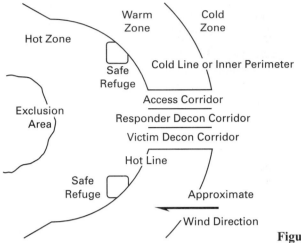

Figure 7-7   The Warm Zone

is dependent on that of the decontamination corridor. Truly, the only reason it exists is so the decontamination corridor has a specified location.

The decontamination corridor or corridors are also subzones of the warm zone in which all decontamination procedures are performed. Routinely a single decon corridor is established to provide decontamination for response personnel. However, a second decon corridor is necessary should contaminated victims be encountered at an incident scene. The actual dimensions of this corridor depend on the specific type and extent of decontamination required. Generally, this corridor ranges up to a maximum of 60 or 70 feet long by 25 to 30 feet wide. (Decontamination and the corridor are covered in more depth later in this chapter.) As a direct result, the warm zone ranges up to a maximum of 60 to 70 feet.

Cold Zone.   The cold zone is located within the incident perimeter but outside the other zones. It functions as a safe operational area. Located within the cold zone are the Incident Command Post, the hazmat command vehicle (forward or field command), medical monitoring, medical rehabilitation (rehab), victim transportation, other operational personnel, fire apparatus, and so on. This zone must be free of any health and safety hazards related to the incident itself or it is too close to the hot zone.

Having said that, it is not meant to indicate that there are no hazards in the cold zone. All the routine response hazards are still present, except for those related to the chemical aspects of the incident. Further, the cold zone can become endangered should incident or meteorologic conditions change, or should terrorism be involved.

The cold zone is often entered by people other than response personnel. Most commonly these individuals are representatives of various governmental agencies, support organizations, industrial representative, technical experts, and so on. In many instances, this may also include representatives from the news media. Before such individuals enter the cold zone, they must be appraised of the scene hazards (i.e., the right to know) from the data found in the site safety plan. Further, they must receive instruction regarding appropriate and inappropriate actions in this zone. It is essential to identify the levels and types of PPE required for entry to each zone. Further, it is highly recommended that the decontamination procedures and set-up be described as well.

This approach not only meets any employee right-to-know mandates, it also helps to eliminate potential problems down the road by establishing the ground rules for the incident scene. The concept is similar to the onsite training mandates required for workers at hazardous waste sites and further makes good sense.

Additional Zoning Considerations

Obviously, every incident is unique and may have multiple variables that impact zoning. However, in this regard, there are several variables that are commonly encountered:

1. Three-dimensional zoning
2. Wind
3. Operations within structures

Three-dimensional Zoning. Unlike the two-dimensional graphics of the zones, zoning is a three-dimensional process simply because contaminants are able to spread in three dimensions. Three-dimensional zoning becomes acutely important when the incident occurs within a multistory structure or in a well developed or urban setting.

Three-dimensional zoning means that zones are spherical in nature. They extend above and below the plane of the incident if there is any way for the contaminant to spread in those directions. The contaminant will spread until it meets a physical barrier. However, just like water, smoke, or heat, contaminants do their best to penetrate the barriers through any available conduit and continue to spread.

Unfortunately, these conduits and the potentially affected areas sometimes cannot be readily identifiable. It requires a concerted effort to identify the potential conduits and the area that they can affect. In some instances, it is easier to identify the potentially vulnerable areas and assess the presence of contaminant than it is to identify the conduit.

One of the first considerations is the physical properties of the product. Is it a solid, liquid, or a gas? Generally, solids are the least likely to spread through conduits unless they are in a fine particulate form such as powders or dusts. Liquids always tend to move downward and seek the lowest level. They also produce vapors that most commonly also seek low levels, although there are some exceptions to this. Vapor density and pressure are critical factors. Likewise, gases may either rise, sink or be neutrally buoyant. Again, vapor density is a critical factor. In most instances, vapors that are heavier than air will not dissipate by natural ventilation alone and thus will remain in low lying areas for a considerable time. The higher the vapor density, the more tenaciously it will linger in the area and resist natural dissipation.

One of the next considerations is locating the potential vulnerable area above and below the level of the incident. If the incident occurs within a structure, areas on floors above and below, HVAC plenums, drop ceilings plenums, utility chases, and so on must be considered potentially hot (contaminated) or threatened with becoming hot, until proven otherwise. Basements, utility vaults, sumps, floor drains, floor trenches, pipe chases, areas below elevated floors (e.g., as found in computer rooms and clean rooms.), storm sewers, sanitary sewers, tunnels, remote catch basins or retentions, and any other potentially contaminated area must be identified.

During one incident involving a pipeline valve failure, a fuel product was found to have flowed into basements, storm sewers, and sanitary sewers as well as an underground electrical vault. The vault was filled with six feet of liquid product.

Sanitary and especially storm sewers are notorious for establishing venturi-like flow patterns that can move vapors and gases in unpredictable directions and patterns. As a result, they can cause vapors to show up in unanticipated places and areas.

Wind.   Wind is a major factor in the zoning process, because it, more than almost any other factor, dictates where airborne contaminants spread. As such, the access and decon corridors as well as the operational areas of the cold zone are to be located upwind from the incident within approximately $\pm 20°$ of the perpendicular (Figure 7-8).

Unfortunately, a multitude of variables can affect this ideal situation. First, access to the incident is often limited. Interstate highways with sound barriers or rail lines are notorious for this problem. In the case where access is limited, it may be necessary to set-up outside the recommended plus or minus 20°. However, since conditions are not ideal, the hot zone must be expanded by moving the hot line further from the incident. The expanded hot zone provides an additional safety buffer to offset the adverse situation. However, the deviation should not exceed approximately 60° at any time. The distance the hot zone and line are expanded depends on variables such as:

1. The physical state and form of the product
2. Its vapor density and pressure
3. How far from the plus or minus the 20° norm is needed; i.e., the further from 20°, the greater the zone expansion
4. The wind speed and constancy of direction

Second, the rule of thumb is to always approach and establish access and decon corridors uphill and upwind. Unfortunately, incidents do not always follow this rule. Uphill and upwind are the ideal but response situations are often far from ideal. It is not uncommon to encounter situations where being uphill puts responders downwind.

Now the question is, "When this situation is encountered, how is it handled?" Since the ideal is not attainable, upwind takes precedence over uphill; zoning almost always follows the upwind precedence. This indicates that the hot zone must be expanded to a greater distance than would be necessary under ideal conditions. The de-

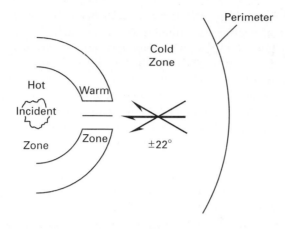

Figure 7-8   Operational Zones

gree of expansion depends on the same considerations identified under the first list of variables and also includes the following:

1. The quantity of product, especially the liquids involved
2. The degree, if any, that the vapor density is above 1
3. If there are physical barriers between responders and the incident that will help limit spread
4. If no spread limiting barriers exist, whether they can be built and how long that will take

Third, large buildings and other man-made structures (e.g., highway sound barriers and levies) produce major variations in wind direction and speed. Routinely, large buildings are found at fixed sites such as industrial facilities as well as in highly developed or urban settings. These structures can easily change wind direction as much as 90° from that of surrounding areas. These buildings produce eddies and vortices that swirl haphazardly over large areas. For example, long, multistory buildings as well as mature tree lines and woods can affect wind speed and direction as far as 2,000 feet downwind. Two or more large, multistory buildings also dramatically increase wind speed by the venturi process as the wind is forced between the buildings.

Operations within Structures.   Operations within structures present many unique and challenging issues. The three-dimensional nature of these operations has been addressed, but there are still other variables. The primary additional variables include air movement, access, and physical layout.

Air movement, like wind, plays an integral role in the zoning process. Fortunately or unfortunately (depending on the situation), structures usually have some type of air handling systems. These systems can be a great ally or a fierce enemy depending on their design and function. Air movement within the structure is independent of conditions outside. These systems can ventilate contaminants to the outside or spread them to unaffected parts of the structure. They may vent contaminants to the outside and thus expose surrounding areas and people, including responders. Further, it is essential to identify the locations of HVAC and other air handlers exhaust discharges so response personnel and apparatus do not stage or go into operation downwind of these discharges.

Zoning within structures also is an issue. As was mentioned previously, three-dimensional zoning is critical when the incident occurs within a structure. However, this is only part of the puzzle. The actual location of zones is often dictated by the physical layout of the building, floor, and so on. Figure 7-9 suggests some alternatives to the zoning process.

The concept of operations is very similar to high-rise firefighting. First, hazmat staging, suiting, medical monitoring, rehab, and similar hazmat functional areas must be remote from the incident location and often the floor. Routinely, these areas are located outside the structure in one story occupancies, or one to two floors below the incident in multistory occupancies. As a result, doorways and stair tower access to the incident floor become critical and must be factored into the zoning process.

Normally, there will be at least two means of access available to the incident floor or area (Figure 7-10). To minimize the potential for cross contamination, a standard access corridor incorporates one of the access door or stair towers and the hallway leading to the incident area. However, an egress corridor is added. The egress corridor leads to the decon corridor and only exists until the first entry team heads to the decon corridor. This corridor is necessary because of the remote location of the decon corridor during structural operations and the associated need to expand the hot zone.

Initially the egress corridor is considered part of the warm zone. However, once the entry team proceeds through this corridor to decon, it becomes an extension of the hot zone. This is due to the spread of contamination all the way to decon. As a result, the need and logic for the use of an egress corridor should become clear.

First, space within stair towers is normally limited to a maximum of four feet, while hallways average 6 to 8 feet. By separating access and decon, their locations and functions are clearly and physically delineated. Further, once personnel are familiar with this set-up, confusion is minimized and efficiency is increased.

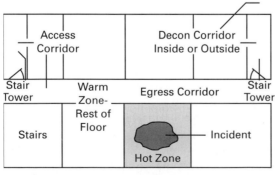

Cold Zone-Area Outside the Structure

Figure 7-9   Zoning Within Structures

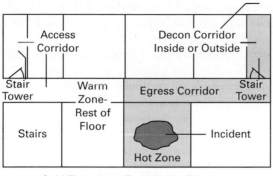

Cold Zone-Area Outside the Structure

Figure 7-10 Zoning Within Structures After First Entry

ISOLATION—INITIAL PUBLIC PROTECTION

Initial public protection tactics involve actions necessary to minimize or prevent injury or death to the public. The two primary alternatives for public protection are in-place protection and evacuation. Both options require considerable interagency coordination, communication, and cooperation for situations that require anything other than limited scale measures. In other words, an effective public protection system is required. Most commonly, the lead and coordinating agency within this system is emergency management. As a result, many emergency responders consider initial public protection someone else's problem.

However, emergency responders must understand what is involved in both options, when each is appropriate and inappropriate, and how to activate the system. They also need to understand what is required for the system to function effectively so their own capabilities can be assessed. Finally, they need to understand the relationship with and functioning of the emergency management agency and the emergency operations center (EOC) that serves their jurisdiction.

In-Place Protection

In-place protection, sometimes called shelter-in-place, involves keeping members of the public within a sealed location to protect them from a toxic atmosphere that exists outside. Although this options has not been widely used in this country, it is a viable approach under the right conditions and may be the only alternative. Further, actual case studies of various incidents have shown its effectiveness.

For example during the mid 1970s, an MC 331 transporting anhydrous ammonia fell from a highway overpass. When the cargo tank impacted on the pavement below, it ruptured completely and released all of its contents almost instantly. A massive vapor cloud ensued, covering the highway clover leaf and several surrounding buildings. In this incident, as well as others, it was found that people who remained within some type of shelter were exposed to lower concentrations of the contaminant than those who tried to evacuate.

In this particular incident, those within the area shrouded in ammonia who remained in their cars survived, although they received exposures. Those who exited their cars and tried to evacuate received lethal exposures.

For in-place protection to work, occupants of a structure use the building and the air within to protect themselves from high concentrations of contaminant. There are several ways to accomplish this type of protection. These methods vary with the type of occupancy but generally include the following:

1. Shut down air-handling, ventilation, and air-conditioning equipment to prevent inflow of outside air.
2. Seal or "button up" the structure or areas within the structure.
3. In high rises and multistory structures, move people from the lower floors to higher floors. This is the case with most airborne contaminants because they have vapor densities greater than 1.

Anyone who remembers the Gulf War also remembers how the Israelis prepared for the potential chemical and biological attacks by the Iraqi the Scud missiles by sealing their homes and businesses. These preparations were similar to those needed for in-place protection. Beside shutting down air handling systems, sealing is especially helpful in dwellings.

Sealable spaces are usually located in or toward the center of the structure since they have no doors or windows to the outside. It is essential for occupants to prevent movement of the contaminant from the outside to the inside. As such, all doors and windows must be closed, preferably with multiple layers such as plastic sheeting, storm windows, and doors with the glass pane in place. The placement of damp towels at the base of the doors and the taping of cracks are also helpful but not essential.

It is important to note that the use of these tactical methods is also applicable as spill control methods for responses involving hospitals or similar occupancies where evacuation is not possible. The use of in-place protection is ideally suited for the following situations:

1. A release resulting in the rapid coverage of the areas with contaminant (including areas where personnel performing evacuation would require respiratory protection)
2. High rise, urban, and similar high density locations where evacuation cannot be accomplished rapidly
3. Occupancies such as hospitals, nursing homes, and prisons where the inhabitants have limited to no mobility
4. Areas anticipated to have relatively low contaminant levels (most commonly involving concentrations at or below the TLV or areas surrounding the downwind protection area)
5. A release that will finish before evacuation is completed

Besides the situations mentioned above, actual scene conditions must always be considered. Three primary considerations for the use of in-place protection include:

1. How long the release will continue
2. The extent to which the existing shelters can protect their occupants
3. A harm-benefit assessment comparing in-place protection versus evacuation (which will produce less harm?)

Short duration releases are the most suitable for in-place protection. Releases lasting one to two hours are optimum. It has been found that, especially in high concentration situations, after about two hours interior concentrations are similar to those found on the exterior. Further, the concentrations linger within the structure. As a result, after the in-place protection period is over, the structures must be opened up and aired out.

There are situations where in-place protection and evacuation can be coupled to provide maximum protection. The ideal situation for this type of combined protection is one where the release has a rapid onset that results in initially high contaminant concentrations. This type of release is generally a large volume flow situation that either depletes the supply of product or is partially controlled. The depletion or partial control allows concentrations to decrease after the initial spike and thus allows a follow-up evacuation. Several incident situations indicate that in-place protection is not a good option. These situations include:

1. The potential for major explosions or BLEVE
2. The potential for mass fire
3. A vapor cloud consisting of flammable gas or vapor

Evacuation

Evacuation is the process of moving people out of the area that is or may be affected by the contaminant. This is the tried and true method of public protection. Many responders believe that when they are faced with a situation requiring evacuation, like Captain Picard of the Star Ship *Enterprise,* all they need do is tell someone, "Make it so."

Unfortunately, it is not nearly that easy. There are many issues and considerations that impact evacuation. The following are examples of some of such potentials:

1. Who has the authority to call for an evacuation?
2. Can evacuation be accomplished and who will perform the functions?
3. How long will it take and is there the potential for public exposure?
4. How will the public be alerted and provided with emergency public information?

The most basic consideration is who is authorized to order an evacuation within the jurisdiction. Most commonly, the answer to that question is found in state laws. In most states, the only individual authorized to order a mandatory evacuation is the governor. Normally, this step cannot be taken until there is an emergency declaration at various levels of government.

Further, what authority is granted to the incident commander (IC) by the local SARA Title III plan? Must the IC request permission? What role does the emergency management agency or its director play? What agencies are to be involved in the evacuation? What are the agency roles and responsibilities, and how is coordination accomplished? Failure to perform in accordance with these plan requirements and statutes can lead to serious legal entanglements for the IC and the jurisdiction.

Additionally, there are many other considerations that come into play. One of the first is whether the threatened area can be evacuated. As mentioned, highly urbanized and developed areas require extended time to perform evacuation and subsequent sheltering. Nursing homes, hospitals and similar facilities are extremely difficult to evacuate. Commonly patients die as a result. Consider critical care sections of a hospital such as intensive care units (ICU), coronary or critical care units (CCU), trauma units, burn units, and operating rooms. These areas cannot be evacuated without extensive time and preparation.

How will the evacuation be performed and by whom? Door to door evacuations are only effective for small scale evacuations. Such personnel intensive evacuations during hazmat incidents are generally impossible when the areas extend more than a quarter mile, other than in places with extremely low population density.

This leads to the next consideration. How will the public be alerted to the problem and then receive emergency public information? The old tried and true answer given by many is to activate the Civil Defense sirens (emergency broadcast system). In some jurisdictions this will work; in others it will not. Areas around fixed nuclear plants and those in the tornado alley of the Midwest are most likely to have the greatest success with this method. The reason for this is that the public has been educated to what the system means and what their role in the system is. Other parts of the country are not as well prepared.

Consider the following examples:

Example 1.   During the mid 1980s, a state emergency management agency was "testing" their civil defense warning systems internally. "Accidentally," the warning to take-cover was transmitted to the Emergency Operations Centers (EOC) of half of the counties in the state. One county's communications center received the message and, following their procedures, activate the county-wide Emergency Alerting System (EAS) sirens.* Unfortunately, when the sirens went off, nothing happened. Sirens were commonly used for volunteer fire and ambulance squads. Even in career departments, these sirens were disregarded. One deputy chief even called the communications center and requested maintenance personnel be notified of a siren malfunction.

---

*NOTE: EAS is the new replacement system for the older Emergency Broadcast System (EBS).

Example 2.   A line of severe thunderstorms was approaching a community. The EAS siren system was activated because of a tornado alert. People in the community had seen the nasty looking skies, heard the sirens, and many went directly to their basements. Fortunately, the main body of the storms passed outside the city limits. Several hours later, a second line of thunderstorms approached. The siren system was again activated. Many people returned to their basements. Unfortunately, this siren activation was not for a tornado but flash flooding. In this case, several people drowned in their basements as the flood waters inundated their homes.

Both of these examples pinpoint the need to educate the public on the options available to help protect them as well as their role in an emergency situation and what is expected of them. This requires ICS Coordination with Emergency Management and the EOC.

One approach that has been found very effective in jurisdictions where it has been tried, is the response of Emergency Management Agency (EMA) personnel to the scene of the incident. There, they usually report to the Incident Command Post (ICP) or one of the operational commands such as Hazmat Command.

Some jurisdictions refer to this process as field or partial activation of the EOC. In this capacity, emergency management personnel assist emergency responders with the implementation and management of agency coordination, EOC communication, evacuation, in-place protection, and, often most importantly, emergency public information.

Although field activation requires some coordination and practice, it has been found to be an effective and efficient tool to assist with many aspects of public protection. When done correctly, it removes a tremendous burden from hazmat personnel as well as the IC and Staff. The reasons are primarily two-fold.

First, EMA personnel are supposed to be coordinators. They know the ins and outs of most plans because they write and update them. EMA personnel know who is responsible for what, what resources and support are available to them, and, most importantly, how to access it. Routinely they have considerable additional communications capability available as well as public information capabilities and personnel.

Second, by having a representative on the scene, communications are streamlined. Requests and needs are directly communicated to the representative. If additional information or clarification is needed, it is requested immediately. Most importantly, communications with the EOC and other appropriate agencies is delegated to someone other than an emergency responder. This relieves the IC or a subcommand of that responsibility and often streamlines the process.

ISOLATION—WITHDRAWAL

The final tactical objective option for isolation is withdrawal. Withdrawal, as the name implies, is a fully defensive process. It is called for when the incident is of such a size, intensity, or complexity that there is little to be gained by intervening. However, this does not imply that there is nothing for responders to do. Five primary tactical objectives and methods must still be performed. These tactics include:

1. Establishing a perimeter
2. Evacuation
3. Public alerting and notification
4. Emergency public information
5. Denial of entry beyond the perimeter to all except those who have been cleared

Although this tactic is simple and straightforward, anyone who has experienced an incident of this magnitude will attest to the pressure and fevered pitch of this type of incident.

NOTIFICATION

The process of alerting and communicating with mandated or needed resources and the public	**Primary Tactical Objectives:** Incident Level Identification Establish Communications Links Request Assistance Public Alert and Notification Emergency Public Information

Notification involves tactics needed to establish vital communications links for the purpose of informing other parties of the incident, obtaining information, and requesting needed assistance.

NOTIFICATION—INCIDENT LEVEL IDENTIFICATION

Not all incidents are of the same size, complexity, or magnitude. Not all incidents require the same resources or notifications. Further, not all jurisdictions have the same resources or capabilities. Some jurisdictions need mutual aid to consider handling a hazmat incident while others do not. Most jurisdictions have their own notification requirements and criteria based on their response plans, systems, and needs. Additionally, jurisdictions within a given state must follow state notification protocols. This means that there is a wide variety of possible notification needs and requirements that are dependent on the jurisdiction and the incident itself.

Such situations are ripe for confusion and haphazard application. So the question is, "What is the most effective way to handle this grab-bag of options?" An answer can be gleaned from the management approaches to handle other types of potential or actual emergency situations such as fires, EMS, and so on.

First, consider fire operations. The most basic and fundamental resource allocation and notification systems revolve around engine and truck (ladders) companies and
their allocation. Allocation is handled through "alarms" and alarm patterns. A vehicle
fire receives a different allocation and notification of resources than does a structural
fire than does a rescue.

A standard structural alarm possibly provides three engines and two trucks or two
engines and a truck—whatever is determined appropriate by the jurisdiction. When command officers arrive on the scene, they determine, based on their knowledge of the job,
procedures, and experience, when to strike additional alarms or special call units. When
additional alarms are struck, many jurisdictions provide automatic notifications to additional staff and command officers, emergency management personnel, public works,
politicians, and so on. Thus there is uniform application of resources and notification.

Second, consider EMS operations. The allocation and notification procedures
vary depending on the number of victims. Most jurisdictions have a basic run that consists of a Basic Life Support (BLS) or Advanced Life Support (ALS) system. This may
be incorporated with enhanced 911 first aid protocols, first responder engine and truck
companies, and various other options.

However, once it is identified that there are more than five or ten victims, the mass
casualty system is often employed. When more than 20 or 30 victims are involved, the
incident normally triggers a higher level mass casualty response. As the number of victims increases, so does the mass casualty level. Predesignated ALS and BLS units as
well as hospital facilities, EMS coordinators, emergency management personnel, and
political leaders are notified. Again, this provides a uniform allocation of resources, and
notification is achieved.

In both examples, the greater the magnitude of the incident, the more resources
are needed and the more complex the notification process. However, each example incorporates a tiered response system based on incident needs required by the incident's
magnitude. This categorization of incidents allows the development of a uniform tiered
notification and response system.

In the NFPA Standard 471, *Recommended Practices for Responding to Hazardous Materials Incidents,* three incident levels are identified. Level I is the least severe, while level III is the most severe. Further, NFPA identifies a series of criteria to
differentiate between each level.

On the basis of these categories, many jurisdictions, ranging from local munici-
palities to state governments, have developed their own criteria for identifying incident
levels based on their needs and response systems. Further, each level incorporates its
own notification protocol for types and numbers of emergency response personnel and
equipment as well as mandatory or procedural ancillary notification. Such notifications
may include but is not limited to:

1. Public health or environmental agencies
2. Medical and EMS systems and providers
3. Public works and utilities
4. Jurisdictional managers such as chief executives, department heads, etc.
5. Politicians such as councils, legislative bodies, mayors, executives, etc.

Unfortunately, it is impossible to describe each level in terms of other than generalities due to local diversity. Again, a big incident in one locality may be routine in another. Further, the protocols designated in the SARA plan and other emergency management plans vary dramatically. As a result, the following are general descriptions of the three levels and are provided in terms of the incident's hazard severity, magnitude, and complexity.

Level I

A Level I incident is a relatively small and minor incident that can be handled through the defensive actions of operations level responders. There is little or no threat to the public or environment and no evacuation or other public protection is required outside the immediate site of the incident. The involved substance is identified and is a liquid or gaseous fuel or another hazardous substance with low toxicity and high stability. As a result of the limited hazards, full structural turnouts and SCBA are appropriate PPE.

For the fuel type liquids, quantities are limited. Some jurisdictions use arbitrary quantities ranging from 5 to 50 gallons. However, environmental or health agencies should be contacted to determine if they have established any specific criteria. For substances other than fuels, the quantities will vary depending on the associated hazards. Further, criteria for minimal container damage are often identified, as is the potential for explosion or fire.

Level II

Level II incidents are of moderate size and potential impact. They may require defensive or offensive actions to control. As a result of the increased hazard, the possible need for special thermal or chemical protective clothing and the possible need for offensive actions, a hazmat team is normally required for full control of this level incident. There is a moderate threat to the public or environment and the incident may require localized or limited public protective actions.

These incidents involve moderate hazards, larger quantities, or multiple substances. They may affect sensitive environmental areas such as ground water, surface water, or water sheds. The containers involved have received damage but are not in danger of failing further. Also, the release is not controllable without special equipment and materials. There may be the threat of potential explosions.

Level III

Level III incidents are larger in size, either by substance quantity or vulnerable area, and have severe impact potentials. They are a severe threat to the public or environment, or they are a major threat over a smaller area but will require extended operations (campaign mode). Thus these incidents may affect either large geographic areas or require major public protection actions or both. The incident may require offensive operations of such a magnitude that initial withdrawal may be appropriate. Further, it may be required to allow these incidents to run their course with little or no outside intervention.

There may be one or more known or unknown substances involved. The substances may possess high hazards, including Extremely Hazardous Substances (EHS), or be present in large quantities. Containers are often large or damaged severely. The potential for catastrophic failure may exist. There may be the potential for large or mass explosions.

The Use of Incident Levels

Incident leveling presents a host of advantages. First and foremost, it identifies a predetermined series of notifications for each incident level. As long as the IC identifies the level correctly, the system works very well. Some of the agencies that may require notification include:

1. Mutual aid emergency response agencies or teams (e.g., first responders, hazmat teams, EMS, and law enforcement)
2. Municipal agencies (e.g., public health, environmental, public works, water and sewer authorities, and public utilities)
3. State "one call" notification numbers and systems

Second, incident leveling provides a "heads-up" for potentially affected individuals, agencies, and groups. For example, the chief executive of a jurisdiction may find it quite helpful to know a major incident is underway and have some basic information about the incident. Such higher-ups do not like being caught off guard. Should other branches of government, the state, the media, or anyone else ask about what's going on, it is extremely unpleasant for higher-ups to say they have no clue. (It then usually becomes unpleasant for those who did not provide the higher-up with information.)

Third, incident leveling can allow for the massing of resources, including personnel. For example, units respond to a Level III incident half an hour before shift change or right at shift change. It may be advantageous to hold the outgoing shift to provide additional personnel or resources and limit the need for call backs.

Finally, incident leveling helps communications in general because when an incident is assigned a level, it lets everyone know its approximate magnitude. A good analogy is the initial status report from units or the IC. When command states that, "all units are committed", it provides a rough magnitude and indicates that additional resources may be needed.

NOTIFICATION—ESTABLISH COMMUNICATIONS LINKS

The sophistication of communications links obviously varies from jurisdiction to jurisdiction. However, at the most basic level, two-way radio communications must be established and maintained. Ideally, such communications have the availability of multiple frequencies and possibly bands or identifiers (800 MHz). These options are needed for the designation of frequencies for command, tactical, suit, and other levels of communications. If mutual aid is involved, hopefully there is more than just one mutual aid frequency available.

Additional communications links often include cellular phones and fax machines, computer modem/cellular links, and so on. The ability to tap into hard lines is very helpful in long duration operations (campaign operations). Electronic communications through computers is extremely effective and efficient due to the nature of the communications. Faxes provide hard copy data in the street that is useful for analysis, review, and documentation.

Even more advanced communications nets are available in some jurisdictions. Such systems include satellite positioning to satellite up- and down-links. Most commonly, such systems are employed for communications between various levels of government.

NOTIFICATION—REQUEST ASSISTANCE

Possible assisting agencies, groups, and organizations are unique to each jurisdictions. However, the following list identifies the most common types and groups encountered:

1. Local agencies e.g., police, EMS, public health, public works, emergency management, water treatment, and hazmat coordinator
2. Local government (e.g., mayor, council, supervisors, and engineers)
3. Utilities (e.g., phone, gas, and electric)
4. Local industry specialists
5. Local public and private groups (e.g., Red Cross, ham radio, aviation, media, Salvation Army, and medical facilities)
6. State agencies (e.g., environmental, highway, police, fire marshal, enforcement, public works, and health)
7. National groups and agencies (e.g., CHEMTREC, National Response Center, EPA, and the Coast Guard)

NOTIFICATION—PUBLIC ALERT AND NOTIFICATION

As mentioned earlier, public alert and notification is an integral part of a successful public protection effort. Responders must be familiar with the local systems used for this purpose. This is not to imply that responders are responsible for setting up, accessing or maintaining the system. Rather, they must understand how the system is activated, whom they must contact, the need for clear and concise initial emergency public information and how to put all of these pieces together at a moments notice.

Consider an incident that requires evacuation for one mile downwind and in-place sheltering for another two miles. The estimated population in the evacuation zone is 5,000, while the population in the in-place protection zone is 15,000. Obviously, this is a decent-sized public protection action.

However, to properly protect the public, they must be first alerted and notified that there is a problem. Next, they must receive information as to the nature and extent of the problem. Finally, they must be advised of the appropriate actions they need to take. Those who are to evacuate must be differentiated from those who are to shelter in-place. Information on evacuation routes and shelter must also be provided. Obviously, it is impossible to meet these demands of a sizable public protection action by simply going door to door. This size incident requires more.

There are many potential options available to address these needs. However, siren systems and the activation of the Emergency Broadcast System (EBS), now called the Emergency Access System (EAS) are the two most commonly employed means. (The extent of EAS coverage ranges from local radio and television stations to complete control of cable system channels.) Routinely, EAS forms the cornerstone because it fulfills multiple purposes. First, it helps alert and notify the public of an unusual event. Second, it is a readily accessible medium to provide available and, just as important, current information. Finally, it is an effective means for providing ongoing and updated information. As such, EAS is a primary component of the Emergency Public Information system.

NOTIFICATION—EMERGENCY PUBLIC INFORMATION

Once the public has been alerted to the existence of a problem, they must be provided with information. The information must be clear, concise, and accurate. This often requires taking highly technical data and reworking it into a more understandable form. Unfortunately, experience has all too often shown that this is beyond the technical scope of many Public Information Officers (PIO). Routinely, hazmat personnel must function as "interpreters" of the information.

Further, emergency public information requires rapid access to the electronic media and EAS to provide timely dissemination of information and instructions. Live, on-air reporting of events and interviews are often involved. This is not to suggest that the hazmat team is responsible for establishing this system. Rather, the team needs to understand its role in the system and the tasks it may be required to perform.

These situations are ideally suited for partial or field activation of the EOC with emergency management personnel physically on the scene. Routinely they can coordinate efforts in the field with those in the EOC. They can often provide the assistance of public information specialists. They can help manage reporters, news briefings, interviews, and so on to relieve command of and the PIO of an extensive task. This is not to imply that the PIO has nothing to do. Rather, this allows the PIO to coordinate with emergency management personnel and focus on the collection of information and preparation of releases.

Media Relations

Unfortunately, many emergency responders have formed an adversarial relationship with the media. These responders view the media as the enemy. However, in the case of emergency public information, the media must be a close ally for the system to work. Obviously, if an adversarial relationship exists, the release of emergency public information will be difficult or impossible to accomplish.

When dealing with the media, several points must be remembered. First, reporters have a job to do. That job is to get "the story." Reporters have bosses who expect them to do just that. If they do not get the story, they may not keep their job. As a result, they have a great impetus to get *any* story. If responders do not provide a story, they will get it somewhere else be it the neighbor down the street, the local "expert" in the field or anyone else who offers. The consequences are not good for the public or responders.

Hazmat must be ready to provide accurate, concise information to the incident PIO or possibly the media regarding the technical and control aspects of the incident. In some instances, a representative from hazmat may even have to provide a live interview, depending on the circumstances. Several points to remember in this case are:

1. **Always provide facts.** Sometimes this requires transforming technical data into nontechnical words. Never use acronyms or terms without defining their meaning. If information is not available at that point, say so.
2. **Anticipate their questions.** Routinely these questions include: who, how, what, when, and why.
3. **Always tell the truth.** Provide only accurate information. Never lie. Lies ruin credibility and call all other aspects of the response into question.

The Role of Public Education

A significant factor in the effectiveness of public protection and emergency public information is the basic knowledge level of the public. For public protection options to function as intended, the public must have some basic understanding of the nature of the problem and the actions they are required to take.

For example, if in-place protection is the chosen option, the public must understand what the term means and what they are expected to do. Further, if they are to evacuate, the public needs to understand some realities. If they are on medication, they need to take it with them. If the children are at school they should know what to do. If people are at work, they should also know what to do. These are just some of the issues that need to be addressed.

This is not to imply that public education will eliminate all the potential problems. Rather, such public education programming should seek to minimize misunderstandings and problems when the public needs to be protected. Additionally, these programs help provide the public with an understanding of the potential hazards and vulnerabilities with which they and the community are faced. Furthermore, this education can help remove some of the total lack of understanding and irrational fears often harbored by the public.

Although this arena is not traditionally covered by fire or police department public education programs, it is extremely important. When included in a broad spectrum approach, such public education fit very nicely with flood, earthquake, and similar disaster education and preparedness programs. As such, the LEPC, CAER, or TRANSCAER groups are ideal candidates for the development and release of public education materials and information.

Finally, various levels of emergency management may be able to assist in the development and release of such programming. These same agencies may also have information on grant programs, alternative funding sources, prepared materials, public service announcement, and so on that can be used as part of the program.

IDENTIFICATION

The process of identifying, confirming, and obtaining information about the product, container, and environment	**Primary Tactical Objectives:** Recognition and Identification (R&I) Data Retrieval Interview Reconnaissance Monitoring Sampling

Identification involves tactics that in some fashion help identifying, confirming, and otherwise obtaining information about the product involved in the incident. This process routinely includes the same steps regarding the container and possibly the environment. Further, as the incident evolves, the process involves identifying the location, extent, and spread of contamination.

IDENTIFICATION—RECOGNITION AND IDENTIFICATION (R & I)

The most basic level of identification involves the six basic clues to the presence of hazardous materials. These six clues include:

1. Occupancy and location
2. Container shape and size
3. Colors and markings

4. Papers
5. Placards and labels
6. Senses

These six clues are the most basic and fundamental means of identifying the hazardous substance, container type, and so on. However, these standard R&I clues are not the full answer to the identification. As such, there are additional tactics that are needed.

IDENTIFICATION—DATA RETRIEVAL

Data retrieval, as the name implies, involves all types of information gathering and documenting procedures. Some of the most common ways include:

1. The use of preincident plans and surveys including SARA Plans and CAER Plans
2. Technical reference materials such as reference books and regulations
3. Electronic data from computer programs and data bases, fax information, thermal imaging, and so on

As discussed in previous chapters, once the data is retrieved, it must be separated, documented, analyzed, and assessed. Through the use of data sheets, this process becomes more structured and systematic.

IDENTIFICATION—INTERVIEW

Often, one of the most underutilized identification tools is the interview. Workers or witnesses who were in the area when the incident occurred or are familiar with the product, processing, situation, and so on should be located and questioned regarding the situation and occurrences. These individuals include people such as:

1. Safety personnel
2. Production personnel
3. Area supervisors
4. Facility or emergency response coordinators
5. Drivers
6. Loading dock personnel
7. Personnel working in the immediate area

Unfortunately, most responders consider truck drivers, facility workers, and even safety personnel to be unreliable and even not truthful. In some instances, these reservations are fully warranted and justified. However, these individuals should not be disregarded off-hand without at least talking to them.

There are many situations that indicate a good probability of a reliable source. Normally, the closer the source is to the manufacturer, the more reliable the information. Chemical company drivers as opposed to contractors are usually well trained and informed in emergency procedures. One well-known company requires 12 to 16 weeks of product specific training for its cargo tank drivers. On the other hand contractors normally only meet the minimum DOT training requirements for a hazmat commercial license.

When the incident involves a fixed facility, employees may be able to explain exactly what happened and what is involved, and possibly provide other vital information. This information can eliminate the need to send additional reconnaissance teams in to the hot zone. Further, these individuals can provide insights into potential hazards, problems, and solutions to the problem at hand.

Entry Team Operations

Reconnaissance, monitoring, and sampling all involve entry team operations. In most instances, such operations involve entering the hot zone to perform the assigned task(s). Although entry team operations are not a specific tactic, a brief discussion of such operation is necessary to ensure operational efficiency and efficacy.

The assignment of entry team activities sounds simple and straightforward. However, there are several nitty-gritty operational actions that must occur to assure successful entry operations. These actions include the following:

1. Assignment of the entry team
2. Pre-entry coordination and checks
3. Pre-entry briefing

Assignment of the Entry Team. Routinely, a two-member entry team equipped with some type of chemical protective clothing is assigned for a given entry. However, there are other options. As discussed in the zoning section of this chapter, it is possible to have multiple entry teams working in the hot zone under various situations and conditions. Further, these teams will normally have different tactical assignments that must be coordinated to achieve the overall goal.

Entry team personnel must be assessed for their ability to perform assigned functions in the type of PPE and under the conditions to which they will be exposed. This is especially true if nonhazmat personnel are being deployed. These nonhazmat personnel perform defensive tactics such as spill control, fire control, vapor suppression, and so on.

Additionally, the minimum number of personnel needed to perform the function must always be considered. In 29 CFR 1910.120 the IC is required to limit the number of personnel exposed to hot zone hazards to the minimum needed to perform the assigned function. It also states that the buddy system must be employed. As a result, there is a tendency to use only two personnel to perform the task.

Unfortunately, the safe, effective, and efficient performances of the assigned task may require three or four personnel. As such, the key word is *minimum*. The task must be assessed to assure that adequate personnel are provided to perform it safely and effectively.

Pre-entry Coordination and Checks.   When the entry team(s) is identified, coordination and pre-entry checks are essential. First pre-entry medical monitoring must be performed prior to the entry. Depending on scene conditions, medical monitoring may be performed on decontamination personnel as well. EMS personnel who perform this monitoring must document the results and must perform the post-entry monitoring as well. Pre-entry medical monitoring will be addressed in more detail during the discussion of protection.

Second, the suiting process (donning of PPE) must incorporate other checks and coordination. Checks involve ensuring that the proper PPE and its components are present, in good working order and condition, and being used. Such checks assure that there are no PPE problems or difficulties when the team enters the hot zone. Coordination involves ensuring that all members of the entry team are suiting at the same rate and time. Coordination assures that one member of the entry team is not suited or hooked up to the air source before the other.

Third, all equipment must be checked to assure that it is functioning properly before entry occurs. There is nothing more exasperating than sending an entry team into the hot zone and finding that some piece of equipment is not functioning properly. Not only do such situations slow down the operation, they can compromise the safety of the team. This is particularly true of radio equipment and monitors.

As such, radio and communications checks must be performed routinely to assure problem-free communications. Such checks should include audio tests of all in-suit radios, their mikes, speakers, and radio interface, as well as audio levels and clarity for both incoming and outgoing messages. Further, monitoring equipment should be checked well in advance so personnel do not experience heat stress or use air while waiting for a replacement. In some instances, monitoring equipment requires ten to twenty minutes (in some cases longer) to warm-up and should be turned on well before the entry.

Pre-entry Briefing.   Finally, a thorough pre-entry briefing must be performed and should include as a minimum hazmat command personnel, the entry team members, and the hazmat safety officer. The pre-entry briefing should assure that entry personnel understand the plan of action, the specific tasks that they must accomplish, and optional task they may perform if conditions and time allow. Such understandings are essential to an efficient, effective, and safe entry. Without this knowledge and understanding of their assignments, personnel will waste precious air and work time by performing potentially unnecessary, ill-thought-out, or unsafe acts. Such waste often requires the use of additional entry teams. Additional entry teams require additional PPE and expense as well as more personnel to enter the hot zone than are absolutely necessary.

Additionally, personnel must be fully aware of all potential safety and health hazards and considerations that exist in the hot zone. Locations, conditions, or situations that are precarious must be identified and explained. Operations or locations that must be avoided require identification and explanation. Key indicators of potential problems or hazards, such as visual clues, sounds, positions, etc., also require identification and explanation.

Although any team that enters the hot zone is generically called an entry team, their specific function is often used as part of their name. For example, an entry team whose function is to perform an initial entry to gathering information about the incident's status and exact situation is routinely called a reconnaissance team. An entry team whose function is to gather information through the use of monitoring equipment is often called a monitoring team. An entry team whose function is to gather actual samples of the product(s) involved is often called a sampling team.

Each of these functions is its own tactical option. As such, the following sections will discuss in more detail their specific role and options.

IDENTIFICATION—RECONNAISSANCE

Reconnaissance (recon) is a tactic that involves surveying the incident scene, normally from within the hot zone, to gather additional information. Routinely, the initial team to enter the hot zone performs this function, possibly along with other assignments. The check lists for information gathered about the product, container, and environment along with the estimating sheets help identify additional information that must be gathered. Such information may simply clarify or confirm already gathered information or consist of information that is impossible to acquire without an up-close look.

The data can be specific to the container, the product, or the environment or any such combination. This information allows hazmat personnel to have higher confidence in the plan of action they develop or help identify additional needs. In some instances, the full development of the plan of action is impossible until the recon team clarifies or acquires additional information.

Examples of they type of information collected by a recon team or other entry team include:

1. Product data
   a. Instrumentation reading to determine specific hazards such as flammability, corrosivity, radioactivity, and oxygen deficiency; may include pH, percent LEL, parts per million, and so on
   b. The quantity released and available for release
   c. Product name or UN identification numbers
2. Container data
   a. Exact container type and specification
   b. Type and extent of damage
   c. Physical and structural stability
   d. Number and orientation of containers involved

3. Environmental data
  a. Actual or potential exposures
  b. Soil types
  c. The presence of confinements or conduits
4. Miscellaneous data
  a. Documentation including shipping papers, MSDS, etc.
  b. The presence of victims, their condition, and degree of entrapment.
  c. General site conditions

Finally, one of the foremost functions is the use of instrumentation or sampling to acquire information otherwise not available. The specifics of this function are covered by the next tactic.

IDENTIFICATION—MONITORING

Monitoring is the use of specific devices to gather information about the incident. Most commonly, monitoring involves the use of instruments to gain site-specific information about the product. However, with the proper equipment, information can be gathered about the environment and the container as well. Since there are fewer options available for environmental and container monitoring, they will be addressed first.

Environmental Monitoring

Environmental monitoring almost exclusively involves standard meteorologic instruments, such as the thermometer (temperature), barometer (barometric pressure), anemometer (wind speed and direction) and psychrometer (humidity). This information is often critical for accurate plume projections as well as product movement and behavior. Teams with onboard computers and plume-generating programs often carry weather stations with electronic interface capabilities to minimize the time and effort required for accurate, real-time data entry.

Container Monitoring

The monitoring equipment available for container monitoring is not very common. However, when available, it is extremely useful. One of the most simple instruments is the rail dent gauge described in Chapter 4. Others include infrared video cameras or head sets and ultrasonic thickness gauges that identify fractures and cracks.

The infrared cameras measure heat. Their most common application involves large product tanks. Commonly, there is a temperature lag between a liquid product and its surroundings. As a result, the infrared camera detects the temperature difference above and below the liquid level within nonjacketed or uninsulated tanks. In the case of compressed gas cylinders, infrared cameras identify the temperature drop that occurs when a gas goes from a higher pressure to a lower pressure (that is, if there is a significant release). These cameras have been able to identify the leaking cylinder among the many found in a cylinder storage area or loading dock.

Ultrasonic thickness gauges are highly sensitive devices that allow examination of the container's wall thickness. Not only can this indicate stretching and thinning of the wall, some of these devices will also detect fractures (i.e., cracks that do not extend completely through the container's wall).

Product Monitoring

Product monitoring involves the use of instruments that detect the presence of various substances that become contaminants when they are released into the environment. Additionally, the most useful types of devices enable the user to identify specific hazardous properties of the contaminant and possibly its concentration.

Instrumentation is needed in this context for three primary reasons. First, our senses are often incapable of detecting the presence of various contaminants. Examples of such contaminants include hydrocarbon gases not treated with an odorant (e.g., methane and ethane), ionizing radiation, odorless substances, and colorless gases and liquids.

Second, the use of our senses would result in an exposure to the contaminant. Such exposures could easily cause acute toxicity, long-term negative health impacts, or even death. Further, such actions violate many of the basic response tenets required by the regulations and common sense.

Third, our senses cannot provide the needed data. Specifically, if our senses are able to detect the presence of a contaminant, and the resulting exposure does not desensitize, sicken, or kill, our senses still are unable to accurately qualify or quantify the contaminant.

Such monitoring data is often critical to the actions chosen on the incident scene. Consider an incident involving a creamy, yellow, syrupy substance leaking from the rear of a box trailer. The trailer is parked at a rest stop on an interstate highway. The driver indicates that the substance is a nonregulated material. What should the hazmat team do?

With the right instrumentation, the answer is clear. Responders should check for flammability, corrosivity, oxidizing ability, radioactivity, and possibly oxygen concentration (depending on the exact conditions and situation). If all of these checks come back negative, there should be no major problem. Obviously, the material should not be allowed to continually spread through the environment. Under these circumstances, there is no need for special PPE or other standard hazmat precautions.

However, suppose the pH reading was 1 and the product had a positive indicator for being an oxidizer. This changes the whole completion of the response. Instead of being a minimal hazard release, there is some real potential for serious problems. PPE, decontamination, and all standard hazmat operational approaches are needed. Because of the oxidizing potential, chemical interactions or even fires are possible.

In this example, as well as many actual responses, instrumentation readings provide defining facts. Such readings help identify appropriate and inappropriate strategic and tactical options, provide a more reliable estimate, and so on. Unfortunately, all too often, hazmat teams do not have the needed instruments or fail to use them routinely, thus over- or under-reacting.

Types of Monitoring Data

As previously identified, there are two general types of data that instruments collect: qualitative and quantitative. Qualitative data deals with the quality of the contaminant. In this context, quality refers to the presence and hazards associated with the contaminant. Quantitative data deals with the quantity of the contaminant within the area monitored. Furthermore, in this context, quantity refers to the concentration of the contaminant within the area monitored. Generally, qualitative data is needed first, followed by quantitative. The process of going from general to specific definitely applies to monitoring.

Qualitative Data.   Consider this example of a situation requiring qualitative data. It is early in an incident involving possibly two or more unknown products. Responders need specific information about the hazards of the contaminant to identify appropriate strategies and tactics.

The answers to the following types of questions are essential: Can the contaminant be detected? (Remember, the absence of evidence is not necessarily evidence of absence.) What are its hazards? Is it corrosive, flammable, radioactive, an oxidizer, etc.? Has the contaminant displaced air? Is it heavier, lighter, or neutrally buoyant in air? Is it organic or inorganic? Depending on the types of instruments available, these questions should have answers when the initial monitoring is completed.

In the realm of qualitative monitoring, there are five primary hazards for which monitoring is done routinely:

1. Flammability
2. Corrosivity
3. Radioactivity
4. Oxygen deficiency
5. Toxicity

It is important to note that in a true unknown situation (such as illegal dumping), radiation detection is essential to thoroughly evaluate the site. Even if the presence of radiation is considered remote, radiation monitoring is essential to assure that personnel are not exposed to an unrecognized or unidentified hazard. Such monitoring must be part of a team's routine monitoring protocol, at least when handling unknowns.

Other qualitative data can be derived through monitoring techniques or the use of multiple instruments and the readings they provide. For example, the vapor density of the contaminant can be estimated by the nature of the readings. If higher concentrations are found toward the ceiling, the contaminant is lighter than air (or possibly heated). If higher concentrations are found towards the floor, the contaminant is generally heavier than air (or cold). If the concentrations are relatively uniform at the ceiling, floor, and in between, the contaminant is neutrally buoyant (or HVAC systems are mixing it well).

All the indicators are valid when the contaminant's temperature is approximately the same as the ambient temperature. If the contaminant's temperature is higher than the ambient temperature, its density will be less than at ambient temperatures. As such, the contaminant will rise until it cools. Conversely, if the contaminant's temperature is lower than the ambient temperature, its density will be greater than when it is at ambient temperatures. As such, the contaminant will sink until it warms.

Quantitative Data and Units of Measurement.   Once the general qualitative questions are answered, the need for quantitative data may or may not be needed. Such data would include values such as the percent of lower explosive limit (LEL), pH, and part per million concentrations (ppm).

Percent of LEL.   The specific units of measurement in quantitative monitoring again vary depending on the specific incident situation and conditions, as well as the specific type of instrumentation used. For example, if the only determination needed is whether there is a flammable atmosphere present, the use of a combustible gas indicator (also called a percent LEL meter) may be the primary instrument of choice. The combustible gas indicator (CGI) provides readings based on the percentage of the LEL based on the calibrant gas.

For example (Table 7-1), a CGI is calibrated with the imaginary gas zeethane (the calibrant). When this CGI is used to measure an atmosphere containing zeethane, it provides accurate readings. Now, let's say zeethane has an LEL of 10% in air. When the monitor provides a reading of 10% of the LEL, it indicates that there is actually 10/100ths or 1/10 of the LEL. This means the actual concentration of zeethane in the air is 1.0% [(10%/100%) × 10% = 1.0%, or 0.10 × 10% = 1.0%]. Likewise, a monitor reading of 20% of LEL indicates that there are 20/100ths or 2/10 of the LEL. This means the actual concentration of zeethane in the air is 2.0% [(20%/100%) × 10% = 2.0%, or 0.2 × 10% = 2.0%].

When this information is compared to EPAs recommended action levels (as found in the joint NIOSH/OSHA/USCG/EPA document, *Occupational Safety and Health Guidance Manual for Hazardous Waste Site Activities*), this concentration of flammable contaminant approaches their withdrawal action level or 25% of LEL. (The National Fire Academy recommends consideration of a lower value of 20% of LEL.) As such, personnel are in a potentially very hazardous atmosphere.

Table 7-1   Reading for Zeethane Spill (%)

% LEL reading	% in fractions	% in decimals
10%	10/100	0.1
15%	15/100	0.15
25%	25/100	0.25

However, although this data provides a basis for the evaluation of potential fire hazards, it generally provides little assistance when considering the toxicity of the situation. Again, consider zeethane. Zeethane has a TLV-TWA of 50 parts per million (ppm) and an IDLH of 500 ppm. The question is, how do these percent of LEL readings compare to parts per million?

The answer is that they do not. Specifically, the percent of LEL reading must be converted to ppms. To do this, the LEL must be multiplied by the percent of LEL readings in the form of hundredth values (%/100) as was done earlier. This means a monitor reading of 20% of LEL indicates there is an actual concentration of zeethane in the air of 2.0% [(20%/100%) × 10% = 2.0%]

Table 7-2 provides several examples of converting meter readings from percent of LEL to the actual percent of contaminant in the air.

Parts per Million. Now the question is, "How is the actual percentage of contaminant in air converted to the actual concentration of contaminant in air in parts per million?" To convert percentages of contaminant in air to ppms, multiply the percentage by 10,000 (2% in air × 10,000 ppm = ppm of contaminant in air). The 10,000 comes from dividing 1,000,000 by 100 (1,000,000 ppm equals 100%; thus 1% equals 10,000). Using this formula, we find that there are 20,000 ppm of zeethane in the atmosphere (2% × 10,000 = 20,000 ppm) when there is a reading of 20% of LEL for a CGI calibrated on zeethane. Table 7-3 provides several examples of the conversion process used to convert the actual percent of contaminant in air derived from the percent LEL from the meter to the actual ppm in air.

Since the IDLH for zeethane is 500 ppm, a reading of 20% LEL (20,000 ppm) indicates a concentration 40 times greater than the IDLH. Based on these values, a read-

Table 7-2   Readings for Zeethane Spill (ppm)

Meter reading (% LEL)	LEL	Conversion	Actual % in air
5%	6%	(5%/100%) × 6% = ?	0.3%
20%	6%	(20%/100%) × 6% = ?	1.2%
15%	3%	(15%/100%) × 3% = ?	0.45%
15%	6%	(15%/100%) × 6% = ?	0.9%

Table 7-3   Readings for Zeethane Spill (Actual ppm)

Meter reading (% LEL)	Calculated actual % in air	Conversion	ppm in air
5%	0.3%	0.3% × 10,000 = ?	3,000 ppm
20%	1.2%	1.2% × 10,000 = ?	12,000 ppm
15%	0.45%	0.45% × 10,000 = ?	4,500 ppm
15%	0.9%	0.9% × 10,000 = ?	9,000 ppm

ing of 0.5% on the CGI (indicating 0.5% LEL) equals the IDLH concentration of zeethane [500 ppm/(100 × LEL%) = 0.5%]. Table 7-4 provides several examples that show the conversion of ppm to percent LEL readings.

One last conversion to ppm involves vapor pressure. As discussed in Chapter 5, vapor pressure is a measure of the volatility of a liquid, and as such provides some insight into the amount of vapor produced by a given liquid. The question is, "How can this data be used when considering air monitoring?"

First, the relationship between vapor pressure, boiling point, and ppm must be addressed. Vapor pressure and boiling point are inherently linked. Specifically, the boiling point is defined as the temperature at which the vapor pressure of the liquid equals atmospheric pressure. Vapor pressure values are routinely provided in millimeters of mercury (mm hg or mm). Using these units, atmospheric pressure is 760 mm. When the liquid has reached its boiling point, the area immediately above the liquid's surface is saturated with vapor. In other words, the atmosphere is 100% vapor.

As such, a correlation between mm of mercury and ppm can be made. If the atmosphere above the boiling liquid is assumed to contain 100% vapor, it would have a concentration of 1 million parts per million. Since 760 mm of mercury also is assumed to equal 100%, this means that 760 mm equals 1 million ppm. By dividing 1 million by 760 it is found that each mm of mercury equals 1,300 ppm.

At this point the question is, "How do we use this information?" First of all, several considerations of vapor pressure must be reviewed. Vapor pressure and temperature are directly proportional. This means that as the temperature increases, so does the vapor pressure. To gain anything from a vapor pressure value as provided in various references, one must know the temperature at which the readings were taken. In many instances, the data are corrected to standard temperature and pressure (STP). STP is a temperature of 68°F or 20°C (for gases, 32°F or 0°C) and an atmospheric pressure of 760 mm Hg.

However, be sure to check each reference to ensure they are using STP values. The *Handbook of Chemistry and Physics* provides vapor pressure data at a series of selected temperatures. The CHRIS (Chemical Hazard Response Information System) reference uses the Reid vapor pressure. Reid vapor pressure is taken as standard atmospheric pressure (760 mm) but at a temperature of 100°F. Further, CHRIS provides the data in psi_a.

When the vapor pressure and the specified temperature are identified, they can provide some further data. First, if the contaminant is within a confined space, such as

Table 7-4   Readings for Zeethane Spill (Percent LEL)

ppm value	LEL	Conversion	% LEL reading
50 ppm	6%	50/(100 × 6%) = ?	0.08%
250 ppm	6%	250/(100 × 6%) = ?	0.4%
1,000 ppm	3%	1,000/(100 × 3%) = ?	3.33%
1,000 ppm	6%	1,000/(100 × 6%) = ?	1.66%

a tank or sump, a general idea of the anticipated concentration within the head space can be calculated using the 1,300 rule (multiply the vapor pressure by 1,300). Do not forget to consider thermal layering and its influence as well.

Further, in an open spill, vapor pressure can give an indication of distance from the spill where vapors may be found. For example, a substance with a vapor pressure of 1 mm at 80°F will only produce vapor readings close to the spill's surface when its temperature is 50°F. Further, a substance with a vapor pressure of 450 mm at 80°F and under the same conditions will produce vapor readings a great distance from the spill's surface. Obviously, considerations such as the surface area of the spill, vapor density, humidity, and so on must also be factored into the equation.

Hopefully this discussion has helped to point out one of the major instrumentation considerations when choosing appropriate devices to be used in the site characterization process or the monitoring plan. Specifically, this consideration is sensitivity. Although the CGI is well suited to the gross process of determining or verifying the potential flammability of an atmosphere, its readings are on such a large scale that they are of little value when trying to identify concentration below 5,000 ppm. In other words, CGIs lack the sensitivity needed to measure low concentrations of flammable gases or vapors. To effectively measure such low concentrations, colorimetric tubes, photoionization detectors (PID), or flame ionization detectors (FID) are needed.

Another consideration is selectivity. The CGI as well as the colorimetrics, PIDs and FIDs, and all other instruments, are selective in what they detect. For example, the CGI, FID, and PID are useless in identifying the presence of radioactivity unless the contaminant is flammable or has other properties the instruments can detect. Colorimetric tubes and the others cannot identify the pH of a liquid or gaseous release. As such, the capabilities and limitations of all instruments must be fully understood before they are deployed on an incident scene.

Relative Response.   Thus far, our discussion has revolved around the detection of contaminants for which the instrument has been calibrated. However, what happens if the device is not properly calibrated? Further, what if it is used to monitor a substance for which it has not been calibrated? The answers to these questions are essential. First, if the monitor is not properly calibrated as recommended by the manufacturer, all of its readings must be considered erroneous.

The answer to the second question has two parts and requires a more indepth explanation. The first part of the answer involves the properties of the contaminant itself.

Initially, a key factor is whether or not the monitor can even detect the presence of a contaminant (other than its calibrant). The answer depends on the specific properties of the contaminant. For example, a CGI will not detect a nonflammable contaminant. An FID will also not detect a nonflammable contaminant. A colorimetric tube designed for chlorine will not detect carbon monoxide. A PID will not detect a contaminant with an ionization potential greater than the rating of its lamp, and so on.

Further, there are contaminants known as interferants that will cause a detector to give false positive or negative readings. For example, propane, ethyl alcohol, isopropyl alcohol, and mercaptans (thiols) are interferants for carbon monoxide and hydrogen sulfide detectors. Silicates, metal and sulfur containing compounds, liquids, and dusts are interferants for CGIs.

As a result, it is imperative to identify if the contaminant can be detected by the instrument and whether it is an interferant before monitoring takes place. Additionally, some interferants work by depositing all or part of the contaminant on the sensor. This type of interferant is often said to "poison" the sensor, eventually causing total failure. If circumstances dictate, such destruction of the sensor may be a necessary evil but will require replacement of the sensor before its next use.

An example of this situation is the use of a CGI to monitor a gasoline spill that has a vapor suppression foam blanket applied. Many of the available firefighting foams used to suppress vapors contain silicon compounds that poison the sensor over a period of time. If the situation requires, the CGI should still be used but loss of sensitivity must be noted.

This can be done by using the calibrant as a baseline. If the calibrant is not readily available, a known concentration sample can be established by placing a liquid such as diesel fuel in a small container (e.g., test tube, sample bag, and plastic bottle). The container is placed in a location that will maintain a relatively constant temperature such as in the cab of a unit.

Before the CGI is used to monitor the spill the first time, a baseline reading of the diesel sample is taken. Following each monitoring routine, the constant diesel sample is again monitored and any changes in the reads are noted. When a significant error is identified, the unit should be replaced with an unaffected monitor. Upon completion of the incident, each monitor must have its sensors replaced.

The second part of the answer deals with contaminants that can be detected by the monitor but are not the calibrant gas. The CGI again provides a good example of this.

Common calibrants for CGIs include methane, propane, pentane, or hexane. The calibrant is used in the calibration process. The calibration process simply assures that the readings provided by the meter are accurate. In the case of the CGI, the calibrant gas is burned within the meter, and its heat output is measured (a more indepth discussion of a CGI's function is found later in this section). The heat output of the various substances is the central issue.

Specifically, each calibrant and flammable contaminant has its own unique heat output signature. Following the general rules of chemistry, the greater the substances molecular weight, the wider its flammable range; and the more complete its combustion profile, the greater its heat output. This means that methane produces somewhat less heat than does hexane when considered molecule for molecule. Further, this means that a concentration of 20,000 ppm of methane produces less heat than a concentration of 20,000 ppm of hexane.

As such, when a CGI is calibrated for methane and used to monitor hexane, the readings (% LEL) will indicate a concentration higher than is actually present in the air. In other words, it will produce readings that are artificially high.

Conversely, when a CGI is calibrated for hexane and then used to monitor methane, the readings (% LEL) will indicate a concentration lower than is actually present in the air. In other words, it will produce readings that are artificially low. Of the two possibilities, artificially low or high, the artificially low readings are the more dangerous. Specifically, the artificially low readings could indicate a safe atmosphere when in fact the atmosphere is dangerous.

Next, consider the following graphic that compares meter readings, heat output, and actual concentrations, of gas in the air. The first graph in Figure 7-11 shows that the meter readings correspond exactly (or very closely) to the actual concentration of the calibrant present in the atmosphere. For example, when the meter reads 25% LEL, the actual concentration is also 25% LEL. When the meter reads 50% LEL, the actual concentration is also 50% LEL.

However, when gases that burn hotter or cooler than the calibrant are monitored, discrepancies arise. First consider a gas that burns hotter than the calibrant. In this example, when the meter yields a reading of 75% LEL, there is actually only about 35% to 40% LEL in the atmosphere.

On the other hand, a gas that burns cooler than the calibrant produces meter readings of about 40% LEL even though the actual concentration is 100% LEL. As a result such readings can be extremely dangerous because they provide a false sense of security even when the atmosphere is potentially lethal.

The manufacturers of many devices have found that the amount of the error in measuring any individual noncalibrant gases or vapors is proportional to the actual concentrations of the substance (Figure 7-12). This means that through the use of conversion factors, the meter readings can be converted to indicate the actual concentration of contaminant in the atmosphere monitored. These conversion factors are commonly called relative response or conversion factors. This relative response data is normally contained in a relative response chart developed by the manufacturer.

When considering the purchase of a CGI, PID, FID, or similar instruments that will be used for more than one contaminant, request a relative response chart. If this information is not available or only addresses a handful of substances, the versatility and value of the instrument is greatly diminished.

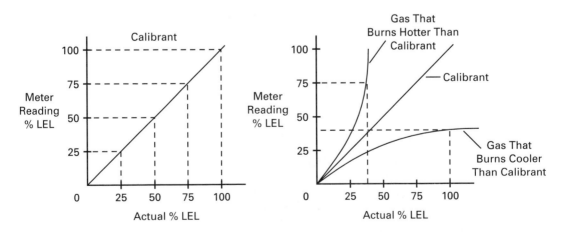

Figure 7-11   Concentration and Heat Output Comparison

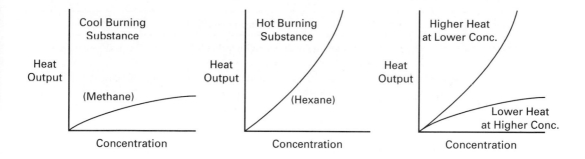

Figure 7-12   Heat Output, Meter Readings, and Actual Concentration Comparison

The most common form in which relative response factors are found is a decimal number (e.g., 0.6, 1.2, 2.5, etc.). In this case, to convert the meter reading of the non-calibrant gas to the actual concentration one simply multiplies the reading by the relative response factor. However, this is not a universal system and various manufacturers use various systems.

Consider a storm sewer that is suspected of containing propane and must be monitored to determine if an explosion is possible. The only available monitor is calibrated to pentane. The manufacturer's relative response chart indicates a conversion factor of 1.4. Actual meter readings taken within the storm sewer system indicate a concentration averaging 80% of LEL.

To identify the actual concentration within the storm sewers, the actual reading is multiplied by the relative response factor (80% LEL × 1.4 = 112% LEL). In this case, the conversion yields a reading that indicates the existence of a flammable atmosphere within the storm sewer as opposed to a potentially flammable atmosphere identified by the raw meter readings.

Again, not all manufacturers use the decimal type of relative response conversion factor. It is essential to follow the manufacturer's recommendations and procedures when performing this type of conversion.

Class, Division, and Group.   Many of the various instruments used for monitoring contain some form of electronics. Further, these devices are often used to monitor flammable or potentially flammable atmospheres. Atmospheres become potentially or actually flammable when they are contaminated by gases, vapors, powders, dusts, or fibers. If monitoring devices are not appropriately designed and constructed for a given atmospheric contaminant and situation, the monitor could act as an ignition source, resulting in a fire or explosion.

To assure the safety of monitors and their use for proper situations, a classification system has been developed. This system is based on the National Electric Code published by the National Fire Protection Association (NFPA).

There are three ways to construct electrical equipment to prevent them from igniting a flammable atmosphere. Table 7-5 identifies each and the how the protection is gained.

Table 7-5   Protecting Electrical Equipment

Construction	Protection method
Explosion proof	Use of explosion and flame suppression technologies within the device (e.g., flame arrestors, and exhaust cooling).
Intrinsically safe	Encasing components with solid insulation, thus preventing contact with the flammable atmosphere
Purged	Shielding components with an inert atmosphere, thus preventing contact with the flammable atmosphere

It is often found that direct reading, air monitoring equipment employs two or more of these construction methods to ensure the safety of the device. Now the question is, "How does the user know what type of atmosphere can safely be monitored by any given device?" A system, using the terms *class, group,* and *division,* has been developed to address this question.

Class.   The class of an instrument indicates the physical form of the atmospheric contaminant for which it is designed to monitor. The following table provides this data.

Class	Type of atmosphere
I	Flammable vapors and gases
II	Combustible dusts
III	Ignitable fibers

In most instances, hazmat teams will require Class I instruments. The only time other classes would be needed is when the specific hazard of one or both of the other classes is identified. Even in that situation, the facility or operation may have all the instrumentation needed.

Group.   There are seven groups used to identify specific substances with similar hazards regarding electricity and ignition. The source for these groups is the NFPA/ANSI Standard 497M in the *Classification of Gases, Vapors and Dusts for Electrical Equipment in Hazardous Locations.* They use alphabetical classifications A through G (e.g., Group A, Group B, etc.). Further, specific groups correspond with each class. The following table provides a correlation of the classes and groups.

Class	Groups
I (gases and vapors)	A, B, C, and D
II (dusts and powders)	E, F, and G
III (fibers)	No groups identified

Division.   Division identifies the type of release scenario involved. Division 1 releases are continuous, intermittent, or periodic and into an open, unconfined area. Division 2 releases are from ruptures, leaks or other container failures or closed systems. Most commonly, a device classified as Division 1 also meets Division 2.

What all of this boils down to is that most hazmat teams need electronic monitoring equipment that is classified as Class I; Group A, B, C, and D; and Division 1. This is also seen as Class I, Division 1.

Specific Types of Instrumentation.   This section will discuss specific types of instrumentation available and recommended for emergency-response use. If a specific manufacturer's device is identified, it is used as an example and does not imply it is the only such device available.

The primary types of devices used in emergency response are the combustible gas indicator (CGI), various colorimetric devices, oxygen and other specific substance electrochemical detectors, radiation detectors (not discussed), photoionization detectors (PID), flame ionization detectors (FID), and FID portable gas chromatographs. It is important to remember that there are many other types of instruments available. However, these are the most common and widely used in the emergency-response field.

Before continuing, there is a terminology clarification needed for organic vapor analyzer (OVA). As the name implies, OVAs are used to analyze organic vapors. However, they are not a single instrument or even a single type of instrument. Rather, OVA refers to a group of various instrument types. This group includes flame ionization detectors, gas chromatographs, infrared spectrometers, and many others.

Combustible Gas Indicators (CGI).   Combustible gas indicators are one of the most basic and fundamental types of monitors needed for emergency response. They identify both quantitative and qualitative data, and they detect a broad variety of flammable gases and vapors.

However, they require the use of relative response data for substances other than the calibrant if quantitative data is desired. Some CGIs have the relative response factors built into their electronics. This means that the user need only identify the specific gas or vapor involved, push the appropriate button(s) for the substance, and the conversion is automatically made.

There are three basic types of CGIs, classified by the types of readings they provide. The following table provides more specifics.

Type	Function
% LEL	Identifies the percent of the lower explosive limit
% GAS	Identifies total percent gas above 100% LEL
PPM	Converts percent LEL readings to ppm

There are three types of sensors used by these devices (Table 7-6). They include the catalytic filament or bead, the semiconductor (solid state), and the thermal conductivity sensor. Some instruments may use a combination.

Catalytic devices use a similar technology to produce readings. This technology is called a Wheatstone Bridge electric circuit (Figure 7-13). In the case of the catalytic and thermal conductivity sensors, a sample of the atmosphere is drawn into the device either by an electric or manual pump. This allows the atmosphere to contact the sensor.

Table 7-6 Types of Sensors

Sensor type	Functioning
Catalytic	Filament coated with catalyst that helps burn the contaminant; may also have second filament without catalyst coating
Semiconductor	Semiconducting silicon chip with metal oxide coating (metal oxide sensor—MOS)
Thermal conductivity	Catalytic filament plus a sealed noncatalyst filament

If there is a flammable gas or vapor in the atmosphere, its molecules burn when they contact the heated filament.

The combustion process generates heat that increases the electrical resistance of the filament. This change in resistance causes a reading on either the analog (needle gauge) or digital display. The more contaminant there is in the atmosphere, the more heat is produced when it burns, the more resistance is created in the filament, and the greater the readout is on the display.

Many of the modern % LEL meters use a compensating filament (not coated with catalyst) as a constant. In this case the readout shows the difference between catalytic and compensating filaments. The use of the second filament dramatically limits the impact of lowering electrical battery output as the instrument works.

Likewise, the % GAS meters use a compensating filament. This compensating filament is not only free of catalyst, it is sealed from any exposure to the atmosphere. When the meter is reading % LEL, the device works like a standard % LEL meter.

However, when the LEL is reached, a switch is thrown and the thermal conductive (TC) filament, which has no catalyst, is activated. Both the TC and compensating filaments are heated to high temperatures that are precisely controlled. As combustible molecules hit the TC filament, they cool it in comparison to the compensating filament. As the TC cools, its resistance decreases and provides the display.

Finally, the semiconductor's metal oxide sensors (MOS) are sensitive to many contaminants. When the contaminant contacts the MOS, it causes a change in the elec-

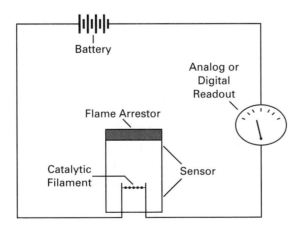

Figure 7-13 Wheatstone Bridge Circuit

tron flow through the semiconductor, producing a change in its electrical conductivity. The same type of electrical conductivity variation is caused by resistance in the % LEL and % GAS systems. There is no relative response for MOS.

All CGIs have interferants that will affect the accuracy of their readings. The following are some of the substances and conditions that potentially impact their readings. However, for the behavior of any individual instrument, the manufacturer's specifications and recommendation must be referenced.

1. High humidity can act to artificially cool either the % LEL or % GAS sensors. Further, it can also change the electron flow of the MOS. Although most manufacturers provide filters, this may still be a concern unless they are down to the 1 micron level.

2. Aspiration of liquid into the sensor will poison all sensors. Again, although manufacturers may provide traps, filter, or similar devices, most sensors will be damaged or destroyed.

3. High or low oxygen levels in the atmosphere can affect the various devices. This is one reason oxygen level monitoring should always accompany CGI monitoring.

4. Atmospheres warmer than the sampling tube and the device can lead to condensation of water or contaminant vapors within the instrument. This can poison the sensor or damage the device.

5. Corrosive atmospheres will damage most CGIs as well as many other detectors through corrosion.

6. Dusts, fibers, and similar particulates will poison sensors and possibly damage the detector. Again, most devices are equipped with filters but consult manufacturer for specifics.

7. Temperature extremes can be detrimental. Routinely, operating ranges are from between about 32°F to about 100°F.

8. Electronic interference is a routine problem with most electronic detectors. The most common type of interference is radio waves produced by mobile and portable radios, cellular phone, etc., that can produce widely erroneous readings.

9. Sensor poisons, including organic heavy metals (organometallics), silicones, silicates, some sulfur containing compounds, and so on, deposit solid fume particulates.

10. Chlorinated hydrocarbons can indicate flammable atmospheres when none exists.

11. Oxygen-acetylene atmospheres require special detectors.

Oxygen Sensors. Oxygen sensors are substance specific detectors that use electrochemical sensors which use the principles of a galvanic cell (see Chapter 4, Pipelines). In this cell are two electrodes connected by an electrolyte solution that contains ions. The electrolyte can be either acidic or alkaline. Since the electrodes and the electrolyte conduct electricity, a circuit can be formed when a battery is added to the system. Oxygen monitors draw a sample of the atmosphere into

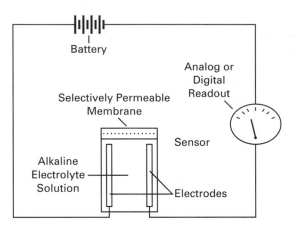

Figure 7-14   Oxygen Meter

the detector where the sample enters the sensor through a selectively permeable membrane (a membrane that allow some substances through while keeping others out). Oxygen changes the ion flow within the electrolyte, thus producing meter readings (Figure 7-14).

A particularly important factor to remember is the presence of the electrolyte. Specifically in the case of the oxygen sensor, the electrolyte is alkaline. If this electrolyte comes in contact with an acidic substance, a standard acid-base neutralization will occur. Once the electrolyte is neutralized, a circuit will no longer exist, and the sensor will not function.

Since the oxygen sensor is alkaline, any acidic atmosphere will neutralize and poison it. One common example of such a substance is carbon dioxide (CO_2). It forms weak carbonic acid (H_2CO_3) when it contacts the electrolyte solution, and thus neutralizes the electrolyte over time. (So much for checking the oxygen meter by exhaling into the sensor.)

Because of this behavior, most monitors that employ electrochemical technologies have limited sensor lifetimes. For example, oxygen sensors routinely require replacement every 12 to 24 months.

Other monitors that use electrochemical sensors include carbon monoxide, hydrogen sulfide, and ammonia. Dependent on the specific target substance of the sensor, the electrolyte is either acidic or alkaline.

Colorimetric Indicators.   The category of colorimetric detectors is wide and varied. However, they all have two things in common. First, each uses some type of color change to provide qualitative and quantitative data. Second, colorimetrics are highly substance or property specific.

The two most well known colorimetric detectors are pH paper and colorimetric tubes (e.g., Dreager, MSA, and Sensidyne). Both provide qualitative and quantitative data.

Colorimetric—pH Paper.   pH paper identifies the hydrogen ion concentration of a solution. pH is the negative log of the hydrogen ion concentration. The term *nega-*

tive in this context indicates an inverse relationship. This means that as the number gets large, the H^+ concentration gets smaller. Conversely, as the number gets smaller, the H^+ concentration gets larger. As such, a pH of 0 or 1 indicates a very high concentration of H^+, while a pH of 13 or 14 indicates a very low H^+ concentration.

The term *log* indicates that the scale is logarithmic, not arithmetic. In an arithmetic scale when the number changes by one whole digit (e.g., 2 to 3 or 5 to 6), the concentration increases or decreases by one. However, in a logarithmic scale, when the number changes by one whole digit, the concentration increases or decreases ten fold. For example, when the pH changes from 7 to 6, the H^+ concentration changes by a factor of ten. When the pH changes from 7 to 5, the H^+ concentration changes by a factor of 100 (10×10). When the pH changes from 7 to 3, the H^+ concentration changes by a factor of 10,000 ($10 \times 10 \times 10 \times 10$).

The most common and useful types of pH paper for emergency response use span the pH scale from 0 to 14. This paper has been treated with indicators that change to specified colors at specified pH ranges. As such, pH paper is inexpensive, generally sensitive, and accurate enough for emergency response activities.

pH paper must be used in aqueous systems (water-based solutions). One problem that routinely arises is when very low concentration acids or bases are being tested for their pH. In the range of about 6 to 5.5 to about 8.5, pH paper can provide erroneously high or low readings. This is especially true in bodies of water such as streams, rivers, and ponds due to the presence of extraneous ions. In these situations, a pH pen or liquid pH indicators are preferable.

Another problem with the requirement for an aqueous system is that pH paper alone cannot be used to monitor for acidic (corrosive) or basic (caustic) atmospheres (unless it is present in extremely high concentrations). For the paper to work in this application, it must be wet. However, plain tap or hydrant water must not be used because it also contains dissolved minerals and ions. These dissolved minerals and ions affect the ability of the indicator dyes to function properly the same as in a stream or low concentration corrosives. As a result, the pH paper must be wetted with distilled water. Since corrosive or caustic atmospheres can damage many monitoring instruments, it is highly recommended that the air be checked for pH whenever there is a possibility or probability that such destructive atmospheres exist.

Colorimetric—Tubes.   Colorimetric tubes (Figure 7-15) are simply glass tubes filled with a chemical reagent that changes color when it contacts a given substance. The glass tubes are sealed at both ends to prevent accidental contamination, and are equipped with graduated markings to assist in determining the concentration of the contaminant in the atmosphere tested. Before their use, the sealed ends of the glass tube must be broken to allow atmosphere to flow through the tube. Most commonly, air flow through the tube is generated by a negative pressure created by a bellows, syringe, or thumb (bulb) pump attached to one end of the tube.

These tubes provide both qualitative and quantitative data, dependent on how they are used. To obtain accurate quantitative data, the manufacturer specifies a given volume of air that must be drawn through the reagent. This exact volume of air is controlled

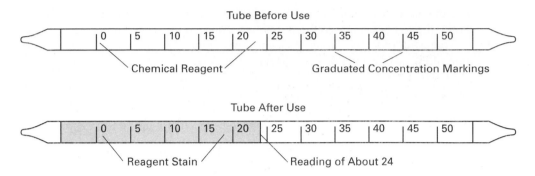

Figure 7-15   Typical Colorimetric Tube

by the number of compressions on the bellows pump, pulls on the syringe, or compression on the thumb bulb.

The reagent within each tube is quite specific regarding the exact substance with which it will react. For example, a chlorine tube will not stain if it contacts xylene or ammonia. However, it will stain if exposed to vapors coming from a sodium hypochlorite or calcium hypochlorite solution. So, although the tubes are substance specific, close relatives can also cause staining.

This situation has both a positive and negative side. On the positive side, a given tube could be used for a qualitative yes/no reading for a closely related substance. On the negative side, if the close relative were present, it could produce a false positive and most definitely will skew any quantitative readings.

Because of the tube's specificity, a large number may be required for all-around monitoring capabilities. Commonly, a packet of tubes contains eight to ten tubes that each have a shelf-life of about two years as a general rule. As a result, maintaining a large number of tubes for many different substances can become a rather expensive endeavor due to the number of tubes needed and their shelf-life limitations.

Refrigeration of the tubes can extend their usefulness. However, cold tubes can cause condensation and are less responsive than those adjusted to the ambient conditions to which they will be exposed. Additionally, there are quite a few conditions and situations that can adversely affect readings, as follows:

1. Most tubes are designed to function accurately between 32°F and about 104°F.
2. High humidity can artificially raise or lower quantitative readings, depending on the specific substance and reagent involved.
3. Light and elevated temperatures have a negative effect on most of the reagents. Exposure of the tubes to rapid heating and ultraviolet light can cause the reagent to become useless in a matter of hours.
4. The existence of interferants for each tube (close relatives) can produce false readings.

A new wrinkle being added to the colorimetric arena is a new technology. In this case it employs a small electronic chip in a hand-held pump. Again, the chips are specific like the tubes and have a shelf-life. However, they are not read by the user's eye but rather by the electronics in the device that provides a digital readout.

Additional Colorimetrics—Dip Sticks. Dip sticks function similarly to pH paper except they can identify a wide range of various qualitative data. One of the more common dip sticks tests for multiple types of substances at one time. Each stick is made of a plastic strip that is about 1/2 inch wide and anywhere from four to twelve inches long. On the surface of the plastic are one or more colorimetric patches. These patches change color when they contact things such as hydrocarbons, oxidizers, acids, bases, gasoline. In some instances they are designed to identify the presence of single substances its general concentration range.

Dip sticks can be a very useful tool when dealing with unknowns or unknown sources. For example, consider a storm sewer that discharges into a stream. An assortment of odors is reported coming from the pipe, ranging from raw sewage to gasoline to diesel fuel to acids. By testing the effluent with a broad-spectrum dip stick, the field can be narrowed and possible identified.

Additional Colorimetrics—Water Dyes. Water dyes are environmentally safe, water-soluble dyes. Although they are not routinely needed during emergency response, they can be a valuable tool under the right circumstances. For example, a contaminant shows up in a stream or some other body of water but its source is not readily identifiable. The problem may be a chronic or acute situation. Responders who have been involved in such incidents will attest to the frustration they can cause.

Dyes can sometimes be of great value in such situations. After some investigation, a possible source is identified. At that point, the dye is placed in the possible source location and the body of water is visually inspected to see if and when the dye appears in the water. A warning: Do not rush the process. It may require considerable time for the dye to work its way from the source to the environment. In some instances, the source may involve a sump pump or similar situation that requires a minimum level be reached before the discharge occurs.

Additional Colorimetrics—Liquid Tests. Liquid-based colorimetric tests are not the most common type of test but do have their place. They are normally product specific and are often quantitative as well as qualitative in nature. Normally, a small sample of a suspect liquid or solid is gathered. To this sample, a reagent is added. If the specific contaminant is present the reagent changes color (in some cases acquires a color). The darkness or shade of the color may vary if the test has quantitative abilities. Some of the most common examples of this test are for PCBs and asbestos.

Photoionization Detectors. Photoionization detectors (PID) employ ultraviolet (UV) energy to ionize contaminants drawn into the sensor. UV is a form of electromagnetic (EM) energy or radiation. As such, it is the same type of energy found in the standard electromagnetic spectrum that includes radio, television, radar, microwaves, infrared visible light, x-rays, and gamma rays. This energy is found in the form of individual packets of energy known as photons, hence the photo in photoionization.

Although it is closely related to visible light, ultraviolet has a shorter wavelength and higher frequency. Consequently, ultraviolet is located on the opposite end of the visible spectrum from infrared (longer wavelength and lower frequency than the lower or red end). This means UV is located just above the upper or blue end of the visible light spectrum.

UV is invisible and is able to ionize certain substances depending on its chemistry as well as the intensity of the UV. In PIDs (Figure 7-16), ultraviolet radiation is generated by the lamp. The lamp is nothing more than a glorified light bulb that emits radiation in the ultraviolet spectrum instead of the visible spectrum. Further, just like a light bulb, UV lamps have varying intensities. In light bulbs, the unit of measurement used to identify intensity is the watt. In UV lamps, the unit of measurement used to identify intensity is the electron volt (eV). The most commonly found intensities are 9.5 eV, 10.2 eV, 10.6 eV, and 11.7 eV.

A PID works by drawing a sample of the atmospheres into its sensor. Within the sensor are two electrodes that flank the lamp. The lamp emits a tight, cohesive beam of ultraviolet energy. When a contaminant enters the beam of UV, it is ionized if the UV is of sufficient intensity. If ions are produced, they will allow an electrical current to pass between the electrodes (going from the bias to the collecting). This electrical current is measured and thus produces a readout. Most commonly PIDs provide readings in one of three ranges: 0.1 to 20 ppm, 1.0 to 200 ppm, or 1.0 to 2,000 ppm. As a result, most PIDs have an effective range of 0.1 to 2,000 ppm. (Newer models have higher ranges.)

The big question is how does one determine if the UV lamp can ionize a given contaminant. The answer is by identifying a specific chemical property of the contaminant known as ionization potential. Ionization potential is the amount of energy required to cause an atom or molecule to release an electron. In other words, it is an indication of how easily a substance can be ionized. Further, ionization potential (IP) is expressed in electron volts, the same as the lamp intensity.

To determine if a substance can be detected by a PID, one must know the electron voltage of the lamp and must use a reference to find the IP of the substance. Such references include manufacturer's information, the NIOSH *Pocket Guide to Chemical Hazards,* among others. If the contaminant has an IP less than the electron voltage of

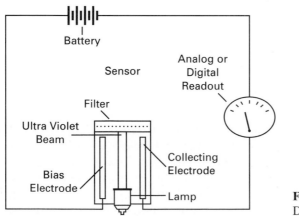

Figure 7-16   Photoionization Detector

the lamp, it will be ionized by the ultraviolet radiation. The greater the concentration of the contaminant is in the air, the greater the number of ions produced, the higher the reading on the meter.

PIDs are unique among almost all of other instruments in that the contaminant is not chemically altered as it passes through the device. Further, instead of having multiple calibrant options, PIDs are calibrated with isobutylene. (The original calibrant was high-hazard benzene.)

PIDs are also unique because they detect substances based on their ionization potentials and not their hazard or chemical classifications. For example, a PID with a 10.6-eV lamp detects various elements including:

1. Simple inorganic compounds including those of sulfur and nitrogen
2. Saturated and unsaturated hydrocarbons
3. Aromatic hydrocarbons and derivatives
4. Hydrocarbon derivatives including halogenated hydrocarbons, aldehydes and ketones, alcohols, ethers, and organic acids

This wide variety of detectable substances is a great advantage and disadvantage, depending on the situation. On the plus side, if the substance is known and has a low enough IP, the PID can be used in a large number of situations to provide essential monitoring information. However, if the substance is not known or a second unidentified contaminant is also present, the reading can be distorted.

One option for the use of the PID is the determination of a contaminant's possible presence. To perform this function, a zero air reading is taken. This means, after the unit is calibrated, a reading of air well away from any potential exposure to the contaminant (i.e., well back in the cold zone) is taken. The device is then zeroed (providing a reading of zero). Through this process, the instrument removes any background ions from subsequent readings. As a result, the appearance of any readings indicates the presence of some contaminant.

However, the absence of readings above the zero air setting does not mean there is no contaminant present. The contaminant simply could have an IP greater than the lamp's intensity. To help minimize misinterpretation when dealing with unspecified contaminants, the PID (or any other instrument for that matter) should be teamed with other instruments that read different substances, properties, or conditions.

For example, consider a situation where the PID provides no readings above zero air. The following instruments may provide valuable information:

1. An oxygen detector produces a reading below its initial 21%, indicating that something has displaced or depleted oxygen from the atmosphere. If the area is not tightly sealed, the readings must result from displacement not depletion (which only takes place in areas that receive no air exchange).
2. The flame ionization detector will detect many hydrocarbons and hydrocarbon derivatives that have IPs above standard lamp intensities. This includes substances

such as methane, ethane, propane, methanol, ethanol, acetylene, acrylonitrile, and methyl acrylate.

3. The radiation monitors will detect radioactive materials.
4. CGI will detect high concentrations of flammables that indicate withdrawal.
5. pH paper wetted with distilled water will identify corrosive atmospheres.

Granted, to some this may seem like overkill. However, in a true unknown situation or when there is the potential for multiple contaminants, such a monitoring protocol is the only appropriate approach for the situation. Even if the data is only used for qualitative purposes, it is better than having no data.

PIDs also have relative response considerations similar to those described for CGIs. For specific information and tables, consult the manufacturer's literature. Further, the ppm readings only apply for monitoring the calibrant. When monitoring other substances, the relative response provides an estimate or the relative concentration and not the actual concentration (as is true for all devices).

Additionally, there are quite a few conditions and situations that can adversely affect readings of the PID, as follows:

1. Humidity is one of the most common interferants for PIDs. Humidity within the PID scatters the ultraviolet beam in the same way fog scatters the light from a car's headlights. This scattering reduces the intensity of the UV beam and thus its ability to ionize contaminants. One means employed by manufactures to minimize this effect is the use of filters. These filters are metallic and restrict movement of humidity and other contaminants. The finest filters are as small as one micron (one millionth of a meter, or 10,000 angstroms).
2. Dust and fine particulates produce the same effect as humidity.
3. In high concentrations, nonionizing gases and vapor can also produce an effect similar to humidity.
4. The lamp condition plays a big role in the intensity of the UV beam. If the lamp is dirty, contaminated with condensate of contaminants or humidity, its intensity will decrease. The decreased intensity will affect its ability to ionize, thus the validity of the readings. The lamp should be cleaned following the manufacturer's recommendations.
5. The 11.7- or 11.8-eV lamp has a UV window made of lithium fluoride. Lithium fluoride is hygroscopic thus absorbing moisture from the air. This causes the window to swell and reduce the intensity of the UV. Additionally, UV also degrades the lithium fluoride and reduces its intensity even further. As such, the lamps have a relatively short service life that ranges from about 3 to 6 months.
6. PIDs should be field calibrated before each use or as specified by the manufacturer. When personnel are properly trained, calibration requires only a matter of minutes. If this recommendation is not followed, the data provided by the instrument is questionable.
7. Depending on the exact make and model of the instrument, radio transmissions and other electronic noise will affect reading from these devices.

Unfortunately, not many responders are familiar with the capabilities of the PID, let alone have one available for their use. These devices are very versatile and are required to provide data involving low concentration releases. Once a team has one and learns how to use it, they cannot understand how they operated without it.

Flame Ionization Detectors. Flame ionization detectors (FID) are a type of organic vapor analyzer (OVA). FIDs function in a fashion similar to PIDs except they use a hydrogen flame to generate ions instead of UV. They detect a type of carbon ion produced when the contaminant is ionized through the combustion process. The intensity of the hydrogen flame is enough to ionize organic compounds with an IP of up to 15.4 eV. Their effective range is from 0.1 to 100 ppm, 1 to 100 ppm, and 10 to 1,000 ppm, depending on the scale chosen. (Some models go significantly higher.)

As an OVA, FIDs (Figure 7-17) are capable of detection substances containing organic carbon. They do not detect inorganic compounds (e.g., silanes, hydrogen sulfide, and nitrogen oxides) even if the compound contains carbon (e.g., carbon dioxide or carbon monoxide). Further the organic compound must contain multiple carbon-hydrogen or carbon-carbon bonds to be detected or detected effectively. As such, substances such as hydrogen cyanide and formaldehyde are not detected, while substances like tetrachloromethane (carbon tetrachloride), acetaldehyde, and methanol are not detected efficiently.

The efficiency of the ionization process is important when it comes to the relative response of the instrument. Generally, the lower the efficiency, the smaller the relative response value and the greater the relative concentration of contaminant in comparison to the reading. Conversely, the greater the efficiency, the greater the relative response value and the lower the relative concentration of the contaminant in comparison to the reading.

For example, consider an FID calibrated on methane (as with all FIDs). Methane's ionization efficiency is considered to be high and thus has a response value of 100 (100%). That means when the meter reads 100 ppm, the actual concentration is 100 ppm. Compare this to propylene which has moderate ionization efficiency. Propylene's

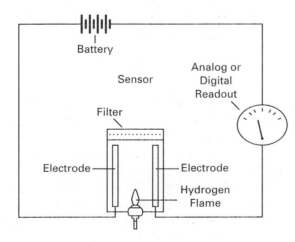

Figure 7-17   Flame Ionization Detector

relative response value is 75, meaning when the meter reads 100 pm there is actually a significantly higher concentration present. Based on this data, there is about a third greater concentration than the meter indicates.

Conversely, consider acetylene whose ionization efficiency is considered very high. Its relative response value is 225, meaning that when the meter reads 100 ppm there is actually a significant lower concentration present. Based on this data, there is less than half the concentration indicated by the reading.

Unfortunately, FIDs intimidate many responders simply by their appearance. One of the most common FIDs is a box about one foot square and about six inches deep. When the lid is removed, one finds two pressure gauges, three rotary valves, several switches, a coil of tubing, and various other doodads that make the instrument seem impossibly complex. The reality is the device is extremely simple to use once it is understood. Some teams have even found that the device is one of the most commonly used instruments because it is so user-friendly.

One drawback to FIDs is that they need time to warm-up before they are used. FIDs require a warm-up process ranging from about 2 to as long as 15 minutes before the flame can be ignited. Once the flame is ignited, it must burn for an additional ten to fifteen minutes before it is used.

Although this sounds like a long time, it can be ready by the time suiting and other preparatory work is completed. However, if its need is only identified immediately prior to the entry, delays will result. As such, if there is any possibility it will be needed, the FID should at least be turned on and warmed-up in advance.

Obviously, since the FID uses a hydrogen flame to ionize the organic it must have a hydrogen source. Routinely, when response teams first purchase FIDs, they also purchase hydrogen supplies. Most commonly, the purchasing agent buys a low purity hydrogen. Further, they often purchase a cylinder large enough to last for twenty years. As a result, a lot of hydrogen is wasted.

For an FID to function at all, it requires ultra high pure hydrogen sometimes known as research grade or as five nine hydrogen (99.999% pure). This hydrogen should have no more than 1 ppm hydrocarbon contamination and no more than 1 ppm water content. The reason for these stringent purity requirements is one of contamination. The slightly less pure grade of four nine hydrogen (99.99% pure) can have 0.01% contamination which translates to 100 ppm of contaminant. On the other hand, five nine hydrogen can have only 0.001% contamination which translates to 10 ppm of contaminant. Since FIDs are routinely calibrated to 100 ppm concentrations of methane, a 100 ppm contamination level of the hydrogen makes the device useless. A telltale indicator of contaminated hydrogen is the inability to calibrate the instrument. The maximum of 10 ppm contaminant found in the five nine hydrogen can be compensated during calibration.

Furthermore, the large compressed gas cylinders are commonly constructed of high carbon, HSLA steels. Under the high pressure found within the cylinders, the hydrogen and even minute concentrations of water vapor will react with the carbon in the steel. The resulting product is methane. Routinely, the concentration of methane will approach 100 ppm from one to two years after the hydrogen is purchased. Again, this problem is most commonly detected when calibration runs are performed after several refills of the instrument's onboard hydrogen cylinder (made of stainless steel). Each time the instrument is filled, it becomes more difficult if not impossible to calibrate.

When this occurs, it is best to purchase new, ultra pure hydrogen less than 1 ppm total hydrocarbon contamination and less than 1 ppm water, in the smallest cylinder available and return the original hydrogen. Commonly, the smallest cylinder will be the D or D1 type. Some teams routinely, every 18 to 24 months, purchase new hydrogen and return any old hydrogen. This helps minimize any potential problems.

Additionally, a few conditions and situations can adversely affect readings of the FID, as follows:

1. FIDs are sensitive to oxygen levels in the atmosphere they are monitoring. If the oxygen level drops, the flame intensity will also drop. This produces artificially low readings. Further, if the atmosphere becomes too oxygen deficient, the hydrogen flame will go out. Obviously, this will result in no readings. Most of these instruments are equipped with a flame-out alarm.

2. Dust and other particulates are potential concerns. All FIDs are equipped with filters to remove most external particulates. However, FIDs can develop particulate deposits within the combustion chamber if high concentrations of organic compounds are drawn into the device. Also, these particulates can absorb vapors and then release them over time. This is noted by a slow recovery time from the onset of the meter reading.

3. Temperature ranges are from approximately 40°F and 105°F. Below the lower end of the range, water vapor produced by the combustion process can condense or freeze in and around the exhaust port. This interferes with response and recovery time and can lead to flameout. Users have found that it can be used for short periods (10 to 15 minutes) when temperatures are below freezing, although it is not recommended.

4. Since FIDs are designed for use in flammable atmospheres, they must be intrinsically safe. Nonintrinsically safe FIDs must not be used in Division 1 or Division 2 locations.

Although the FID is not one of the most commonly carried instruments by response teams it should be. Its flexibility and versatility make it an essential piece of instrumentation.

Finally, some FIDs have an optional gas chromatograph (GC) accessory. Here, we will discuss a highly simplified explanation of the GC accessory. It allows the monitoring of an atmosphere of a pure substance or a mixture by breaking it into its constituent components. The process produces a graph that can be compared to known samples to help provide identification or potential constituent information. This accessory is not needed by all response agencies. However, it may provide assistance to those who wish to take advantage of its capabilities.

This discussion of instrumentation and air monitoring has been brief and only covered the high points. However, it is hoped that it has identified the need for thorough and detailed monitoring plans and programs during emergency-response operations. When a cadre of the devices discussed becomes available and is used on the emergency scene, responders will truly wonder how they did the job before. Instrumentation is the crucial link between what we expect and what is really the case. In many situations, air monitoring and the data it provides enables responders to perform a more safe and effective response while minimizing over- or under-reaction due to lack of information.

To learn more about the vital area of air monitoring for emergency responders and others, this author strongly recommends a text written by Carol and Steve Maslansky, entitled *Air Monitoring Instrumentation—A Manual for Emergency, Investigatory and Remedial Responders* (Van Nostrand Reinhold, 1993). It provides a simple, easy to understand discussion of not only the general types of instrumentation, but also describes the use of individual devices. It is a must for all who are serious about safe and effective emergency-response operations.

IDENTIFICATION—SAMPLING

Sampling is often mistaken for air monitoring. Although both are often performed during an incident, they are two unique and distinct operations. Unlike air monitoring, sampling involves gathering a given quantity of the contaminant from the hot zone. The contaminant can be a solid, liquid, or gas.

Sampling is done for a variety of reasons. In some instances, a sample is required to identify the contaminant through the use of a HAZCAT Chemical Identification System (Trade Name of San Francisco: Haztech System, Inc.), gas chromatograph, infrared spectrometer, and so on. In other instances, the sample may be needed to document non-compliant or illegal activities or handling of chemicals. Regardless of the reason, all sampling involves removing a substance from the hot zone and taking it into the cold zone.

Samples can be drawn using an instrument such as a syringe, drum thief, Coliwasas water column sampler, or drum core samplers. Routinely, liquid or solid samples are placed in test tubes, bottles, small cans, or even plastic bags. Gases are gathered in large syringes or gas bags. Since these samples are taken in the hot zone and involve the contaminant, they must be properly packaged and handled or cross contamination will result. Further, the sample should consist of the minimum amount needed to meet the purpose of collection in the first place. In other words, if 2 milliliters (ml) or cubic centimeters (cc) will do (since, 1 ml = 1 cc), do not take 250 ml (approximately a cup).

Most commonly, proper handling involves the collection of the sample and its placement within a suitable, preferably new, container (e.g., vial, bottle, bag, or tube). This is intended to secure and properly hold the sample. Next the sample proceeds to the hot line where, using strict sterile technique, it is placed within some type of clean and appropriate overpack. Decontamination of the actual sample containers is generally not the best option because it introduces the possibility of sample contamination, spillage, or cross contamination of later handlers.

The overpacked sample is then taken to an assigned location to undergo chemical monitoring, analysis, logging, and other procedures. An essential requirement for any sample(s) that may be used in any type of enforcement action (regulatory, civil, or criminal) is logging and the establishment of a chain of custody.

The logging process involves establishing a paper trail that documents who, when, where, why, and how the samples (now the evidence) has been seen, tested, handled, and so on. This is done to ensure that only authorized and known individuals have contact with the sample. The chain starts with the individual who originally collects the sample. The overpack should be marked to indicate at a minimum, who took the sample, the location where it was taken, and the date and time when it was taken. If there are multiple samples taken from various locations around the scene, exact locations must be listed.

Following these procedures dramatically minimizes the possibility of tampering by outside parties. As a result, a chain of custody needed for enforcement or trial situations is established. However, the subsequent handling of the sample is also important. As pointed out by recent, high visibility trials, improper handling of evidence can be disastrous. The following examples are just a few of the possible improper handling techniques that can cause evidence to be thrown out by a judge or arbitrator, or not believed by a jury.

1. Samples placed in the trunk of a chief's car and allowed to stay there until the next shift
2. Samples placed in the cab, jump seats, or compartment of apparatus and allowed to remain there after the response is complete
3. Samples placed in the common areas of a fire or police station that have open access to anyone in the station

As soon as practical after gathering, any sample to be used as part of an enforcement action should be logged into a secured evidence area. Such an area should contain an appropriately designed and protected chemical storage cabinet or closet. It is possible an arson evidence area may be so equipped (again, this is for very small and limited quantities of the contaminant, not 55-gallon drums).

Any fire, health, or safety code implications must be addressed beforehand. If there is any leakage or potential that the sample or overpack may open, one should repackage it in a more secure container. Other areas that may be suitable include chemical laboratories at municipal water, sewer, or health facilities.

When an environmental or law enforcement agency is involved and will handle the samples, one should make sure they establish the chain of custody. If they arrive late in the operation and a log has been established, note in the log when, where, and to whom the samples were turned over.

Although, sampling is not the most commonly used identification tactic, it can be very important and valuable when performed correctly. Further, there is a growing trend toward the formation of environmental crime units at all levels of government. As such, every team should have identified procedures for the acquisition and handling of samples that will be used in enforcement actions.

PROTECTION

The process of assuring the safety of the public and response personnel	**Primary Tactical Objectives:** Secondary Public Protection Reassess Zones Personal Protective Equipment Decontamination EMS and First Aid Preentry Briefing Preentry Medical Monitoring Safety Assessment Rescue

Protection is the process of assuring the safety of the public and response personnel. Although a small portion of this goal addresses the public, the majority targets response personnel operating on the incident scene.

PROTECTION—SECONDARY PUBLIC PROTECTION

Secondary public protection occurs after initial public protection is identified and underway. It may or may not be needed, depending on the efficacy of the initial protective measures taken.

In many situations, initial public protection performed as part of isolation will satisfactorily address public protection. However, for larger incidents or those that require large scale evacuation or inplace protection or a combination of both, it is often advantageous to perform staged or staggered evacuations.

The approach involved is similar to that employed when rescuing or evacuating trapped building occupants during a structural fire. Initial efforts focus on those individuals in the most threatened areas (i.e., those in the fire area, above, and on either side). In a staged evacuation, the most threatened areas are evacuated first followed by evacuations of successively less threatened areas (i.e., those immediately downwind, those at successively greater distances downwind, and those on either side). Furthermore, areas may initially only need inplace protection but later require evacuation. Also it is often not feasible to move everyone at one time. As a result, some areas are protected inplace until it is their turn to be evacuated.

PROTECTION—REASSESSMENT OF ZONES

Reassessment of zones normally occurs when a hazmat team first arrives on an incident scene where zones (or perimeters) have been established by first responders. The process is simply a double check to ensure that the zones and perimeter are appropriate for the incident situation and conditions encountered.

Further, reassessment of zones is needed when the following types of situations or conditions occur on the incident scene:

1. The incident has been going on for a relatively long time.
2. Meteorologic conditions change or are expected to change; includes wind direction, wind speed, humidity, temperature, as well as when the dew point is reached or precipitation occurs and so on.
3. Release conditions or rate change. These include additional container failure, greater rate of release, lesser rate of release, physical or structural instability of the container, and so on.
4. Accomplishment of strategic goals such as spill control, leak control, or fire control.
5. Whenever any unexpected or major event occurs on the scene.
6. When the incident moves from the emergency to the recovery phase.
7. When clean-up, product transfer, container righting, or similar activities are set to occur.

Reassessment of zones is often a result of the evaluation phase of the seven-step process. As the efficiency and efficacy of the operations being performed is evaluated, zone reassessment is a common result.

PROTECTION—PERSONAL PROTECTIVE EQUIPMENT

Personal protective equipment (PPE) is one of the most important aspects of personnel safety and thus the strategic goal of protection. PPE is any type of equipment worn by personnel to provide them with some form of protection. As such, work gloves, steel tip boots, safety glasses, and so on, are part of this large group of objects known as PPE. The real challenge is to identify the exact nature of the hazard(s) that personnel will face and identify appropriate PPE to protect them.

NFPA identifies six hot zone hazards and uses the acronym TRACEM to help remember them. TRACEM stands for:

1. **T**hermal
2. **R**adiation
3. **A**sphyxiation
4. **C**hemical
5. **E**tiologic (biohazards)
6. **M**echanical

These six hazards are, of course, not the only hazards encountered during hazmat response. A very common one, especially within fixed facilities, is electricity. Other hazards include guard dogs, controlled access points, and air locks. However, TRACEM works when discussing PPE.

When examined closely, these six hazards, with the exception of chemical and etiologic, require their own unique PPE or in some instances, procedural safeguard. As a result, our examination of PPE will use these hazards as its focus.

Mechanical Hazards

Mechanical hazards involve blunt force of some type or form. Slips, falls, pinches, cuts, impacts, and punctures are all examples of situations that involve mechanical hazards. The primary PPE available to protect personnel from these hazards include outer gloves (worn over chemical gloves), steel tip and shank boots, turnouts, and helmets.

Although this type of PPE is recommended and necessary, it has substantial limitations and must be incorporated with safe operational procedures. For example, consider an entry team member who releases the latch on a rear, double, swing door of a box trailer involved in an accident. This responder will receive only nominal protection from any of his or her PPE should a full 55-gallon drum fall from the door as the door is opened.

This particular operation always holds the potential for danger when it is necessitated by an accident or simply a cargo shift. The reason is simply that cargo may have shifted and now is resting against the door. It is unreasonable to expect steel tip boots, leather gloves, or a hard hat to provide any real protection from a falling 500- to 800-pound weight. As a result, personnel must also be protected by the use of appropriate procedures.

In this particular example, the potential hazards can be blunted by using straps, cables, or a come-along to control the opening of the door. The use of a remote opening device such as ropes attached to the latch handle do nothing to prevent the load from flying from the door and impacting on the pavement. (Something the hazmat team really does not want to have happen either.)

The focus of PPE for mechanical hazards is short and sweet. Although there is equipment available, it only protects against minimal hazards. The primary way to protect personnel from mechanical hazards is to identify their presence and institute procedures that prevent or minimize any potential encounters.

Radiation Hazards

Radiation hazards are not the most common nor the most challenging hazards with which to deal. The primary reason is that instrumentation provides good information about the nature and extent of the hazard faced. (That is assuming the instrumentation is used.) Respiratory protection and disposable suits, when used according to accepted procedures, will prevent internal and direct external contamination.

However, this PPE will not prevent exposure to the associated ionizing radiation (except for alpha particles). Again, PPE provides only part of the answer. Appropriate procedure provides even greater protection than the PPE itself in most instances. The old tried and true TDS (time, distance, shielding) are still the best protection available. Time refers to limiting the duration of any exposure. Distance refers to increasing the distance from the source of the radiation to decrease the exposure. This tenet is based on the inverse square law that states that every time the distance from a gamma source is doubled, the intensity of the radiation is decreased by 75%. Shielding refers to the placement of mass (soil, water, steel, etc.) between the source and the individual. Again, procedures are just as essential as PPE to provide protection.

Asphyxiation Hazard

Although NFPA specifies asphyxiation, this hazard really encompasses all potential respiratory implications. As such, there are three primary types of respiratory PPE available to protect the respiratory tract. They include air purifying respirators (APR), positive pressure self-contained breathing apparatus (SCBA), and in-line air systems.

The primary and defining difference between these three types of respiratory protection is their source of air. APRs condition ambient air in several ways, while both SCBA and in-line systems use cleaned, dried, and filtered compressed air.

Air Purifying Respirators.   APRs consist of either a full or half face piece. For use in emergency response, only full face mask APRs should even be considered. Full face APRs have a mask that is very similar to a standard SCBA mask except there is no connection to a bottled air supply. Instead, air from the environment passes through a filtration cartridge or canister.

Cartridges screw directly on to the inhalation ports of the face mask, while canisters (sometimes called a Type N gas mask) attach to a belt and are then connected to the face mask through an inhalation tube. The primary difference is the larger filtering capacity of the canister. Most commonly, APRs equipped with cartridges are used for emergency and nonemergency applications. Canisters are most commonly found at fixed facilities and are generally used for higher concentration applications.

A new twist to the straight canister is the PAPR that stands for pressurized air purifying respirator. In this configuration, a canister type APR is equipped with a fan that pushes air into the mask to provide a slight positive pressure. Obviously, the positive pressure has definite safety advantages.

Since all APRs rely on the filtration of ambient air to supply the wearer, they have many usage limitations. First and foremost, SCBAs must be worn until the ambient air has been tested by direct reading instrumentation and these tests have determined that a lesser level of respiratory protection is acceptable. The primary conditions that must be met for a lesser level of protection provided by the APRs to be acceptable include the following:

1. The atmospheric oxygen level is above 19.5%
2. The contaminant and its concentration are known

3. There is a cartridge available for the contaminant, and the concentration of contaminant will not exceed the maximum rate concentration of the cartridge.

4. The contaminant has good warning characteristics (e.g., odor or sensation) so it can be detected should the cartridge fail.

Although many take exception to using APRs during an emergency response at all, they do have their place. They are always used as part of a Level C chemical protection system. One example of a situation when they can be used quite safely and effectively (as long as prerequisites are met) is during decontamination in many but the most severe incident situations. Further, they can also be used when control efforts involve substances having low or no volatility, or when very low atmospheric concentrations of contaminant exist.

To indicate that APRs can never be used during emergency response is to take one of the valuable tools away from responders. However, to safely and effectively use this equipment, responders must fully understand the criteria for its use as well as its limitations.

In-line Air Systems.   In-line air systems are identical to SCBA in function except the user does not carry a cylinder of air on his or her back. Instead, an umbilical hose runs from a cylinder bank to a regulator most commonly worn on the hip. From there the inhalation tube provides positive pressure to the mask.

Although in-line systems have the advantage of an almost limitless air supply, they do have limitations. First, the umbilical cannot extend more than 300 feet from the air supply due to friction loss considerations. Second, this umbilical hose has a nasty tendency to become entangled with all sorts of things found on an incident scene. Third, to perform tactical operations, the umbilical often becomes contaminated. Finally, since a failure of this hose would totally interrupt the user's breathing air, an escape pack with a minimum five-minute air supply is required.

There is a hybrid in-line and SCBA system used by some teams. In this configuration the hose is attached to a fitting on the outside of the chemical suit. Another hose then attaches to the regulator of the SCBA. As a result, entry personnel use the umbilical air while they approach the problem and then disconnect. At that point, they start breathing from the SCBA cylinder. Some teams will then reattach the umbilical for exiting.

Obviously, this approach has some of the limitations of a straight in-line system, but it also has the advantages of an SCBA. One area of concern is that if the umbilical is reconnected for the return to decontamination, would it not be possible to force contaminant through the fitting as well? This would result in contamination of the respirator and air supply.

SCBA.   Present day SCBAs are all positive pressure to prevent any infiltration of contaminant into the face piece. If anyone has any old demand masks, they must contact their supplier and determine what needs to be done to make the masks compliant with today's standards.

Although SCBAs have many advantages, they too have their limitations. For one, the actual usage time varies substantially from the rated time (e.g., 30 minutes, 45 minutes, or 60 minutes). Most teams have found the minimum time rating necessary to perform entry operations safely and effectively is 45 minutes, with the 60-minute units being the best choice of all. These packs provide sufficient work time yet provide a safety margin so personnel can go through decontamination without running out of air.

In other words, this does not imply that personnel work for 45 or 60 minutes. The amount of work time is identified as the operational work interval. Routinely, this work interval is 20 to 30 minutes, depending on meteorologic and work conditions. In extremely hot and humid or cold to frigid conditions, the interval may be cut to 15 minutes.

Thermal Hazards

Thermal hazards consist of hot or cold conditions. The primary hot thermal hazards include those associated with fires and related heat, as well as process and transportation heat. Process heat is the heating required for part of the processing of the substance, while transportation heat is the heating required for transporting the substance.

The primary cold thermal hazards include those associated with cryogenic liquids, liquefied compressed gases, and compressed gases. Cryogenic hazards result from the cold temperature of the product itself, while liquefied compressed gas and compressed gas hazards result from the dramatic cooling that occurs when they undergo decompression.

There are three primary types of effective thermal PPE. These three types are structural turnouts, proximity suits and entry suits. Further, each type provides body protection and respiratory protection. The respiratory protection is the same, consisting of SCBA. The primary difference between the three types is the degree and type of body protection afforded.

Structural Turnouts. Standard full structural turnouts, as specified by NFPA 1970 and 1980 series standards, are designed to protect the wearer from the typical hazards associated with the structural fire fighting environment. OSHA also has a turnout standard that is found in 29 CFR 1910.156. (Although this equipment is acceptable, it provides a much lesser degree of protection.)

Regarding body protection, turnouts consist of a series of individual components that combine to provide a formidable array of protection against thermal as well as moderate mechanical and electrical hazards. In the realm of thermal hazards, the typical turnout will allow the wearer to function effectively in dry heat where the temperature ranges from 170° to 180°F. However, this effective temperature range decreases to between 130° and 140°F when confronting moist heat (as develops when hose lines are placed in service).

Turnouts also provide a reasonable degree of protection against moderately intense radiant heat. However, their protective qualities dramatically decrease as the intensity and magnitude of the radiant heat increases.

They are not designed to sustain direct flame contact. Although they will provide protection from direct flame contact, primarily through the thermal mass of the liner and flame retardency of the shell, the duration is extremely limited. Direct flame contact for more than 5 to 10 seconds, depending on the design and construction of equipment as specified by the manufacturer, will result in injury. As the duration of the contact becomes greater, the protective value of the equipment becomes less. Generally, direct flame contact for 30 seconds or more will result in severe injury.

Yet with all these seeming limitations, turnouts generally function superbly for structural firefighting, assuming they are worn and maintained properly. However, ask it to do what it is not intended to do or modify the ensemble, and all bets are off. For example, years ago, it used to be common to see firefighters remove the liners from coats during the summer. The result of this unwise action was often a major burn (most commonly from steam, radiant heat, or direct contact) to the upper arm, shoulders, or back while performing rather routine structural firefighting operations. The same was true for the hands when the rubber-coated cotton (Red Ball) gloves were worn inside.

On the cold end of the thermal spectrum, turnouts provide protection to about –20° to –30°F. Below that range, not much keeps anyone warm. If liquids are involved, temperatures in the 30° to 40°F range can prove fatal in minutes. Further, parts of this protective ensemble are particularly susceptible to cold. Boots, inhalation tubes, and other equipment constructed of elastic polymers (elastomers) will become brittle in temperatures between –30° and –100°F, depending on the specific material involved. As a result, they will crack, split, and fail.

Now consider liquid oxygen (LOX) at a temperature of –297°F and liquid nitrogen (LIN) at a temperature of –320°F. Not only is the liquid that cold, the vapor, especially close to the liquid, is close to that temperature as well. At these temperatures, as well as those considerably higher, boots will shatter and inhalation tubes will break.

As such, turnouts provide extremely limited protection and even become dangerous when any piece or the ensemble as a whole is asked to perform outside its operational range. Further, turnouts and any other thermal protection will provide only limited protection to splashes of cryogenic liquid. Immersion will result in severe injury and possible death. To bear this out, ask representatives of a cryogenic liquid manufacturer what type of PPE they use for liquid spills. The answer they will give, if they are honest, is none because there is none that is effective.

Proximity Suits.   Proximity suits, found in NFPA 1976, are designed to provide protection from high-level radiant heat. These suits are similar to structural turnouts in basic design but have several very unique features. First, they are equipped with a hooded helmet to protect the face and head from the radiant heat. Second, the SCBA is located within the coat so it is also protected from the heat. Third, and probably most strikingly, the outer shell of the hood, coat, pants, and boot covers each have an aluminum surface layer.

The aluminized surface is the key protective factor for the entire ensemble. Specifically, the aluminum reflects as opposed to absorbs the radiant heat generated by the fire. This allows the wearer to approach more closely and withstand greater intensity.

Proximity suits do not do much better than turnouts when it comes to direct flame contact or exposure to extreme cold. Although their liners are somewhat thicker, they only increase the duration of protection by seconds.

Entry Suits.   Entry suits are designed to protect the wearer from direct flame contacts. They are extremely thick (3/4 to 1 inch) multiple piece suits and are equipped with a hood, coat, pants, boots, and mittens. The mittens are needed because gloves would allow excessive heat to reach the hand. The face piece in the hood is covered with a gold mist layer to reflect the heat that would otherwise sear the face (the same is true for proximity suits).

Entry suits can be exposed to direct flame contact for a limited duration (some manufacturer's literature indicates 30 to 60 seconds over the lifetime of the suit). Their primary use is in fixed facilities for shutting down valves, tightening flanges, and other actions in the event of fire. They are not rescue tools because if the situation is so extreme that they are needed for the rescue, the likelihood for success of the rescue is nil. They can also be helpful in situations involving process or transportation heating.

Flash Protection for Chemical Protective Clothing.   This is one of those murky and turbulent areas where technologies and hazards have not quite kept pace with each other. Further, this is one example of an attempt at a hybrid form of PPE (i.e., thermal and chemical). There are many grave questions about the attempt as well as its value. Although flash protection is available and required by NFPA 1991 for chemical protective clothing, many believe it is the type of equipment that merely instills dangerous overconfidence in the ill-informed user. As a result, this topic is one of the most controversial in the realm of thermal protection.

At the heart of this problem is a protection incompatibility that exists between chemical and thermal hazards and chemical and thermal protective equipment. Each hazard and its corresponding type of PPE provide little if any protection against the other. Specifically, thermal protective equipment provides limited protection against anything but minor chemical exposure. Furthermore, chemical protective clothing provides no protection against thermal hazards. With the existing design feature, the old comic analogy of "overgrown ziplocks" or "shrink wrap bags" holds some validity when it comes to their ability to provide thermal protection.

Now enter the flash protection outer envelope or the newer integral design, and we are on our way to controversy. The concept sounds intriguing until the details are examined.

This flash protection envelope or layer is constructed of a fire retardant fabric such as PBI or Nomex with an outer aluminized covering. The thickness of the fabric is similar to a turnout coat's fire retardant outer shell. Based on manufacturer's information, flash protection is designed for just that, a one to three second exposure to radiant heat from a vapor flash. It is not designed to protect against, and will not provide protection against, direct flame contact. Direct flame contact will produce wearer injury similar to that anticipated by a firefighter wearing only the outer shell of the turnouts.

Table 7-7   Levels of Chemical Protection

PPE	Body protection	Respiratory protection
Level A	Maximum—vapor and gas tight, totally encapsulating	Maximum—SCBA or in-line
Level B	Moderate to low—splash protection, not vapor or gas tight	Maximum—SCBA or in-line
Level C	Low—minimal splash	Minimum—APR
Level D	Ordinary work place	None

In addition, several field tests have shown varying results. Some of these results have been disturbing at best. It has been purported that the only reason flash protection is part of the standard is so some manufacturers can meet the abrasion test requirements. There are still other purports that NFPA response to such questions was that they are not responsible for the lack of technical expertise of the end user of their standards.

The bottom line is that if responders follow the 20% to 25% LEL withdrawal recommendation, most of the controversy becomes moot simply because personnel are not being placed in harm's way. In the event ignition occurs, the flash protection will not be exposed to direct flame contact so it can function as designed. Further, plain chemical PPE should receive minimal damage as well.

Chemical Hazards

Chemical protective equipment consists of two primary components similar to thermal protection. These components are body (or skin) protection and respiratory protection. To accommodate a wide variety of potential hazard levels, there are four levels of protection identified. In descending order of the protection they provide (highest to lowest), these four levels are designated A, B, C, and D (Table 7-7).

Although EPA and most other sources list these four levels of PPE, it is essential to understand that, in essence, Level D provides no protection against health or safety hazards beyond the most rudimentary hazards found at a normal work site. OSHA describes Level D protection as, "A work uniform affording minimal protection, used for nuisance contamination only." In other words, this PPE can only be worn when contaminant levels are at or below the PEL or TLV concentrations. As such, many agree that this level is really not a true level of protection, rather simply a work uniform as OSHA states.

Level A.   As specified in the table above, Level A PPE is used when maximum body protection and maximum respiratory protection is needed due to the chemical hazards of the substance involved. Level A PPE provides the highest degree of chemical protection available for the eyes, skin and respiratory system.

Level A is to be used in situations where there is little or no thermal hazard and a high degree of chemical hazards. Such conditions include the following:

1. There is little or no fire hazard associated with the substance or incident situation
2. There is little or no potential for the substance, its surrounding atmosphere, or equipment to sufficiently cool the suit's protective membrane to cause embrittlement and suit failure
3. A contaminant is skin absorptive, and a skin exposure possesses a high probability of impairing escape, immediate serious illness or injury, or immediate death
4. There is the potential for high atmospheric concentrations of gases, vapors, or particulates that are tissue destructive or highly toxic through skin absorption
5. The site operation and work functions involve the potential for splash, immersion, or exposure to unexpected vapors, gases, or particulates that are tissue destructive or highly toxic through skin absorption
6. The known or suspected presence of substances possessing a high degree of hazard to the skin and contact is probable
7. Operations are conducted in poorly ventilated confined locations, or in confined spaces where the need for Level A equipment has not yet been ruled out
8. Substances with high toxicity through inhalation, skin absorption, or contact are present
9. Atmospheric oxygen concentrations are below 19.5%

OSHA specifies in 1910.120 that Level A PPE is, at a minimum, composed of the following equipment:

1. Maximum respiratory protective equipment provided by SCBA or in-line air system; SCBA must be worn inside the protective suit
2. Maximum body protection provided by a total encapsulating chemical protective suit (TECP), which is a suit covering the entire body and the respirator
3. A minimum of two layers of chemical protective gloves, including an inner chemical resistant glove and an outer chemical resistant glove
4. Chemical-resistant boots equipped with steel toes and shanks

There are various options also identified by OSHA as follows:

1. Coveralls, long underwear, and hard hat to be worn under the TECP
2. An additional outer glove
3. Disposable outer boot covers
4. Disposable outer suit

One thing that OSHA does not list and really should is the use of in-suit, two-way radio communications. For the sake of safety and efficiency, such capability is essential to communicate such things as information, conditions, unexpected situations, status reports, or briefings for back-up or next entry teams.

Level B.   Level B PPE is used when a lesser level of body protection yet the maximum respiratory protection is needed due to the chemical hazards of the substance involved. Level B PPE provides a lower degree of chemical protection for the skin but the highest level of protection for the eyes and respiratory system. Level B is designed for use in situations where there is a determined degree of chemical hazards and little or no thermal hazard, as follows:

1. There is little or no fire hazard associated with the substance or incident situation
2. There is little or no potential for the substance, its surrounding atmosphere, or equipment to sufficiently cool the suit's protective membrane to cause embrittlement and suit failure
3. The type and atmospheric concentrations of contaminants is known and requires a high degree of respiratory protection, but a lesser degree of skin protection
4. Atmospheric oxygen concentrations may be less than 19.5%
5. The presence of incompletely identified contaminants is indicated by direct reading organic vapor instrumentation, but are not suspected of containing high levels of chemicals harmful to, or absorbed through, the skin

OSHA specifies in 1910.120 that Level B PPE is, at a minimum, composed of the following equipment:

1. Maximum respiratory protective equipment provided by SCBA or an in-line air system; SCBA worn inside or outside the protective suit
2. A lower level body protection provided by a semi-encapsulating chemical protective suit (i.e., a suit with hood and integral face piece and non-integral gloves) or hooded chemical resistant clothing (i.e., one piece hooded coveralls, one or two piece chemical splash suits, or chemical resistant overalls)
3. A minimum of two layers of chemical protective gloves, an inner chemical resistant glove, and an outer chemical-resistant glove
4. Chemical resistant boots equipped with steel toes and shanks

There are various options identified by OSHA and they include:

1. Coveralls or long underwear to be worn under the chemical-resistant clothing
2. Hard hat
3. Disposable, outer boot covers

Again, OSHA does not specify in-suit communications. However, it is highly recommended in this level as well.

Level C.   Level C PPE is used when a lesser level of body protection and only a low level of respiratory protection is needed due to the chemical hazards of the substance involved. Level C PPE is used only when the substance or substances are known, their airborne concentration is also known, and all criteria for the safe use of an APR are met.

OSHA specifies in 1910.120 that Level C PPE is, at a minimum, composed of the following equipment:

1. Full or half face APR with appropriate cartridge installed
2. Body protection provided by hooded chemical-resistant clothing (i.e., one piece hooded coveralls, one or two piece chemical splash suits, or chemical resistant overalls)
3. A minimum of chemical protective gloves; an inner chemical resistant and outer chemical resistant
4. Chemical-resistant boots equipped with steel toes and shanks

Optional equipment includes:

1. Coveralls or long underwear to be worn under the chemical resistant clothing
2. Hard hat
3. Disposable, outer boot covers
4. Additional full face shield
5. In-suit, two-way radio communications

Inspections and Testing.   In the three levels of chemical protective clothing just listed, the chemical suit, like the SCBA or APR, is an integral part of the entire protective system. Both the respirators (SCBA or APR) and the chemical suit require special care and consideration due to their critical role in protecting the wearer. As a result, this section examines some of the basic care issues concerning both.

A big factor in the exact care, handling, inspection and testing of PPE is whether the equipment is disposable or reusable. Both types of equipment require an initial inspection when they are received from the vendor and before each use.

As with any reusable respirator, an individual file must be established for each SCBA. The SCBA must reflect all maintenance, tests, chemical exposures, and so on for that unit.

For suits, gloves, boots, and other equipment, the initial inspection is to ensure they meet the specification desired in the order. Further, they must be inspected for any visible signs of damage or other defects, as well as the presence of any options requested. Suits and gloves must be checked for penetration failures.

Penetration failures are openings within the equipment as a result of manufacturing (e.g., sewing needle holes, zipper closures, minor cuts, and pinholes). Suits and gloves can be held up to the light or a light can be placed within for a visual inspection. Further, gloves are twisted shut, submerged in water and squeezed as a further check.

Reusable suits require the establishment of an individual file to document the history of its use and any testing that is required. Most commonly, totally encapsulating chemical protective suits (TECP) comprise reusable suits (although some organizations also have reusable splash suits as well). In order to establish a file, each suit must be assigned a unique number that is recorded in the file. Often, the manufacturer's serial number is used for this purpose. Further, the file must contain the brand name, date of purchase, and material of construction, and it must identify any unique fit features.

Totally encapsulating (TE) suits must meet the requirements established in 29 CFR 1910.120 (g)(4)(ii) and (iii). These sections indicate that the employer is responsible for providing TE suits that can maintain a positive pressure and also prevent inward leakage of a test gas. See Appendix A of 1910.120 for testing options.

There are two test methods listed. The first method is identified as a qualitative test and is the most widely used. The second method is identified as a quantitative test and is rarely used. The primary reason is that this method requires the use of a live chemical, i.e., ammonia from ammonium hydroxide, at a level over twice the IDLH. (This is definitely not one of the most intelligent test methods dreamed up.) All test results must be documented in the suit's file.

After inspection and testing, the suits have some type of indication that the inspection or test was performed. For suits that come in plastic bags, one method is to apply a tape that must be broken or leaves some indication it has been opened. This way, if the tape is broken, the suit is not used until it has been reinspected or retested.

After Use.   After each use, reusable TE suits must be thoroughly decontaminated before they can be reinspected and then retested. This retest is necessary to assure that requirements of (g)(4) are met. Further, the file must be updated to reflect the substance(s) to which the suit was subjected. In comparison, disposable or limited usage suits are routinely thrown out after each use. As such, no file must be generated or updated.

Failure Mechanisms for Suits.   Chemical protective suits can fail in three ways: penetration (already discussed), degradation, and permeation.

Degradation is the chemical or physical breakdown of the suit that will allow the movement of the contaminant through the protective membrane. Physical degradation can take the form of punctures, tears, cuts, abrasions, and so on. Chemical degradation involves a chemical incompatibility between the contaminant and suit material. The reactions can range from chemical degradation to dissolution (dissolving). Five primary signs indicate chemical degradation of suit material:

1. Color changes
2. Cracking and flaking
3. Swelling and thickening
4. Loss of flexibility
5. Stickiness

Permeation (Figure 7-18) is a physical process where a contaminant moves through the suit's protective membrane at the molecular level. An example of this process is the placement of a cut onion in a tightly closed plastic bag and then placing the bag in a refrigerator. The next morning when the refrigerator is opened, one smells onion even though the bag is still tightly closed. Permeation of the chemical responsible for the onion's odor is the reason.

To understand this process, it is important to understand two other physical processes. They are diffusion and osmosis. Diffusion occurs when a substance moves from an area of greater concentration to an area of lesser concentration. The suit material is a membrane that allows only certain substances through. As such, it is known as selectively permeable. Under these conditions, permeation is an osmotic process (i.e., related to osmosis). Osmosis is the movement of a substance from an area of greater concentration (the outside of the suit) to an area of lesser concentration (the inside of the suit) through a selectively permeable membrane (the suit itself).

There are two terms used to quantify permeation after a sample of the suit material undergoes testing prescribed in NFPA 1991, 1992, or 1993 (depending on the specific type of suit). The first is breakthrough time, which is the amount of time required for a contaminant to be detected on the inside of the suit. The second is permeation rate, which is the quantity (in milligrams per square meter per minute or micrograms per square centimeter per minute) of contaminant that passes through the membrane once breakthrough occurs.

Obviously, permeation has the potential to cause contamination to the wearer of a suit when the contaminant passes through the membrane. However, this is only part of the story. There are two other major concerns associated with permeation that involve multiple substance exposures and their affects.

The first multiple substance concern involves exposing the membrane to two or more different contaminants at the same time. Routinely, the testing data provided by the manufacturers as well as required by the NFPA certification process involves only single chemical testing.

However, actual emergencies often involve the release and possible mixing of multiple contaminants. The single substance data generally provide no insight into how

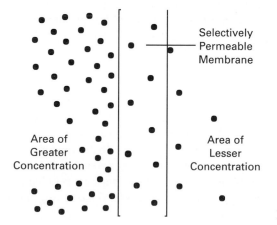

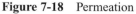

Figure 7-18   Permeation

the membrane behaves in contact with mixtures. Some very disturbing data were generated a few years ago when one manufacturer provided information on multiple versus single substance breakthrough times. In a single substance test they found a breakthrough time of 140 minutes. In a different single substance test they found a breakthrough time of 38 minutes. Next they took a 50/50 mixture of both substances and found a breakthrough time of only 18 minutes.

This accelerated breakthrough time is an example of synergism, which is a multiplicative as opposed to an arithmetic response (the substances enhance each other's effect). Furthermore, if synergism occurred in this case, it could easily occur in others and that would have major implications on the safety of the suit's wearer.

In response to this concern, manufacturers indicate that there are innumerable potential mixtures that could be encountered during an emergency. As a result, meaningful testing is impossible. However, the same could be said for individual substance testing, yet a test procedure was developed and is part of the NFPA certification process. A solution to this issue could be accomplished quite simply by requiring round-robin testing with the substances specified in NFPA 1991, 1992, and 1993. At least that information would assist responders in making informed decisions.

The second multiple substance contact concern involves exposure of the membrane to a different contaminant after some degree of permeation has occurred. This concern primarily involves reusable suits since disposables are not subjected to multiple uses. Specifically, every time the suit membrane is exposed to a contaminant, a certain amount of permeation may take place. In some instances, the contaminant may only contact the surface and only superficially enter small intermolecular gaps. In other instances, it will enter and start to migrate through the membrane.

If the substance permeates into the membrane more than superficially, it actually enters the body of the suit material. When it enters the body of the membrane, the membrane has become impregnated with the contaminant. The question is, "What will happen to the impregnated contaminant?"

The answer depends on the specific contaminant involved. The two options are that it will either eventually permeate out of the membrane or remain impregnated. Some data from various manufacturers indicate that minute quantities of various contaminants remain impregnated within the membrane. How long they remain is not clear.

The next question is, "What effect will these impregnated contaminants have on the breakthrough time and permeation rates of other contaminants during subsequent usage of the suit?" The answer is not clear. The only assured way to determine an answer would be to follow the testing methods found in NFPA using their battery of test chemicals for the specified time. After this test run is completed, the samples would be held for varying amounts of time, such as 48 hours, 96 hours, and then 2 weeks. After this time, these same samples are then exposed to the other substances on the list to determine if there is any change in their permeation characteristics. The results of such round-robin testing would be invaluable to answering this very important question.

Failures of PPE

There are three primary reasons for PPE to fail the user. The first reason is because the PPE is not properly maintained, stored, or otherwise handled. The second reason is because the user is not properly trained and knowledgeable in its use and limitations. These two are simple enough to address. They require some knowledge and effort by the user. According to OSHA, 29 CFR Subpart I (starting with 1910.132) requires that employees be properly trained in the safe use, storage, inspection, and limitations of all PPE.

The final reason, and the most probable to result in injury or death, is because the PPE is asked to protect the wearer from a hazard or a magnitude of hazard for which it was not designed. In this case, when the PPE fails, there is no protection against the hazard for the wearer. Unfortunately, this is often the most difficult reason to address, especially when it involves emergency response.

The reason is the users rely on linear thinking. An example of this problem is the thermal protection/chemical protection incompatibility. Consider an incident where a skin absorptive mixture of flammable and combustible liquids is spilled on the loading dock of a fixed facility. The temperature is 78°F, the humidity is 55%, and the winds are calm. It is found that the mixture contains toluene, xylene, methyl ethyl ketone (MEK), several acetates, and acetone. All of these are listed as skin absorptive toxins. What is appropriate PPE to handle an incident involving this mixture?

Routinely, the answer to that question is Level B. That answer is wrong. In fact, that answer is potentially *lethal!* That answer can lead to fatal burns for both members of an entry team during an actual incident!

The substance described is lacquer thinner, which is a flammable liquid. Yes, it is skin absorptive, but it is only slightly less toxic than gasoline by this route. In this situation, the thermal hazard is the primary and over-riding hazard—flammability. As such, only the proper PPE will provide thermal protection, that is full turnouts including SCBA, not Level B. Asking personnel to wear Level B in this situation is like asking them to wear Level B for a similar gasoline spill.

Another fundamental question is, "Does the use of turnouts eliminate the hazards to personnel?" The answer is not by themselves. If personnel walk through this spill and it ignites, they will be injured and possibly killed.

So what is the answer? The answer is tactical operations that address the hazards. One of the primary tactical approaches when handling flammable liquids is suppression of the vapor by using foam, adsorbents or some other type of blanketing. Before, during and after the application of vapor suppressant, air monitoring is used to identify the exact hazard level based on % LEL or other applicable readings.

This is followed by diking to minimize product spread. If a foam or some other type of blanket is established, the flammability hazard is greatly diminished. (It is not eliminated when foam is used because the foam can break down or otherwise lose its effectiveness.) Now turnouts provide much more assured protection, again, unless personnel tramp through the blanket and spilled product.

There are two general rules to follow. **Rule #1:** Never walk through flammable or combustible liquid spills. **Rule #2:** Never disturb the foam blanket unless there is some other vapor-suppressing tactic ongoing in the immediate area of the disturbance. Such vapor-suppressing tactics include continual application of foam or the use of an adsorbent (see Spill Control—Liquid Surface later in this chapter for additional information).

Suppose the liquid is tissue destructive or highly toxic through skin absorption, as well as flammable. Again, suppress the vapor and the flammability hazard is also suppressed. Now it may be possible to wear a disposable Level B suit over a set of turnouts (obviously with SCBA) if personnel must approach the spill. Again, air monitoring is essential to identify the magnitude of the hazard once the vapors are suppressed.

Suppose it is a tissue destructive gas that is flammable. Again, suppressing the flammability hazard is the priority. If the release is within a building, one must ventilate the area. If the product is water soluble, consider using water fog to dissolve the gas. If neither are possible, it may be necessary to go fully defensive. To have personnel enter a flammable atmosphere is an imminent danger situation and must be stopped by the safety officer. The bottom line is there are times when nothing will be effective.

A final generalization on this topic of thermal-chemical incompatibility; Most commonly, there is a greater ability to suppress the flammability hazard than the chemical hazard. If the flammability hazard cannot be suppressed, the situation is a good candidate for fully defensive operations.

Once appropriate PPE is identified, personnel can enter the hot zone. However, before they can return to the cold zone after the entry, they must go through decontamination.

PROTECTION—DECONTAMINATION

Decontamination is one of the more talked about yet least well performed or understood tactical objectives for protection. One of the reasons is that no one can provide definitive answers or definitions for the process. One thing everyone can agree on is that decontamination is desirable and necessary. This is quite simply because contaminants can cause negative impacts on living organisms and the environment.

Further, contamination occurs through two mechanisms. The first is the direct contamination that occurs when product escapes its container. This released product is then able to contact and thus contaminate people, objects, and the environment. The second is cross contamination that occurs when someone or something contacts an object or individual that received direct contamination. In either case, if decontamination does not occur, the contaminant is free to produce the most negative impacts possible.

The entire focus of hazmat emergency response is the prevention of additional contaminant spread through the environment. Decontamination, in particular, is a critical and final step in accomplishing this goal. Specifically, before the entry team, other people or equipment leave the hot zone they must be decontaminated to prevent additional spread.

OSHA defines decontamination as the removal of hazardous substance from employees and their equipment to the extent necessary to preclude foreseeable health effects. Exactly what does this definition say? How clean must it be to preclude foreseeable health effects?

Further, NFPA defines decontamination as the physical and/or chemical process of reducing or preventing the spread of contamination from persons and equipment. Again, what does this really say? How much of a reduction is acceptable? Who makes that judgment?

In reality, it is better to describe decontamination rather than define it. From this perspective, decontamination is a process that involves chemical and physical actions used to prevent the spread of contaminants beyond the warm zone in low enough concentrations so they cannot produce negative impacts on people, other living organisms or the environment.

Gross and Secondary Decontamination

There are two phases of decontamination. Gross decontamination is the process of rapidly removing or altering the majority of the contaminant. Inherent to this process is residual contaminant remaining on the object. Secondary decontamination is the subsequent process of removing as much residual contaminant remaining after gross decontamination as necessary to protect people and the environment.

There are some differences between protected responder and victim decon. For example, both gross and secondary decontamination of contaminated victims occur within the decontamination corridor. Secondary decon for protected responders may occur just outside the decon corridor. In either case, decontamination must be performed by personnel who have received appropriate training and are wearing appropriate PPE.

For a better understanding, it is helpful to examine the decontamination of protected responders and contaminated victims separately.

Protected Responder Decon Phases

In the case of appropriately protected response personnel, gross decon involves removing enough contaminant from the PPE to allow the wearer to safely exit the equipment. It does not involve completely cleaning the PPE. Consequently, gross decon is *always* considered to leave residual contaminant.

Once the wearer exits the PPE, they proceed to a secondary decon location. This personal form of secondary decon normally involves, as a minimum, the washing of one's hands and then face. However, if conditions warrant, this secondary decon may include the total removal of clothing and a full body wash.

However, the decon process is still not complete. Any nondisposable or reusable equipment must now also undergo secondary decon. Contaminated suits should be overpacked for later evaluation and secondary decontamination. (Some agencies do not perform decon on their suits but rather rely on a contractor). Equipment such as SCBAs worn underneath the suit should be examined to determine the need for any additional decon (normally none is required) beyond routine sanitation and cleaning. One option to identify the need for further decon is sealing equipment in a large, clean plastic bag or drum. After some time has passed, the head space within can be checked with monitoring equipment to determine the presence of possible contaminant.

Contaminated Victim Decon Phases

Contaminated victims (including improperly protected responders) also normally undergo at least gross and secondary decon. Gross decon of victims involves the removal of all contaminated or potentially contaminated clothing, sometimes requiring the removal of all clothes. The removal of clothing may or may not be accompanied by the application of water. Obviously, the victim's modesty must be considered in light of the severity and extent of contamination and injury.

Furthermore, an additional decon corridor must be established to manage contaminated victims. The reason is simple, the ability to decontaminate responders must always exist once they have entered the hot zone. Should a problem arise, decon personnel must be able to decon victims and responders at the same time. This is especially true for nonambulatory victims. Since these victims cannot assist in the decontamination process, it becomes labor and time intensive. If all decon capabilities are committed to the victim's predicament, who will decon the entry team? The answer is the second decon corridor.

Secondary decontamination of victims then involves flushing the exposed skin with water for a minimum of 15 minutes. This process also occurs within the decon corridor but more toward the cold zone than gross decon. In the case of acids and especially bases, flushing should continue as long as practical or until the pain has stopped. This flow should be of an appropriate volume to adequately flush the contaminated area thoroughly. There is an additional step to secondary decon, sometimes called medical decon, that involves decontaminating orifices such as the mouth, nose, and ears. Further, it may include flushing the eyes or irrigating a wound.

In some instances, victims will require tertiary decontamination. Tertiary decon occurs at a medical facility. It involves the indepth cleaning and debridement of wounds and possibly surgical intervention. In one case, a worker was contaminated by an organophosphate pesticide. It was later learned that the contaminant permeated and impregnated his fingernails and toenails thus acting as time release patches. This situation necessitated the surgical removal of all ten of the victim's fingernails and toenails.

Decontamination Mechanisms

NFPA lists eight mechanisms available for decontamination:

1. Absorption
2. Adsorption
3. Chemical degradation
4. Dilution

5. Disposal
6. Isolation
7. Neutralization
8. Solidification

However, these mechanisms require closer examination. Four of them are realistically only applicable to environmental decontamination (also known as clean-up or remediation). These include absorption, adsorption, isolation, and solidification. Chemical degradation as defined by NFPA is really two separate mechanisms; one is chemical while the other is strictly physical and has nothing to do with degradation. Further, the list completely ignores disinfection (an increasingly important mechanism and need).

The remainder of this discussion will focus on the decontamination of properly protected emergency responders and contaminated (or potentially contaminated) victims. When considered from this context, there are seven mechanisms. These seven human decontamination mechanisms in descending order of importance include:

1. Emulsification
2. Chemical reaction
3. Disinfection
4. Dilution
5. Disposal
6. Removal
7. Absorption

It is important to note that actual decontamination procedures routinely incorporate multiple mechanisms to be most effective. Further, determination of the exact methods and mechanisms varies depending on the contaminant and its properties. At this point, it is helpful to discuss the mechanisms in more detail.

Emulsification.   Emulsification is one of the most widely used decontamination mechanisms. It involves the use of an emulsifier (e.g., surfactant, soap, or detergent) to produce a physical suspension of two or more immiscible (insoluble) materials.

Most commonly, an emulsifier such as liquid laundry detergent is used to physically break nonpolar liquids, highly viscous liquids, or insoluble solids into small packets (droplets) that become surrounded by a foam bubble. When emulsification is complete, the application of additional solution or water then rinses the emulsified contaminant from the surface. As such, this incorporates both dilution and physical removal.

One of the most convenient and effective ways to mix and apply an emulsification solution is through the use of a 2½ gallon garden pump sprayer. Water is placed into the sprayer and liquid laundry detergent or similar liquid detergent is added and mixed through. Liquid is far preferable to powders because the liquid mixes more easily with the water.

For many years, EPA's emulsifier solution incorporated trisodium phosphate (TSP). Today this solution is frowned upon by most responders due to its tendency to produce a solution with a pH close to 12. Not only is this harsh on any inadvertently exposed skin, but it has a tendency to attack and break down many elastomers used in PPE. Liquid laundry detergents normally have a pH of 8 or 9 (also an advantage when neutralization is needed).

Finally, emulsification is appropriate when the contaminant is a viscous, miscible liquid. The syrupy nature of some viscous miscible contaminants makes them extremely difficult to remove (think of trying to decon someone contaminated with corn syrup or honey). The addition of an emulsifier is very effective in this type of situation.

Chemical Reaction.   Chemical reaction is a mechanism that involves reacting a contaminant with a substance or solution. This chemical reaction causes a decomposition process that reduces the contaminant's overall hazard. These chemical reactions would include the NFPA's neutralization and part of its chemical degradation.

In any case, *the chemical reaction mechanism must NEVER be used on contaminated skin, but only on PPE or other objects.* The reason for this sweeping statement is simple. Both neutralization and degradation are chemical reactions that release energy, most commonly in the form of heat. In many instances the reaction is "vigorous." Many of the chemicals used, especially in degradation, are quite hazardous all by themselves. As such, the reactions and even some of the agents are quite capable of producing additional and potentially severe injury.

Chemical degradation agents cause the contaminant to decompose. These agents are contaminant specific and include, but are not limited to, the following types of substances:

1. Hydrocarbon liquids
2. Halogenated hydrocarbons
3. Ketones, aldehydes, alcohols, and other derivatives
4. Strong bases
5. Strong acids

Furthermore, many of these agents and their degradation reactions are so strong that they routinely lead to failure of all but the most resistant materials. Their use on skin and in some cases PPE being worn by personnel is out of the question.

Neutralizing agents react with either the excessive H^+ ions or OH^- ions and tend to raise the pH of acids or decrease the pH of bases. Unfortunately, such reactions almost inevitably cause wild pH swings from one end of the scale to the other. For example, soda ash, one of the more common acid neutralizers, invariably will swing the pH from 0 or 1 to between 10 and 12 almost instantaneously. Such swings must be considered and monitored as needed.

Neutralization does nothing for the other hazards of the substance. For example, many acids, especially spent (i.e., used) acids, are contaminated with substances including metals, inorganic salts, and organics (hydrocarbons or hydrocarbon derivatives). Neutralization will not reduce the hazards of such secondary contaminants. Additionally, neutralization has the nasty habit of causing many of these secondary contaminants to precipitate (forming a semi-solid substance similar to soap scum) from the solution. These precipitates stick tenaciously to any surface they contact.

Pump sprayers are excellent mixing and application tools for neutralizing agents as they are for other solutions. The specific agent used depends on whether the contaminant to be neutralized is an acid or a base.

Acid neutralization routinely involves the use of a weak base or buffer: The most commonly used weak base is soda ash, also called sodium carbonate (Na_2CO_3). When mixed with water, sodium carbonate partially dissolves and produces a solution with a pH of 11 to 12. The most commonly used buffer is baking soda called sodium *bi*-carbonate ($NaHCO_3$). When mixed with water, sodium bicarbonate (*bi-* signifies the presence of the H) again partially dissolves. However, it produces a lower pH more in the range of 8 to 9. There is an advantage to this fact. Specifically, the H^+ present in the HCO_3 radical prevents the violent pH swings associated with the straight carbonate. As the pH decreases, the H^+ reacts with the OH^- to produce water (H_2O), thus neutralizing the accumulating base. The ability to react with both acids and bases, thus preventing wild pH swings, is why sodium bicarbonate is known as a buffer.

It is also important to remember that various detergents and soaps are also basic. As such, they also will act as neutralizers to some degree. Unfortunately, the degree of neutralization varies from detergent to detergent.

On the other hand, base neutralization routinely involves the use of weak acids. The most common weak acid used is acetic acid in the form of vinegar (which is also the cheapest available). Most commonly, the vinegar is placed in the sprayer and applied at full strength. It can be diluted with water if so desired. Another option includes critic acid. Unfortunately, it is not very common to find this weak acid in large quantities.

Disinfection.   The decontamination of potential or actual biological hazards is taking on ever increasing importance as more and more pathogenic organisms are routinely encountered. Specifically, the increase is not due to greater handling or transportation of etiologic agents (biohazards). Rather it is due to the increasing frequency and number of incident involving bloodborne pathogens.

Every major airline crash in the past several years has become a hazmat operation involving at least biohazard decontamination in the early stages of the incident. Likewise, bloodborne pathogen control and its associated hazmat involvement were critical factors in the Oklahoma City bombing and many other mass casualty incidents. In fact, in the Commonwealth of Pennsylvania, the certified hazmat teams have been designated second only to the State Department of Health for the control biohazards at incidents involving mass fatalities. As a result of this ever increasing public health issue, hazmat responders must be able to perform decontamination involving disinfection.

The recommended agent for disinfection of equipment is still liquid laundry bleach [sodium hypochlorite solution ($NaClO$)]. In some instances, especially those involving direct skin contact or turnouts, bleach should be replaced with disinfectant soaps such as Lysol.

The liquid bleach can be used straight out of the bottle or diluted. The undiluted bleach should be used for direct blood or body fluid contact. This may occur during victim or body-recovery operations. After the initial application of bleach, it is recommended that the solution be allowed to remain on the equipment (i.e., let stand) for about 10 minutes.

Following this initial waiting period, a second application should be made and allowed to stand for another 10 minutes. After that time, thoroughly rinse off the solution. It should be noted that this treatment, when done repetitively, may lead to equipment damage. Additionally, this treatment may also alter the flame retardant characteristics of turnouts, so consult the manufacturer and consider disinfectant soap.

For less virulent situations involving medical wastes, use a mixture of 1 cup bleach for every gallon of water. The same 20-minute total treatment should occur. In either case, the pump sprayers are ideal mixing containers and applicators.

For example, in the USAir Flight 427 crash outside Pittsburgh, a pit was excavated and lined with plastic. As pieces of the plane, its cargo, and other wreckage were removed, they were sprayed with hypochlorite solution as a precaution. This process continued for between 5 to 6 days.

Dilution.   Dilution is the process of reducing the concentration of the contaminant (solute) through the application of some other substance (solvent), most commonly a liquid. In emergency response, the most common solvent is water and the solute is a miscible liquid or solid. It is also routinely used after immiscible liquids or solids have been emulsified.

Disposal.   One of the increasingly common decontamination mechanism is disposal. Disposal is the process of decontaminating PPE, clothing, equipment, and so on just enough to allow its safe removal from the wearer. Once removed, it is then disposed of along with the rest of the wastes on the incident scene. Such disposal often includes chemical protective clothing, gloves, boots, decon containments, plastic sheeting, decon solutions, and victim's clothing.

One of the factors leading to the popularity of this mechanism is the use of relatively low cost disposable and limited use suits. This approach kills two birds with one stone. First, it provides enhanced safety for team personnel because the suit is used only once and then thrown away. Second, it eliminates the logistical problems of individual suit files as well as after-use inspections and testing, and its associated documentation. Obviously, disposal is routinely used in conjunction with other mechanisms.

Removal.   Removal is the process of physically removing the contaminant from the contaminated surface. It always involves the use of pressure or a vacuum. In most instances, the safest and most effective means to accomplish physical removal involve the flow of water. This is especially true for solid contaminants. When the stream from a pump sprayer, spray wand, or other device is applied to the contaminated surface, the force of the water acts to physically remove the contaminant. Routinely, physical removal also involves dilution when miscible contaminants are involved.

Another type of removal commonly mentioned is the use of a brush or similar tool to push the contaminant from the surface. Should such an option be required, exercise extreme caution because it will almost always cause extensive cross contamination. The reason is simply that the brushing action causes pieces of the contaminant to be propelled substantial distances in *all* directions. Furthermore, if the contaminant is in the particulate form, this process can spread tremendous amounts of contaminant over relatively large areas with little or no control over where they will deposit themselves.

Although the following statements contradict many accepted protocols, their merit must be closely considered. This airborne spread of contaminant is *especially dangerous* if the removal is being performed on a contaminated victim or improperly protected responder. These individuals generally have on little, if any, appropriate PPE. The victims and often the responders have on no face, eye, or respiratory protection. As a result, these airborne contaminants can easily and rapidly produce contamination of the mucous membranes as well as internal (respiratory and digestive) cross contamination.

In such situations, it is far superior to use an adhesive roller (such as a lint roller) or a vacuum system such as a certified hazardous substance vacuum cleaner. Both options provide superior contaminant control and cross contamination prevention. They also work for substances that are explosive in contact with water.

Absorption.   Absorption is not commonly used for decontaminating personnel or victims. However, it does have its place. The placement of absorbent pads in the decon containments can minimize the cross contamination of boots due to the presence of immiscible substances floating on the surface of the decon solution.

Decontamination Involving Water Reactive Substances

Decontamination involving water reactive substances is a highly controversial subject that often involves contradictory recommendations. However, a logical approach helps provide sound answers. First and foremost, it is essential to identify the type of water reactivity the contaminant exhibits. The three primary types of this reactivity are:

1. Generation of heat
2. Generation of toxic or flammable gases
3. Ignition or explosion

Contaminants that generate heat are the most common types of water reactive contaminants encountered. The heat is often released by the dissolution or ionization process. Substances such as acids, acid anyhdrides, hydroxides, metal and nonmetal oxides, and hydrides undergo this type of heating. In addition, when they dissolve or ionize, they produce corrosive substances that attack and destroy tissue.

When they contaminate chemical protective clothing, their hazard is relatively minimal. If physical removal, involving even the minimal but continuous flow of a pump sprayer, is employed, there is sufficient heat absorbing ability of the water to minimize the potential for injury or damage. As such, successful decontamination is achieved with little or no adverse effect.

However, when these substances contaminate exposed skin, there is another scenario. First, all skin has a small degree of water on its surface. When these substances contact the skin, they start to dissolve or ionize, creating a rather viscous, corrosive liquid. This liquid, in turn, begins to break down the skin cells and results in the release of interstitial fluid that is primarily composed of water. The additional water becomes available to the contaminant so dissolution or ionization reactions continue.

The continuation of the reaction results in the generation of even greater amounts of heat, further resulting in greater tissue damage and the subsequent release of more fluids. Consequently, the victim receives progressive chemical and thermal burns from the exposure. Thus continues this destructive chain of events.

The conventional protocol states that any solid contaminant within this group should first be brushed from the surface of the skin followed by flushing with copious amounts of water (most EMTs and medics interpret this to mean two 1-liter bags of normal saline instead of one). However, this process is usually counter productive. The brushing process fragments the solid contaminant and spreads a sticky viscous mixture of the contaminant and its resultant corrosive liquid over a greater surface area of the skin. This increases the amount of damaged tissue as well as causing a more difficult decontamination process.

After discussions with burn specialists who are highly familiar with these types of chemical and thermal burns, the recommendation is to skip the brushing (which turns out to be more like smearing) and proceed to flushing. The flushing should involve a water flow with a volume similar to a garden hose for multiple exposure locations or a single location greater than about a square inch.

Eyes can be flushed with multiple bags, but the flow must be from the bridge of the nose outward to prevent cross contaminating the other eye. If both eyes are involved, a nasal cannula can be attached to the IV line to provide flushing for both eyes. In either case, however, EMS personnel must contact medical oversight personnel and should consider contacting the nearest fire units to provide additional assistance and water flow capabilities.

Contaminants that generate toxic or flammable gases are far less common than those that generate heat. Furthermore, for appropriately protected responders, decontamination is handled in the same fashion as for those contaminants that generate heat.

However, for contaminated victims or improperly protected responders, several issues must be addressed. First, many of these contaminants also heat and create corrosive liquids as they react. As a result, they can produce the same problems previously described. Should the tissue be broken, not only will these substances heat, but they will also release gas (again, without any outside water source). Second, the gases present a possible inhalation hazard that must be addressed. Third, if a flammable gas is produced, it could react very poorly to oxygen used on the victim. Finally, the application of appropriate volumes of water may well eliminate the problem. Again, medical control personnel must be contacted for their assistance and advice.

Contaminants that burn or explode on contact with water are a challenge to manage. For properly protected responders, the use of adhesive or vacuum removal techniques are the most appropriate. It simply may be necessary to clean the areas surrounding the entry zipper and other high probability contact locations, followed by a strict, sterile technique to safely remove the PPE. Once it is safely removed, it should be overpacked to prevent potential exposure to moisture.

Direct skin contact becomes a really difficult situation. Again, skin is normally covered by a minute layer of water (and also it contains water), so the reaction is likely to be spontaneously initiated. Explosions tend to propel major portions of the contaminant off the initial contact site but possibly to other locations on the body. To prevent the contaminant from contacting blood and interstitial fluid, mineral oil may be effective, depending on the exact contaminant. Again, contact medical control for further information and guidance.

Decontamination Corridors and Set-up

The process of decontamination begins in the hot zone and finishes within the cold zone. Figure 7-19 provides a typical set-up for a responder as well as a victim decontamination corridor. The exact number of stations, corridor layout, solutions used, and so on varies depending on the following types of considerations:

1. Who or what is to be decontaminated
2. The type of PPE worn (e.g., level B versus level A versus level C)
3. The contaminant involved, including:
   a. Chemical properties, such as reactivity, primary and secondary hazard, toxicity, and routes of exposure
   b. Physical properties, such as physical state, miscibility, and viscosity

Figure 7-19 also provides a model setup for the warm zone. It includes the access corridor, standard responder decon corridor, plus a victim decon corridor. In addition, the graphic shows a maximum layout for the responder corridor that includes two wash stations and two rinse stations. Also, the setup is not to scale, so the distances between various stations would be greater for an actual operation.

An additional consideration is the level of PPE needed to safely perform decontamination. The general rule states that decon personnel should be protected by PPE of the same level or one level below the PPE worn by the entry team. On the surface, that makes sense. However, upon closer examination, this general rule becomes questionable. Consider why personnel wear level A PPE. The primary reason is to protect the skin and respiratory tract from exposure to tissue-destructive and/or highly toxic gas and vapor, or liquid immersion.

Two good examples of such substances are concentrated hydrochloric acid and concentrated sulfuric acid. Hydrochloric acid is a powerful corrosive that fumes. Sulfuric acid is also a powerful corrosive, but it does not fume. In either case, personnel entering the hot zone, where a decent quantity of either of these acids has spilled, probably should be protected by level A PPE (especially for the hydrochloric acid).

Now the question is, "What should decon personnel wear?" It is a good question and can be answered on the basis of the concepts covered under PPE. PPE is to protect

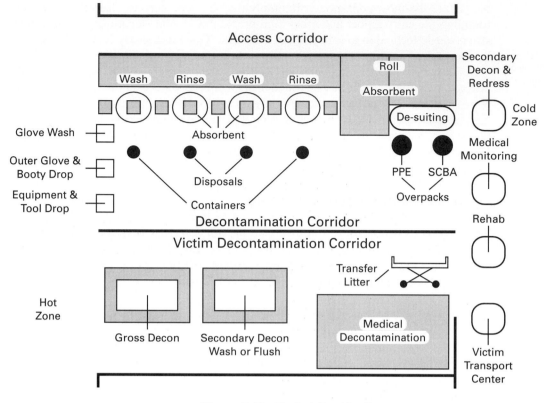

Figure 7-19   Typical Corridor Setup

the wearer from the hazards to which he or she will be or may be exposed. What are the hazards to which decon personnel will be exposed in either case? For sulfuric acid, the primary hazard will be from incidental splashes from the PPE worn by the entry team (no immersion and no vapor exposure). There are no real respiratory hazards in this situation because the liquid has low volatility. Further, the atmosphere in the decon corridor will contain more than 19.5% oxygen (otherwise it is in the wrong location). If there is an appropriate cartridge available, level C should be appropriate and safe. To assure this fact, air monitoring could also be performed.

So, what about the hydrochloric acid? Again, minor splashes of acid are one of the hazards. How about the respiratory hazards since the hydrochloric is a fuming acid? The answer is that there will be no major respiratory hazard. The question is how many teams have ever seen their personnel enter decon and their PPE was emitting fumes? Further, routine air monitoring rarely finds concentrations of any type of contaminant above a few ppm. Additionally, if the team uses a shower as the first station of their decon process, there is almost no chance of any respiratory hazard whatsoever. Again, level C should be appropriate as long as air monitoring indicates no problem.

Hopefully, this brief discussion has raised the reality check issue again. The bottom line is that the PPE used by personnel must be appropriate for the hazards with which they will be faced. Arbitrary rules are thus questionable.

PROTECTION—EMS AND FIRST AID

Not only are EMS personnel necessary for a safe hazmat response, but they are mandated by the regulations. The primary role of EMS during hazmat operations is the performance of pre- and post-entry medical monitoring. Additionally, they must be able to provide first aid if there is injury or accident.

EMS personnel must be qualified to at least the advanced first aid level. Routinely they are detailed to hazmat from the Medical Unit within the Logistics Section of the ICS. In the FIRESCOPE hazmat module, they are shown as reporting to the Medical Unit Leader. As such, it is possible for one set of EMS personnel to perform the pre-entry monitoring while another performs the post-entry monitoring. To most Hazmat Group Supervisors or Branch Directors, this situation is completely unacceptable.

The rationale for this objection is simple. The EMS personnel monitoring and treating entry personnel must be dedicated to only that function. There must be no possibility that they can be replaced by other medics without the full knowledge and concurrence of the Hazmat commander (group supervisor or branch director). It is essential that the same medics perform both pre- and post-entry monitoring so they are fully aware of the condition of entry personnel before as well as after the entry. Subtle changes such as pallor, muscle rigidity, and minor change in alertness can only be identified when the same personnel perform the monitoring. As such, Hazmat EMS personnel must report to the Hazmat commander and be part of the command structure.

NFPA 473 identifies two primary levels of training for EMS personnel involved in a hazmat response. The first basic level is trained to deal with victims or team members once they exit the decontamination corridor. The second level is trained much as a hazardous materials specialist. These personnel provide in-depth analysis and make decisions based on available information and conditions. They can also be trained to enter the warm zone to assist in decon, initial assessment, and so on.

Heat Stress

The most common medical ailment that occurs during hazmat operations is heat stress. As a result, both team and EMS personnel must be fully aware of the symptoms, etiology, and treatment. The four levels of heat stress are:

1. Heat rash
2. Heat cramps
3. Heat exhaustion
4. Heat stroke

Before discussing each of these heat-related disorders, it is important to explain why technicians and specialists need to understand their signs, symptoms, and etiology. As mentioned, heat is the most common and significant medical problem associated with entry personnel, not injury, chemical exposure, or similar situations. As such, each member of the team must be thoroughly familiar with these disorders so they can tell when they, their partner, or other members of the team fall victim to these conditions.

The identification of such problems indicates the need to change the entry procedures. First, additional pre-entry hydration and electrolyte infusion are in order. Second, it is highly probable that the work interval for entry teams should be reduced to minimize potential problems.

Heat Rash.   Heat rash is the first and most basic level of heat stress. This is very uncommon for emergency response personnel. The rash results from chafing and irritation after long exposure to heat and moist clothing. Good personal hygiene and sanitation of PPE go a long way to minimize this problem.

Etiology	Treatment
Chafing and abrasion on wet material	Good personal hygiene and sanitizing of equipment

Heat Cramps.   Heat cramps are the typical muscle cramping experienced by many during periods of physical exertion. It is a result of moderate dehydration and its accompanying electrolyte loss. It primarily occurs in the long muscles of the body, including those of the leg, arms or, in some cases, the abdomen. It is treated by rest and the intake of fluids and electrolytes.

Never use sugar sweetened fluids or electrolytes as the first step in fluid replenishment. The sugar retards the rate of fluid absorption from within the intestines. It is far better to drink plain water or a nonflavored electrolyte solution. Also, never administer salt tablets. They are counterproductive because they can enhance dehydration and elevate blood pressure.

Etiology	Treatment
Fluid and electrolyte loss due to exertion, often over a long period of time; results in uncontrolled long muscle contractions and spasms	Fluid and electrolyte replacement accompanied by rest

Heat Exhaustion.   Heat exhaustion (Table 7-8) is the next level of heat-related disorders. It is typified by profuse sweating and lethargy. The victim's skin is cool, moist, and often pale. Oral temperature may be somewhat low. It is not life threatening but can lead to life threatening heat stroke if it is not properly managed. Heat exhaustion is often later accompanied by cramping of the long muscles as in heat cramps. This is probably the most common type of heat stress experienced by hazmat personnel.

Table 7-8 Heat Exhaustion

Symptoms	Etiology	Treatment
Profuse sweating, lethargy, cool, moist and pale skin; oral temperature may be decreased, deep body core temperature is rising. When advanced, symptoms are similar to hypovolemic shock.	Due to moderate to large loss of fluids and electrolytes, hypovolemic reaction begins. Blood starts pooling in the deep body core. As such, heat dissipation is decreased. This sets the stage for heat stroke.	Remove the victim from the source of heat. Air conditioning is highly recommended. Provide fluids and electrolytes (no salt tablets). Have them recline in feet-up position (feet above the level of the heart).

Heat Stroke.   Heat stroke (Table 7-9) is extremely dangerous and life threatening. Should anyone exhibit these symptoms, they must be treated immediately. In essence, the body has totally lost the ability to control its own temperature. As a result, it rapidly heats to potentially lethal levels. The elevated temperature threatens the brain with permanent damage. These victims will usually exhibit an altered state of consciousness. They may or may not be able to talk and normally cannot move by themselves. They can easily die if not treated immediately and removed from the situation.

Cold Stress

Although less common, cold stress is still a potential danger to entry personnel. Initially in cold stress, the individual is merely cold and shivering, but a precipitous drop in body temperature can follow. This type of temperature drop is most rapid when the cold is accompanied by cool liquids. Additionally, the cold can lead to frost bite or hypothermia.

Many ask, "Where could such conditions be encountered?" One answer is in cold weather operations. Another less obvious situation is any incident involving the decompression of a compressed or liquefied compressed gas. Still another situation involves operations in and around cold storage facilities, especially those using rapid freezing technologies. These technologies include temperatures in the range of $-30°$ to

Table 7-9 Heat Stroke

Symptoms	Etiology	Treatment
Skin is hot, dry, and red. Oral temperature may be 106°F or higher. The state of consciousness may range from disoriented to comatose. Muscle spasms, rigidity, or even convulsions are common.	Severe fluid and electrolyte loss. The body has gone into the equivalent of hypovolemic shock. Blood is pooled deep within the abdomen and thorax. No heat exchange is occurring through the skin.	Immediately remove from the heat and start cooling. If unconscious, start IVs to replace fluids and electrolytes. Place in a head up position. Monitor oral temperature for sudden precipitous drop. Transport victim to hospital.

–50°F, accompanied by high speed, high volume air movement (winds in the range of 40 to 60 miles per hour). In such settings, hypothermia is possible, and frost bite could be probable.

Frost bite is a relatively dangerous condition that can become serious if it occurs over a large area of the body. The most common places for it to occur are on the fingers and toes. However, any skin exposed to temperature below freezing and especially below zero can freeze in a matter of a minute or two. Generally, the colder the cooling media (objects, air, gases, or liquids) the more rapid the onset.

Typically, frost bite causes the skin to take on a white, almost wax-like appearance. If the cooling is rapid and severe, the skin may actually develop frost. In the case of a cryogenic liquid exposure, the skin may develop an indentation.

If not properly treated, frost bite can result in the loss of the affected tissue. It can also lead to gangrene. Any suspected frost bite victim must be transported to an appropriate medical facility.

Hypothermia is a condition where the deep body core cools below 96°F. Commonly, the victim will experience severe shivering, numbness, and an inability to move efficiently. If severe, hypothermia will result in an altered state of consciousness preceded by disorientation and a diminished ability to think clearly and safely.

Hypothermia is treated by the gradual warming of the individual. If this condition is present or even suspected, the individual must be transported to an appropriate medical facility for treatment.

First Aid Considerations

One point that must be made perfectly clear is that first aid is almost impossible within the hot zone. Only the basic ABCs (Airway, Bleeding, Circulation) can be addressed at best. It is impossible to feel a pulse through the layers of gloves. A stethoscope cannot be used because there is no access to one's ears. The use of oxygen or even a defibrillator is contraindicated.

If bleeding is present, it can be stopped, but this should be done with a hand and not a dressing. The reason is simple. Every known protocol states that once a pressure dressing is in place, it cannot be removed outside a hospital. If a dressing is put in place while within the hot zone, there is a high probability it is contaminated. Here is the catch-22. On the basis of decon protocol, the dressing must be removed before the victim can leave the decon corridor. While on the basis of medical protocol, the dressing cannot be removed. This is a major problem. By using the gloved hand, the pressure can be released after most of the decon process is complete and then the dressing put in place. This is but one example of very complex EMS and hazmat protocol conflict.

Once the victim has been decontaminated, EMS personnel can start with first aid. First the nature of the injury must be identified. Is it chemical or trauma? What system or systems are involved and how severely? What is its exact location and is there ABC involvement?

If it is a chemical injury, a series of questions follows. Was it the result of direct or cross contamination? Does it involve external, internal, or both types of contamination? How large a portion of the body's surface area is affected? How long has the exposure gone on? How many contaminants are involved and what is their identity? Was this a splash type of exposure or was the victim in constant contact? If so, how long?

Most commonly, treatment is symptomatic in nature. Since there are few antidotes and those are not normally readily available, EMS personnel have little choice but to treat the symptoms.

Be sure the receiving medical facility is notified as soon as possible that a contaminated victim is involved, as well as the identity of the contaminant. It will take the facility at least 20 to 30 minutes to prepare for such victims.

Some may ask about preparing the transport vehicle. Remember, the victim has already gone through decontamination, the clothing has been removed, and they are clean at least externally. The nature of the contaminant must then be closely examined. Can it be rapidly excreted through exhalation, salivation, perspiration, urination, or defecation? Was any ingested? If so, these victims have the nasty habit of vomiting the contaminant all over the back of the unit. This can result in cross contamination of the attending personnel.

As such, plastic sheeting may be placed on the floor and at least part way up the walls. Personnel should have a minimum of splash and full face protection. Respiratory protection in the form of an APR may be in order.

A full EMS protocol may be developed to handle contaminated victims from the scene through the transport to the medical facility. A hazmat paramedic program could be developed to assist on such incidents. Following this chapter is an example of a Hazmat EMS check sheet. It also contains the pre- and post-entry medical monitoring sheets.

PROTECTION—PRE-ENTRY BRIEFING

As mentioned earlier, a thorough and concise pre-entry briefing is essential to safe and effective operations. All entry personnel must be fully aware of their assignments and what is expected of them. In addition, they must be fully aware of any safety or health hazards that could adversely affect their entry. All assignments should be documented. As these assignments are completed, they must be checked off. If they are not completed, that must also be noted. Following this chapter is an example of an Entry Team Assignment sheet that can help with the briefing and its documentation.

PROTECTION—PRE-ENTRY MEDICAL MONITORING

Pre-entry medical monitoring has become controversial lately. Earlier, it was considered essential in establishing the physical baseline of the individual for that day. As a minimum, this medical monitoring should include:

1. Blood pressure
2. Pulse

3. Respiration
4. Temperature

Additional checks could include naked body weight and EKG.

   Some cutoff values must be identified to help determine when personnel are fit to perform entry work. Beyond temperature and obvious coronary indications there is tremendous controversy about appropriate readings. The vitals one responder feels are just fine are too high or too low for another. These criteria require the input of medical specialists as well as field trials to identify the appropriate vitals for a given jurisdiction. (Even medical specialists are divided over this issue.)

   Following this chapter is an example of a Medical Monitoring check list that can help establish a protocol and provide documentation. At the end of this text in the appendices is a section with each ICS position and its associated checklists.

PROTECTION—SAFETY ASSESSMENT

The safety assessment is just that, a check by the Hazmat Safety Officer to assure that all aspects of safety have been appropriately addressed. Although this assessment process goes on throughout the entire operation, there are various times when it must be documented.

   The first time when safety considerations must be documented is with preparation of the Site Safety Plan. As mentioned previously, the Site Safety Plan identifies specific hazards on the scene and appropriate methods for protecting personnel. Specifically, the site safety plan must address the following as a minimum:

1. Site characterization (i.e., identification of health and safety hazards found on the site)
2. Site work zones (i.e., the location and extent of zones)
3. Site communications and roles (i.e., lines of communications throughout the site operations, such as dedicated radio frequencies, ICS position assignments, etc.)
4. Use of the buddy system (i.e., determination of who is working with whom)
5. Command post or center designation, including an overall ICS structure
6. Safe work practices (e.g., PPE for the zones, unsafe practices, etc)
7. Medical monitoring and triage
8. Decontamination corridor and procedures
9. Air monitoring plan
10. Other relevant issues and areas

   Not all of this information is available at one time. However, the following data should be available early on and stay relatively constant throughout:

1. Site characterization
2. Site work zones
3. Site communications and roles
4. Command post (CP) designation and ICS
5. Decontamination
6. Safe work practices

As such, documentation of this information can occur early in the incident and can be used to provide basic information to all incoming personnel and agencies. For lack of a better term, this can be called the Basic Site Safety Plan because it contains the nitty gritty data upon which all site operations are based. An example of such a Basic Site Safety Plan is provided at the end of this chapter. Copies of the Basic Site Safety Plan can be posted at the command post and the hazmat operational location.

Unfortunately, this Basic Site Safety Plan alone does not assure safe operations or even thorough documentation of the incident. To accomplish this task, each position, at least within hazmat, must perform some part of the documentation. All the check lists identified so far in this text are pieces of the comprehensive Site Safety Plan. Obviously, every incident does not require the completion of every sheet. However, large scale or long duration operations may require extensive use of the sheets.

It is the Hazmat Safety Officer's (Assistant Safety Officer for Hazmat—FIRESCOPE) responsibility to assure that these documents are completed as fully as necessary. Further, before each entry, the Hazmat Safety Officer must take part in the pre-entry briefing to check and make sure all appropriate procedural and safety issues have been addressed. This is a highly structured and well-documented part of the overall safety assessment process. As such, it must be thorough and well documented and should contain the following elements:

1. If the substance(s) is known or unknown
2. Whether known or unknown if hazards have been identified
3. Assurance that PPE is appropriate
4. Assurance that decontamination is appropriate
5. Assurance that decontamination is set-up and ready
6. Assurance that EMS is on scene
7. Assurance that the identity of the entry team is documented
8. Assurance that a back-up team is ready and their identity is documented
9. Assurance that pre-entry medical monitoring has occurred and been documented
10. Assurance that air times are documented (e.g., on air, entry, out, off air)
11. Assurance that the pre-entry briefing occurred
12. Assurance that the safety assessment has been performed

Again, an example of this type of check sheet is included at the end of this chapter.

PROTECTION—RESCUE

As mentioned earlier, rescue of victims is often completed by first responders long before hazmat ever arrives on the scene. However, there are exceptions. For example, the back-up team may have to rescue the entry team if they should have a major problem. As such, this tactic deserves some consideration.

Should the hazmat team be required to perform victim rescue, there are some extremely important questions that must be answered:

1. Is entrapment involved? If so, how severe is it?
2. Is the victim conscious?
3. Is the victim injured or did he or she receive chemical exposure?
4. What is the victim's proximity to the product?
5. Is he or she lying in the product.
6. What type of equipment is needed to perform a rescue?
7. Is there the potential for fire or explosion as a result of rescue efforts?
8. Is this a rescue or a body recovery?
9. To what hazards will personnel be exposed?
10. Is the risk worth the benefit?
11. Are there regulatory or legal implications, and what are they?
12. Is this a confined space? If so, are personnel properly trained and equipped, and are written procedures in place?

These questions may or may not be difficult to answer. However, they must be factored into the decision to attempt a rescue.

One general note needs to be made at this point. Some teach a confined space entry operation that involves removing the SCBA from the wearer's back to allow access through small openings. By performing this action, the user violates OSHA regulations that state that respirators must have and meet NIOSH approval. Specifically, the removal of the SCBA from the back voids the NIOSH approval. As a result, it means the action is not OSHA compliant and thus, Pandora's box has been opened.

When considering the rescue of entry personnel, specific procedures must be in place and part of routine training. The biggest issue is the removal of downed personnel from the hot zone. One of the most effective methods identified involves the use of a stretcher (plastic stokes or orthopedic scoop) or body bag as an excellent tool. The entry team member can be rolled onto or into any of these and then carried without fear of rolling from the device. Backboards are often a problem because the individual must be secured to the board before they are moved. Such securing takes time, is often difficult, and takes additional time to be removed when the decon corridor is reached.

Now that the protection issues and options have been addressed, it is time to move on to control actions, starting with spill control.

SPILL CONTROL

<table>
<tr>
<td>

The process of stopping, limiting or controlling the spread of the product through the environment

</td>
<td>

<u>Specific Tactical Objectives are identified by the exact type of release involved:</u>
Gas/Air
Liquid/Surface
Liquid/Water
and Solid Surface

</td>
</tr>
</table>

Spill control tactics are operations intended to stop, limit, or otherwise control the spread of the product through the environment. As with confinement operations in firefighting, spill control is generally defensive in nature and intended to confine the spread of release product to the physical areas already affected. With some types of spills, notably gas/air releases, it is routinely impossible to confine the product to the initial affected area. As such, spill control acts to limit or possibly control product spread.

Further, spill control tactical objectives are release specific. Although several release types have the same tactical objectives, the methods and resources used to accomplish the objective are quite different. As a result, the spill control tactical objectives and methods are discussed on the basis of the four spill types.

Gas/Air Releases

<table>
<tr>
<td>

Gas/Air Releases include gases, vapors, fumes, and fine particulates released into the air

</td>
<td>

Primary Tactical Objectives:
Ventilation
Dissolution
Dispersion
Diversion
Blanketing

</td>
</tr>
</table>

The first consideration for gas/air releases is the degree of confinement. Is the release inside or outside a structure? If it is inside there are only certain options available to control the spill. If the release is outside, there are other options. In the following section, each option will be discussed.

SPILL CONTROL—VENTILATION

Ventilation is used when releases occur within a structure. It uses a positive or negative flow of air, most commonly developed by mechanical devices, to remove the contaminant from the structure. However, in some instances, the natural buoyancy of the contaminant is also involved.

Natural Ventilation

Natural ventilation relies on the buoyancy of the contaminant to make this tactic work. This means that only substances with vapor densities less than 1 should be considered for this ventilation method. Although substances with a vapor density of approximately 1 may be removed by natural ventilation, the process is slow and quite inefficient. Substances that have a vapor density greater than 1 should not be considered as candidates for natural ventilation because they tend to remain in place rather than disperse.

Further, natural ventilation can easily be foiled by meteorologic conditions. Wind speed, direction, and variability as well as humidity can have tremendous effects on the success of natural ventilation.

Mechanical Ventilation

Mechanical ventilation involves the use of mechanical air-moving devices. Some of these devices may be part of the structure in the form of heating, ventilation, air conditioning (HVAC) systems, or as fume hoods, processing hoods, and ventilators. Other devices include portable fans and blowers (such as smoke ejectors or PPV fans). These devices can accomplish ventilation by using positive or negative pressure (Figure 7-20).

Positive pressure ventilation (PPV) pressurizes the area containing the contaminant and forces it out through any available openings. Normally, PPV is the most effective and efficient way to rid an atmosphere of contaminants because it cleanses the entire space at one time. However, if it is not performed properly, it can also force the contaminant into surrounding areas that are not affected. As a result, in a multi-roomed building, including those containing multiple HVAC or ventilation zone, it may be necessary to positively pressurize surrounding areas and to force contaminant from an area.

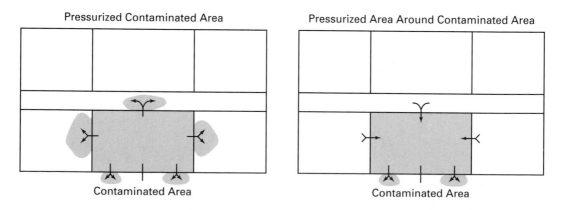

Figure 7-20   Effects of Properly and Improperly Used PPV

Although PPV is more efficient, negative pressure ventilation (NPV) is still a potential option. NPV involves creating a negative pressure within the contaminated space, thus drawing fresh air in through some opening such as a door or window. In some instances, it may be used in conjunction with PPV when surrounding areas have been pressurized.

However, NPV has two primary drawbacks. First, it is less efficient because it "pokes holes" in the contaminant. This means it clears a direct path from the inlet to the outlet where the fan is located. Second, the contaminant must be drawn through the fan. This may or may not be a problem. In the case of flammable gases or vapors, the fan must be intrinsically safe or explosion proof. In the case of fixed ventilation systems or hoods, this may or may not be true. In the case of portable fans, it is often impossible to tell whether this ever was true.

Depending on the nature of the contaminant, ventilation may be put on hold until the surrounding area is readied for the release. This may include having to in-place protect or evacuate the public. Make sure response personnel are also out of the exhausted contaminant and away from HVAC and ventilator discharges. This will prevent accidental contamination of personnel or the public.

SPILL CONTROL—DISSOLUTION

Dissolution is the process of dissolving a water soluble gas or vapor in water. Most commonly, the water comes in a fog stream. Since the contaminant is water soluble, consideration must be given to the spread of the now contaminated water. In most instances, dissolution requires a large fog stream-to-vapor ratio. As such, it is not very effective on large releases.

SPILL CONTROL—DISPERSION

Dispersion is the process of dispersing the gas/air release by spreading it through the atmosphere and diluting its concentration. The most common method involves the use of water fog. In this case, however, the contaminant is not water soluble. Again, there is a rather large fog-stream-to-vapor ratio required. As such, this technique is not very effective on large releases.

SPILL CONTROL—DIVERSION

Diversion tactics always involve changing the contaminant's direction of flow. In gas/air releases, the most common force used to accomplish this change in direction is again the fog stream. The same limitations regarding fog-stream-to-vapor ratio and efficacy that apply to dispersion and dissolution also apply here.

However, there are two other options besides fog streams. One is used within a structure while the other is used outside. Inside, buttoning up is an example of diversion. In this case, doors, windows, and other openings are covered to stop the movement of contaminant into or out of a designated area, thus changing the flow of the contaminant. This covering process normally involves the use of tape and may also include the use of plastic sheeting with the tape. Although it may not seem to be important, this method of diversion is very effective.

Outside, diversion of gas/air releases can sometimes also be accomplished by using plastic sheeting. In this case, the vapors must be heavier than air and topography must be advantageous. In essence, the plastic sheeting is used to form a solid fence to stop or at least minimize the spread of the gas/air release away from its source. If the release is in a small depression, this technique is far more effective.

SPILL CONTROL—BLANKETING

Blanketing is somewhat different from the other spill control techniques because it seeks to stop the gas/air release at its source. Specifically, blanketing is the process of covering the surface of a liquid to suppress the vapors that it is releasing. Most commonly, fire personnel immediately think of fire-fighting foam. However, this is just one of the many potential options.

Again, plastic sheeting can be a very effective blanket as long as it is not damaged by the liquid involved. Once the sheeting is in place, wind may be a problem. However, the placement of a charged hose line, water tube (plastic tube filled with water), sand, and so on will hold the plastic in place.

Sand, soil, and absorbent clay are also effective blankets. One caution is necessary: The greater the depth of the liquid, the less effective these blankets become and the more advisable the use of a foam vapor suppressant becomes.

There are several foam vapor suppressants designed to work on various fuming acids as well. Ansil and Chubb National are two manufacturers that have this type of foam.

Spill Control—Liquid/Surface Release

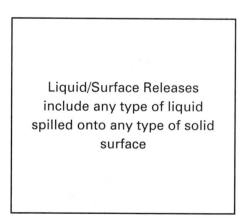

	Primary Tactical Objectives:
Liquid/Surface Releases include any type of liquid spilled onto any type of solid surface	Diking
	Diversion
	Retention
	Adsorption
	Absorption
	Neutralization
	Gelation
	Solidification
	Emulsification

SPILL CONTROL—DIKING

Diking is the process of placing a physical barrier in front of a flowing liquid/surface spill to stop or minimize its spread. Diking can use any solid material ranging from sand to soil to charged hose lines to sand bags to water tubes. Diking is one of the many low-tech control methods available.

When personnel excavate soil to form a dike, it is recommended that a group of four or more personnel stand side by side with their backs to the spill. Obviously, this will require someone to watch the spill while they are digging. They dig on what will be the inside of the dike and place the soil on the downstream side of the hole. This hole increases the capacity of the dike and helps slow its forward movement.

When building dikes, it is important to remember that a series of dikes is preferable to just one. Should one dike fail or is unable to hold all the product, the next dike picks up the slack.

Further, dikes have different shapes and uses. If a product is fast moving, a long tapered, V-shaped dike is recommended. This type of dike will slow the flow of product without being overridden or eroded away rapidly. For a slow moving flow, a U-shaped dike is recommended because it has a greater capacity than the V-shaped dike for the same amount of effort and diking material.

SPILL CONTROL—DIVERSION

Diversion is the placement of a physical barrier to change the direction of product flow. The same materials can be used for diversion as are used for diking. Also, diversions are often teamed with diking to stop the flow.

SPILL CONTROL—RETENTION

Retention is the process of catching the released liquid in some type of holding location. Retentions can include a small can or pail, a five gallon bucket, a decontamination retention pool, a portable water tank or pond, or an excavated hole. Anything that can hold the product can potentially act as a retention.

Should a hole be excavated or an existing depression in the ground be used as retentions, it is often advisable to line the bottom. This lining can be a salvage cover or a piece of plastic sheeting and is used to prevent the product from absorbing into the soil.

SPILL CONTROL—ADSORPTION

Adsorption is a process where molecules of the contaminant adhere (stick) to the surface of the adsorbent material. There is a major advantage to the use of adsorbents over absorbents. Specifically, adsorbents suppress vapors by binding the contaminant to its surface. As a result, adsorbents can help reduce the fire hazard associated with flammable liquid spills by suppressing vapor.

Another advantage to adsorption is disposal. Many of the adsorbents are organic based materials. They range from activated charcoal or carbon to sphagnum moss to polypropylene. As such, if the contaminant is a simple hydrocarbon, nonhalogenated derivative or other low toxicity organic compound, the adsorbent and the contaminant may be authorized for disposal by incineration. Incineration is one of the more cost effective and potentially environmentally friendly disposal options.

Adsorbents have several limitations, however. One is the feasible surface area to which the adsorbent can be applied in a reasonable time. By starting at the edge, applying the adsorbent and then moving forward over the area just covered, large areas can be handled. Another potential limitation is the depth of the spill to which the adsorbent, or absorbent for that matter, is effective and efficient. Routinely, when the liquid is deeper than about 1/4 to 1/2 inch, another option should be considered.

SPILL CONTROL—ABSORPTION

Absorption is the process of physically drawing the contaminant into the inner spaces of the absorbent, like a sponge absorbing water. There are a wide variety of absorbents ranging from vermiculite, absorbent clay, and fly ash, to pads, pillows, pigs, and booms. The mineral based absorbents (e.g., vermiculite, clay, and ash) are simple and straightforward. The synthetic pads, pillows, pigs, and booms are not.

There are two primary types of synthetic absorbents: hydrophilic and hydrophobic. Hydrophilic absorbents absorb water and water-based contaminants. They are often used for acids, bases, emulsions, and other water-based substances. However, if the spill is outside and on a rainy day, they are of little value because they will absorb the rain just as easily and as well as the contaminant.

On the other hand, hydrophobic absorbents absorb hydrocarbons and other immiscible substances. They will not absorb water or water-based substances. They are commonly used when a hydrocarbon or derivative is released onto or into water. It is this type of absorbent used in absorbent water booms.

SPILL CONTROL—NEUTRALIZATION

Neutralization is a chemical reaction that acts to reduce the hazards of the contaminant. By far, the most common emergency response neutralization reactions involve acids and bases. However, neutralization alone is not truly a spill control method in the traditional sense (unless the agent is used for diking or diversion). Truly, it is simply a spill modifier as opposed to a spill controller.

Although neutralization is an effective option, it has several limitations and drawbacks. First, it is generally effective on only relatively small spills (up to a hundred or so gallons). Over that size, its effectiveness is often questionable. One example of a large neutralization operation occurred in Denver in the early 1980s. In this incident, a tank car of 99% nitric acid was punctured on the end, close to the bottom. Almost the entire contents were lost.

It was determined that approximately 700 tons of soda ash (Na_2CO_3) were to be applied by a snow thrower from Stapleton airport. The acid was neutralized, but the soda ash caused a pH swing to 12. Consequently, months of work were needed to counteract the effects of the soda ash. Again, this pH swing is almost inevitable when neutralization is performed.

Further, neutralization does nothing for other contaminants that may be present. Hydrocarbons, metals, and salts are routine contaminants of acids and may precipitate from solution when neutralization is performed. The resulting residue may present an additional hazard.

SPILL CONTROL—GELATION

Gelation is the process of turning a liquid contaminant into a semi-solid gel. One of the more ridiculous and dangerous attempts at this control method occurred in the late 1970s. In this case, a powder was added to a gasoline spill, turning the liquid into a gloppy goo. This goo could then be shoveled into a container.

Unfortunately, the gelation process did nothing to reduce the flammability hazard of the gasoline. In essence, this process had now created napalm. The scooping action of the shovel was an excellent potential source of ignition.* This was another good idea that did not live up to its possibilities.

A reality check is needed at this point. It involves the misconception that non-sparking tools cannot produce sparks. This author experienced one of the most beautiful sparks created by a nonsparking tool while working in a chemical production facility. What was most frightening was that the area in which the spark was generated contained vapors from isopropyl ether. Obviously, the vapors were not within the flammable range, but it still provided an opportunity for one's life to flash before one's eyes.

There are several gelation lab-type kits available for various contaminants. Their primary advantage is that they convert the liquid into a less mobile form that usually can be controlled much more easily.

SPILL CONTROL—SOLIDIFICATION

Solidification is the process of changing a liquid into a solid. The most common example of this type of control is the mercury spill pack. Here a powdery agent is sprinkled on the mercury, leading to a mild chemical reaction. A solid mercury salt (routinely a sulfate) results. Solidification thus suppresses the mercury vapors and binds the liquid in a manner that allows its retrieval. In a recent incident, about 2,500 pounds of mercury

*NOTE: Even plastic and "nonsparking" shovels can produce sparks when used with a scooping action.

were spilled at a loading dock area of a business. After initial diking by responders, a cleanup contractor was called to the scene. They performed a large scale solidification and removal operation. Once the solidification was completed, the entire spill area received an additional layer of blacktop to prevent any further spread of mercury.

SPILL CONTROL—EMULSIFICATION

Emulsification is the process of producing a physical suspension of two or more immiscible (insoluble) materials by using an emulsifier (e.g., surfactant, soap, or detergent). The emulsifier is then applied to the surface to loosen the remaining contaminant.

After the emulsifier is applied, it is usually advantageous to agitate the mixture with a broom to loosen as much contaminant as possible. After this process, another application of absorbent or adsorbent is required. The use of water to flush the emulsion will rapidly spread it over a greater area and thus should not be done.

Further, if the emulsion enters a body of water, it will create tremendous problems. First, since the immiscible contaminant is emulsified, it will form only a very thin membrane over a tremendously large surface area of the water. This makes it extremely difficult and often impossible to control. Second, the emulsion can enter the water column (the water below the surface of the soil). As a result, it can easily produce toxic effects on wildlife.

Liquid/Water Releases

Liquid/Water Releases include any type of liquid released into or onto any water surface	**Primary Tactical Objectives:** Damming Booming Diversion Retention Absorption Dispersion

SPILL CONTROL—DAMMING

Damming is the process of building a physical barrier within a body of water to stop the flow of a contaminant. Diking is the liquid/surface equivalent to damming. There are three types of damming that include underflow dams, overflow dams, and complete dams.

There are two physical properties of the contaminant that dictate which type of damming is appropriate for the substance: water solubility and specific gravity. Substances that are insoluble and have a specific gravity less than 1 will float on the surface of the water. As such, underflow damming is the choice for them. Substances that are insoluble and have a specific gravity greater than 1 sink to the bottom of the water. As such, overflow damming is the choice for them. Finally, substances that are water soluble generally require complete damming.

Underflow Damming

Underflow damming is the process of building a dam and inserting a pipe through the dam (Figure 7-21). The pipe allows water from behind the dam to flow through the dam and downstream. The pipe is placed on an angle so the inlet (upstream end) is well below the surface of the water. The outlet (downstream end) of the pipe is placed higher than the upstream end so water will build up behind the dam to a level higher than the inlet of the pipe. This prevents the contaminant from flowing through the pipe. In this arrangement, the contaminant floats on the surface of the water and remains behind the dam.

An alternative way to make an underflow dam is by using boards (on side) or a ground ladder wrapped in a salvage cover or plastic sheeting. The board or ladder is then extended across the body of water and kept from touching the bottom by placing it on rocks or other objects underneath. As such, the water flows under the dam while the contaminant is maintained.

Overflow Damming

Overflow damming is the process of building a dam that allows water to flow over the top (Figure 7-22). Because the contaminant is heavier than water, it collects at the bottom of the dam and does not spread any further. Overflow damming may also work for some water soluble substances as well. This is particularly true for soluble substances with high viscosity. An example of such a substance is concentrated sulfuric acid.

There is one significant caution about overflow damming. Remember that the breast of the dam must be covered with something similar to a salvage cover or plastic sheeting to prevent it from eroding away. Again, ladders, boards, and other objects can be used as overflow dams.

Complete Damming

Complete Damming is the process of stopping and maintaining the flow of the body of water. Obviously, this will only work for small streams. It is simply impossible to completely dam anything larger.

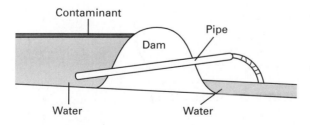

Figure 7-21 Underflow Dam

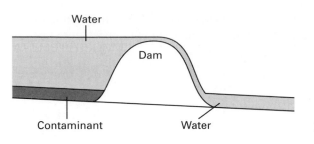

Figure 7-22   Overflow Dam

SPILL CONTROL—BOOMING

Booming is the process of placing floating devices on the surface of a body of water to hold or control the spread of an insoluble contaminant with a specific gravity less than 1. To be most effective, a series of booms is placed to catch and move contaminant toward the shoreline. Booms should never be strung straight across a stream or river. They should be angled to prevent the boom from bellying in the middle (generally the fastest part of the stream) thus causing the contaminant to stay away from the shore (Figure 7-23). There are two types of booms: containment and absorbent.

Containment booms are used to form a physical barrier in the body of water. To be most effective, containment booms are often equipped with skirting. Skirting is a piece of boom that extends down below the body of the boom itself (Figure 7-24). The skirt helps minimize the underflow of contaminant below the boom itself. Skirts of varying lengths can be purchased for different conditions. Generally, longer skirts are used when the water is more turbulent and thus more likely to cause product underflow. In the case of ocean booms, it is not uncommon to find skirts that are up to four feet long. Containment booms can be combined with hydrophobic absorbent pads or pillows to help absorb contaminant as well as control its spread.

Absorbent booms provide a physical barrier to the spread of contaminant as well. However, their primary function is to absorb the contaminant. Normally absorbent booms are used when there is minimal contaminant forming a sheen on a relatively small body of water. They are composed of a hydrophobic absorbent held within a net-like material. They can be purchased in varying diameters ranging from 2 to 12 inches.

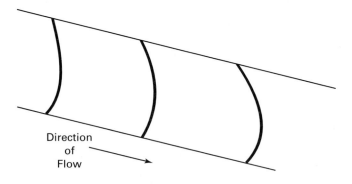

Figure 7-23   Multiple Booms

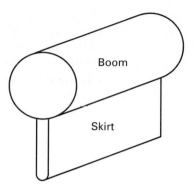

Figure 7-24   Skirt Boom

SPILL CONTROL—DIVERSION

Most commonly, diversion involves the use of booms to move the contaminant to the bank of a body of water. At the bank, the contaminant can be handled most readily and safely. The use of vacuums (skimmers) or continued diverting into a retention are the primary options.

SPILL CONTROL—RETENTION

Retention is accomplished by diverting the contaminant to the bank through the use of booms. If there is a large volume of contaminant, it may be very difficult to effectively contain it simply with booms. In this case, it is often advantageous to excavate an opening in the bank into which the contaminant can be directed. In this fashion, large amounts of contaminant can be contained and recovered.

SPILL CONTROL—ABSORPTION

As previously mentioned, absorption accompanies booms. It can be accomplished with containment booms and absorbent pads or pillows, or with absorbent booms.

SPILL CONTROL—DISPERSION

Dispersion is the process of emulsifying a contaminant when it is floating on the surface of a body of water. It has been used successfully in various major oceanic, crude oil spills. However, it can easily produce very negative impacts when used improperly. Specifically, the emulsification that occurs causes the contaminant to decrease in depth and thus spread over an extremely large area. Further, the emulsification allows the contaminant to mix with and enter the water column. As a result, dispersants make the recovery of the contaminant very difficult, if not impossible, as well as increasing the size of the affected area dramatically.

As a result of these negative impacts, the use of dispersants must be authorized by a Federal On-Scene Coordinator (OSC). If such approval is not granted, the offending party is in violation of the National Oil and Hazardous Substance Contingency Plan (NCP—40 CFR 300) and thus subject to the full force of regulatory enforcement actions. As a result of unauthorized applications, various response agencies have been cited for using dispersants without the required approval.

Additionally, unscrupulous vendors periodically try to sell response agencies dispersants as the cure-all for all liquid hydrocarbon fuel spills (e.g., gasoline, diesel, or fuel oil) wherever and whenever they occur. The use of dispersants in the fashion described by such vendors can lead to very unwanted results.

Solid/Surface Releases

Solid Surface Releases include any form of a solid (i.e., sheets, pieces, prills, dusts, powders etc.) released onto any type of surface	**Primary Tactical Objective:** Blanketing

SPILL CONTROL—BLANKETING

Blanketing in the case of solid/surface spills involves simply covering the spill with some type of plastic sheeting. This is normally only required for fine particulate solids or, in some instances, those that sublime. Once the sheeting is in place, its edges are weighted to keep it in place. When clean-up personnel start the process, they should make sure not to remove the entire sheet. All that is required is for them to cut a small hole in the plastic and to clean up the contaminant directly underneath. When that area is clean, they simply need to cut another small hole and clean that area.

LEAK CONTROL

The process of slowing or preventing the release of product through the breach or other openings in the container	Specific Tactical Objectives are identified by the type of control option chosen: Direct or Indirect

Leak Control is the process of slowing or preventing the release of product through the breach or opening in the container. There are two types of leak control tactical options: direct and indirect.

Direct objectives hold the potential of being one of the most dangerous operations to perform. This is because the performance of direct control options requires entry personnel to be in close proximity to the compromised container. As a result, personnel are almost assured of becoming contaminated with the product and they have no physical safety buffer because they are so close to the container. On the other hand, indirect objectives hold much less potential for danger because, as the name implies, they are performed remote from the container.

Direct Leak Control

Procedures used to control the release of product involving direct control of the container breach	**Possible Tactical Objectives:** Plugging Patching Overpacking Crimping Product Transfer Valve Actuation

There are many options available for each of the direct methods of leak control. Unfortunately, many are unique to specific types of systems or containers involved in the incident. All to often, teams purchase off-the-shelf leak kits that contain items and devices that look good in theory but do not function in practice. In the real world of leak control, what looks good is often not going to work. As such, various tactical options are covered very briefly.

In addition, the internal pressure of the container plays a big role in the potential success of the operation. The reality is that the higher the pressure, the greater the hazard to personnel and the less likely success. Furthermore, in most cases, plugging and patching a breach of any size becomes questionable at pressures of 20 to 40 psi. For small breaches, plugging and patching may be effective at somewhat higher pressures. In a few instances, where there are specifically designed kits, such as the Chlorine A, B, and C Kits, pressures much higher can be addressed. Unfortunately, there are very few of these specific container kits available.

Another major problem with plugging and patching is the type of breach that is involved. Very commonly the breach is the result of trauma, poor maintenance, corrosion, or other stressor to the container. Impacts often result in irregular, awkwardly located breaches. Such situations often make plugging or patching impossible. Corrosion induced breaches are often accompanied by severe thinning of the surrounding area. Such situations can be even more problematic because the use of plugs or patches may in fact simply act to enlarge the breach.

LEAK CONTROL—PLUGGING

Plugging is the process of inserting an object into the breach in the container. This process works best on atmospheric containers where the hole is a uniform shape and size. The further the situation moves from these ideal conditions, the less the likelihood of success.

Plugs can either be driven or screwed into place. Obviously, those that are screwed into place must be solid and are usually metallic. In some instances, screw plugs are fitted with some soft gasket of similar material to assist in stopping the flow.

Those that are designed to be driven must be constructed of materials that are somewhat soft so they can deform and fill uneven locations in the breach. As such, rubber, plastic and wooden plugs are the most common. The wooden plugs should be made of a soft wood like pine, not a hard wood like oak.

Some kits and teams use drift pins or solid metal pins as plugs. The classic example of such a device is the drift pins found in the Chlorine B and C kits. It is important to note that such devices can become lethal projectiles should the pressure within the container spit them out of the breach.

Quite often simple plugs are the most widely used as well as most effective. Such devices include boiler plugs and screws, golf tees, and wooden wedges and cones.

One warning: When the release involves a failed spring loaded pressure relief valve, rupture disk, or valve stem, it is potentially very dangerous to insert a plug into the breach and stop the release of product until the reason for the system failure is identified. Specifically, the failed member of the system (especially in the case of pressure relief devices) may have failed due to an over pressurization and is now venting potentially dangerous pressure. Should this opening be plugged, the pressure relief system will no longer function. As a result, excessive pressure may again build and lead to a more violent failure in some other part of the system.

LEAK CONTROL—PATCHING

Patches are analogous to bandages because they are placed over the outside of the breach and then held in place to stop the flow. Patches range from simple plumbing devices such as pipe sleeve patches to chain thumb screws to complex chlorine bonnets to inflatable drum and container sleeves and bags. One's ingenuity is the only restraining factor in patching.

Unfortunately, it is found that again what looks good in theory does not always work. For example, one of the common training techniques used in leak control is the old MC 306 cargo tank. Usually, this tank is at some training facility and standing on its wheels with the dollies down and supported.

There is a hole in the side of the tank that students are to patch using a banding device over a flexible gasket plate, an inflatable bag system, or some other device. There is water flowing from the tank to simulate some product. As the training evolution unfolds, team after team proceeds to throw straps, cables, or similar restraints over the tank, set the gasket plate, and at least minimize the flow of "product."

Many agree that this was an excellent training evolution. However, was it? First, how realistic was the scenario in question? In the real world, it is very uncommon for hazmat personnel to be called to a cargo tank that is standing upright with a hole in its side. Most commonly they are rolled over and on their side. If, per chance, they are found upright, leakage is most commonly found in valving, piping, or some other appurtenance attached to the cargo tank. Once in a great while, an upright cargo tank will strike some object and develop a hole at one end or the other, but rarely in the middle. As such, the scenario is plausible but quite unlikely.

Will this evolution work in a more probable scenario? Maybe. If the tank is on its side, the evolution normally will not work because bands, straps, and other restraints can either not be placed around the tank or, if they can, they usually cannot be positioned to allow good pressure on the gasket plate or bag. If the tank is upright and the problem is with a valve or piping, the equipment is not appropriate. Further, if the tank is upright and the problem is on one end (or end corner), the equipment may work, but that is dependent on the size of the breach and the amount of deformation associated with it.

This short discussion is not meant to say that patching is not an appropriate option, because it often is. Rather, this discussion is meant to say that the evaluation of patching devices and options must consider the specific conditions and situations that can be encountered. Further, some of the pat answer devices and approaches routinely taught may be of little value, depending on the exact incident situation.

LEAK CONTROL—OVERPACKING

Overpacking is the process of placing a leaking container inside another container to confine any further release of product. For a long time, it was one of the more comon means of handling an atmospheric drum as well as other atmospheric container leaks.

Since then, overpacking has come a long way. New technologies have introduced compressed gas cylinder coffins. Probably the most common cylinder coffin is designed to handle 150-pound chlorine cylinders. It consists of an oversized cylinder with one end equipped with a swinging hatch. When the hatch is opened, a 150-pound chlorine cylinder is slid into the coffin. Next, the hatch is closed and sealed, thus preventing the escape of any additional gas. One warning: These coffins are only made for chlorine and its maximum cylinder pressure. The placement of other or higher pressure cylinders may lead to violent failure of the coffin.

There are also high pressure 2,200-psi coffins available. They are certified for standard compressed gas cylinders. Unfortunately, their cost is approximately $30,000 to $40,000.

All forms of overpacking have several drawbacks. First, overpacking requires moving the container, usually by changing its position. Often the containers and their contents weigh from a minimum of 100 to 200 pounds up to close to 1,000 pounds. As a result, significant strength is needed to perform the task. Further, this weight presents numerous safety hazards ranging from back injury to crushing injuries to severe pinches.

The movement of the container can present other hazards as well. Suppose the container has been exposed to an internal or external chemical stressor. The movement required to accomplish overpacking could easily lead to partial or complete container failure.

Finally, the containers most frequently overpacked are those that are often stored or shipped in groups (drums and cylinders). Routinely, the container with the leak is not the most readily accessible but in the midst of others. This situation requires the movement of unaffected containers and thus additional physical labor for the entry team.

LEAK CONTROL—CRIMPING

Crimping is the process of compressing a tube or pipe to prevent the release of product. Although it is not one of the more commonly used leak control option, it can be very effective should the situation and piping allow. When the release is from an atmospheric container, relatively small diameter (3/4 inch or less) copper, thin walled stainless steel, flexible plastic, or other, piping and tubing can be crimped. Rigid plastic (e.g., PVC and CPVC) is not a candidate because it will crack and break further before crimping will work. Heavy walled steel, stainless steel and other metallic pipe cannot be crimped because of their hardness and lack of give. When pressure enters the picture, plastic tubing and pipe should generally not be crimped because they may fail.

LEAK CONTROL—PRODUCT TRANSFER

Product transfer as part of leak control is an often overlooked option. Most commonly this process is considered for bulk storage containers, after the incident is stabilized, or as part of recovery for smaller containers. However, product transfer can be a safe, easy, and less time-consuming alternative to overpacking.

The recommended procedures for overpacking include plugging or patching the leak before overpacking is performed. The same is true for product transfer. In this procedure, a pick-up tube is inserted into the leaker through a bug, lid, or similar opening and the product is pumped into another container. This process eliminates the need to reposition the leaker, thus reducing the threat of weight related injury or the potential for container failure.

Additionally, should access to the leaker be insufficient for overpacking operations, product transfer may be the ideal answer. Specifically, instead of moving other containers to gain space for repositioning, the pump's pick-up tube is inserted and the product is transferred.

Over recent years, industrial teams have increasingly relied on product transfer to minimize the potential for injury as well as for speed and ease of use.

Some industrial gas emergency-response teams even do product transfer for high-pressure gas cylinders. The process uses an approach similar to an air cascade system. First, a series of three to four empty cylinders and the leaker are manifolded together. Next the leaker is opened, followed by one of the empty cylinders. Gas flows from the leaker into the empty cylinder until their pressures are equalized. If the proper size and capacity cylinders are used, this process drops the pressure of the leaker 50% (e.g., starting at 2,000 psi and winding up at 1,000 psi). The first empty is closed and another is opened until the pressures again equalize. This achieves another 50% drop in pressure (e.g., 1,000 psi to 500 psi). The second empty is closed and the last one is opened. Again, the pressure is dropped another 50% (e.g., 500 psi to 250 psi).

At these pressures, the gas teams, with their specialized equipment, can often control the leak. It is essential to note that this illustration is only given as an example and should only be attempted by properly trained and equipped personnel.

LEAK CONTROL—VALVE ACTUATION

Valve actuation is sometimes an option available to control a leak. In some instances it is a simple task to shut a valve to stop a leak. Many responders have been taught that this is an ideal way to control a leak, which in some instances is true. However, the actuation of valves can be a tricky and potentially dangerous proposition.

To understand some of the considerations better, it is important to understand some basics of product flow. When a liquid or a liquefied compressed gas (and, in some instances, a gas as well) is flowing through a pipe, it tends to want to continue flowing. This is known as inertia. If this flow is stopped abruptly, the trailing product pushes against the leading product, producing the equivalent of a chain reaction accident. The result of this is often a powerful pulse of energy. In water mains and fire hoses, this is known as water hammer. In other cases it is called adiabatic pressure shock or pressure spike.

Unfortunately, when various gases and in some cases liquids are moving through a piping system, a rapid cessation of flow can produce a devastating force. The resulting energy release (similar to water hammer) and heat of compression can cause valves, pipes, and even entire containers to fail violently. Yet, this is exactly what can occur if a valve is actuated too rapidly.

Another potential problem involves trapping liquefied compressed gases in a liquid filled space between two valves or in a cylinder. Gases such as hydrogen chloride, ammonia, and sulfur dioxide, among others, are especially noted for creating violent hydrostatic pressure ruptures in pipes or containers when this occurs. Specifically, since the system is liquid filled, there is no vapor space. Because there is no vapor space and many of these liquefied gases do not compress well, there is no way for the system to absorb a pressure increase should it occur. This problem is especially acute for liquefied gases that are below ambient temperatures within the contaminant system.

As a result should the liquid heat only slightly, the pressure can rapidly spike to extremely high levels. In some instances, this situation has resulted in a pressure spike of thousands of psi. Such spikes will cause almost any container to fail, often violently.

When it has been determined that the closing of a valve is appropriate, remember never use force to open or close a valve. Valves and valve stems have been known to fail if such pressure is applied. In some instances, the stem simply shears off. In other instances, the stem becomes a high speed projectile capable of fatally wounding anyone within range. Further, such a problem will only complicate the leak control process.

Indirect Leak Control

Procedures used to control the release of product without physically controlling the breach	**Possible Tactical Objectives:** Remote Shut offs Emergency Shut-offs Product Transfer Product Displacement

Indirect leak control methods involve some form of leak control where the product is manipulated without manipulating the breach itself. As a result, indirect methods, generally, are substantially safer than direct control operations. The higher degree of safety is primarily due to the lack of close proximity to the breach and the container, thus any escaping product.

LEAK CONTROL—REMOTE OR EMERGENCY SHUT-OFFS

The use of shut-offs, just as the name implies, involves closing some type of product flow-control device to prevent product from reaching the breach. So why consider plugging, patching, and over-packing if all that was needed was to close the valve? There are several reasons.

Before closing the remote valve, try to determine whether the breach was caused by overpressurization of the container. If there is no trauma, why did the pipe or container fail before a relief valve activated and relieved the pressure? Could the piping or container have been damaged before the failure, or does this system lack a relief valve? In any case, if the possible cause involves a pressure increase, the pressure could still be present. As such, shutting the valve could allow the pressure to build until the next weakest link in the system fails.

The use of remote or emergency shut-offs is generally one of the safest means of controlling a leak. Fortunately, as a result of process safety management (PSM) and risk management planning (RMP) requirements established by OSHA and EPA, more and more fixed facilities are reviewing their operating systems to identify where and how problems can occur. A process fault tree analysis is sometime used. Once the analysis is completed, ways of preventing or rapidly stopping the problem are identified. Consequently, remote and emergency shut-offs are becoming increasingly common.

Such devices and their operation should be noted whenever preplanning inspections or visits occur. Further, information regarding their possible existence and locations must be requested when involved in an actual incident.

When considering transportation containers such as highway cargo tanks, many are required to have emergency shut-offs. Unfortunately, accidents, misuse, poor maintenance, and so on can make them nonfunctional.

LEAK CONTROL—PRODUCT TRANSFER

Unlike the product transfer discussed under direct methods, indirect product transfer involves the use of pumps and piping systems that are in place and simply require activation of or realignment of valves. Most commonly, this type of product transfer occurs at fixed sites with bulk storage systems and containers.

LEAK CONTROL—PRODUCT DISPLACEMENT

Displacement is the process of pumping a material, usually water or foam, into a container so the new material replaces the product escaping through the breach. The most common example of such an operation is the placement of water into a leaking vehicle fuel tank so water exits through the breach instead of fuel. Obviously, such an operation will normally not go on for an extended period of time and is not the overall solution to the problem. Rather, displacement is a stop-gap technique used to buy time for some other operation.

Two critical considerations that must be examined are the water solubility and water reactivity of the product if water or foam is to be used in the operation. Obviously, it would be of little value to use water or foam, even polar solvent foam, if the product were water soluble. Just as obvious, it could be lethal to use foam or water on a product that reacts with water.

One specific application of displacement that deserves mention is its use in bulk facility fires involving dike and flange fire or even spill situations. The use of foam or water has been very effective when they were introduced into the piping involved. When the water or foam reaches the piping failure, the fire is extinguished or the product flow is stopped by displacement of the product.

FIRE CONTROL

The process of minimizing the total impact of a fire that involves or threatens to involve chemicals	**Primary Tactical Objectives:** Extinguishment Controlled Burn Exposure Protection Withdrawal

Fire control is the process of minimizing the total impact of a fire involving or threatening to involve chemicals. The primary considerations involve the safety of responders, the public and the environment.

In some instances it is better to allow the fire to burn while in others it is better to extinguish the flames. Some considerations in making an appropriate decision include identifying the following:

1. What is burning: product, container, or environment
2. The potential harm and impact of allowing the fire to burn versus suppression activities
3. The potential for explosion or BLEVE
4. The potential magnitude of explosions or BLEVE
5. If there would be less harm if the fire were allowed to burn

FIRE CONTROL—EXTINGUISHMENT

If the estimate of the course and harm of the fire indicate suppression activities are advisable, there are multiple considerations that must be addressed. First, an appropriate extinguishing agent must be identified. The options include foam, dry chemical, dry powder, carbon dioxide, and some rather unusual options including liquid nitrogen, sodium chloride, soda ash, slurry mixes, and so on. By far, the most commonly used agent is fire-fighting foam, simply because flammable and combustible liquids are the most common fuels.

For situations and substances other than flammable or combustible liquids, foam or other agents may be required (assuming the product and the agent are compatible with each other). On the other hand, there are substances and situations that render foam, commercial dry chemical agents, CO_2, as well as all the other agents totally ineffective or violently reactive.

For example, carbon dioxide, or other chemical agents propelled by carbon dioxide, must *never* be used on flammable metal fires because the CO_2 is an oxidizer to the fuel. The hypergolic combustion of hydrazine and nitrogen tetroxide is immune to extinguishing effects of all the agents. Foam (or water in any form) will react violently with all flammable metals. Consequently, it is essential to research the potential interactions between the agent and the fuel before any extinguishment effort begins.

Since there are so many different potential fuels and agents, the remainder of this discussion will focus on fire-fighting foam.

To use foam effectively and efficiently, some specific data must be determined before proceeding. First, what foam type and expansion ratio are most appropriate for the situations? Second, how much is needed to extinguish the fire? Third, is a sufficient water supply available? Finally, are the right type and capacity appliances available for the effort? Before these questions can be answered, it is necessary to understand some basic aspects of fire-fighting foam. These aspects include how foam is made, expansion ratio options, and its basic properties.

How Foam is Made

Present day foams are known as mechanical foams. Earlier foams were made by combining several chemicals and were known as chemical foam. Mechanical foam is made by creating a precise mix of water and foam concentrate to produce a foam solution. Foam concentrates are identified by percentages. These percentages identify how much foam concentrate is mixed with how much water to make the solution. For example, a 3% concentrate requires that three gallons of concentrate be mixed with 97 gallons of water to make 100 gallons of solution. A 6% concentrate requires six gallons of foam concentrate be mixed with 94 gallons of water to make 100 gallons of solution.

The mixing process is done by an in-line eductor or some type of proportioner. These devices assure that proper mix of concentrate and water.

After the foam solution is mixed, it flows to the applicator or appliance. At the appliance, the solution is mixed with air to produce foam. The amount of air mixed with the solution determines the expansion ratio of the foam.

Expansion Ratio

The expansion ratio is a comparison of the solution volume to the foam volume produced. There are three expansion ratio categories used when discussing foams: low expansion, medium expansion, and high expansion (Table 7-10). As with most things, each ratio has its strengths and its weaknesses and thus has appropriate and inappropriate applications. Also, there are trade-offs: a quality that helps one problem creates another in some aspect.

Low expansion foam's expansion ratio ranges from 5:1 to 10:1. This means that every gallon of solution produces between five to ten gallons of foam. Low expansion foam is generally most suited for fire extinguishment and is generated by an automatic or foam nozzle. Automatic nozzles tend to produce low ration foam (5:1) while foam nozzles produce more toward the high end (10:1). The lower expansion of the automatic nozzle has the advantage of being heavier and thus is less affected by the thermal column produced by the fire. Also since it contains more water, it is less likely to evaporate as readily as the higher expansions. However, the more highly expanded foam produced by a foam nozzle creates a lighter and airier foam that is more suited for vapor suppression.

Medium expansion foam's expansion ratio ranges from 50:1 to 100:1 and is generated by screen or multihead spray cones. Medium expansion foam is often ideal for

Table 7-10 Expansion Ratios

Expansion	Expansion ratio	Appliance
Low	5:1 to 10:1	Automatic or foam nozzle
Medium	50:1 to 100:1	Screen or multihead cone
High	300:1 and up	Fan-powered blower

vapor suppression activities because it covers a large area very quickly due to its greater expansion. Further, the greater expansion means there is less total fluid produced, thus limiting the spread of product over a larger area. Although medium expansion foam will extinguish fires, it is not as quick or efficient as low expansion. Also, it does not behave well in high wind conditions, because large pillows of it tend to blow away, or high heat, which causes it to evaporate.

High expansion foam's expansion ratio ranges from 300:1 to over 1,000:1 and is generated by fan-powered blowers. Its primary use has been within structures. Although it can extinguish fire under the right conditions, it has also failed many times. It is also extremely sensitive to even light wind conditions. Possibly its best use is in limiting the impact of explosives. When it is used to fill the space around explosive devices, the millions of bubbles act as shock absorbers, dissipate the force of the blast, and reduce its damage.

Foam Properties

To identify what type of foam is appropriate, a series of four foam properties must be examined. These properties are drain time, viscosity, flow rate, and polar solvent compatibility.

Drain time is the rate at which the foam solution seeps from the bubbles. The longer the drain time of the foam, the longer the foam blanket will last and work effectively.

Viscosity is the relative fluidity of the foam blanket produced. High viscosity foam tends to stick together and not flow well, while low viscosity foam tends to spread and flow rapidly.

Flow rate is the speed with which the foam blanket spreads across the surface of the product. This should not be confused with the application rate that is the gallons per minute (gpm) of foam needed to extinguish the fire. High flow rate foam rapidly moves to cover the surface of the product, while low flow rate foam does not.

Polar solvent (alcohol resistance) compatibility is the ability of a foam blanket to withstand application to a miscible (water soluble) liquid. Nonpolar solvent compatible foams are destroyed or at least will not suppress vapor or extinguish fire involving polar solvents.

Types of Foam

There are four primary types of foam: protein, fluoroprotein, aqueous film forming foam (AFFF), and AFFF polar solvent. In recent years an additional foam has been introduced: film forming fluoroprotein (FFFP), which also has a polar solvent version available. In either case, FFFP is similar to AFFF in most respects except it usually has a slightly longer drain time. Table 7-11 summarizes their properties.

Protein foam is produced from animal protein. It has a long drain time, high viscosity, and low flat rate. It is not compatible with polar solvents. Due to long drain time, it is excellent for vapor suppression unless the product is polar. Protein extinguishes rather slowly because of its viscosity and its associated low flow rate.

Fluroprotein foam is protein based but has been fluorinated. It has a shorter drain time than protein foam, low viscosity, and thus an excellent flow rate. It is not compat-

Table 7-11   Foam Properties

Foam Type	Approximate drain time	Viscosity	Flow rate	Polar solvent compatibility
Protein	20 to 30 minutes	high	low	none
Fluoroprotein	15 to 25 minutes	low	moderate to high	none
AFFF	10 to 15 minutes	low	high	none
AFFF polar solvent	8 to 15 minutes	low	high	compatible

ible with polar solvents. It is reasonably good at vapor suppression, as well as an excellent extinguishing agent unless the product is polar. Some newer types meet AFFF and polar solvent standards as well.

AFFF is a synthetic surfactant type material. It has a short drain time, low viscosity, and thus excellent flow rate. It also produces a film of solution that covers the surface. However, the film readily sinks in flammable liquids. Any openings in the blanket must be considered to have no film in place. It is not compatible with polar solvents but is a reasonably good vapor suppressant and is an excellent extinguishing agent unless the product is polar.

AFFF polar solvent is basically the same as standard AFFF, but it has a slightly shorter drain time and is compatible with polar solvents.

Quantity of Agent Needed

Bulk Storage Tanks.   Before any firefighting with foam is started, the application rate and total concentrate need must be calculated. The basis for these calculations is found in NFPA 11. The examples used are for fixed bulk storage tanks unless otherwise indicated. All the values listed are based on standard data and may vary from manufacturer to manufacturer.

The first step is to identify the needed application rate of foam solution. This is based on the surface area in square feet. The calculation varies depending on the shape of the surface. If it is square or rectangular, the formula is

$$\text{length} \times \text{width} = \text{surface area}$$

For example, an area that is 20 feet wide and 50 feet long has a surface area of 1,000 square feet.

There are two ways to calculate the surface area of a circle. The first is

$$\text{pi} \times \text{the radius squared} = \text{surface area}$$

Pi(π)has a value of about 3.14, and the radius is one half the diameter of the circle. This means that for a tank with a diameter of 100 feet, the calculation is:

$$3.14 \times (50 \times 50) = 3.14 \times 2{,}500 = 7{,}850 \text{ ft}^2$$

The second way is to square the diameter and multiply it by 0.785. This means that for a tank with a diameter of 100 feet, the calculation is:

$$(100 \times 100) \times 0.785 = 10{,}000 \times 0.785 = 7{,}850 \text{ ft}^2$$

The actual application rate depends on how the foam will be delivered. If the foam is applied through a fixed foam system, the application rate is 0.1 gallons per minute (gpm) of foam per square foot. If the foam must be applied by portable appliances, the application rate is 0.16 gpm per square foot.

Using the 100 foot diameter tank, the calculation for a fixed system fire fighting attack is:

$$7,850 \times 0.1 = 785 \text{ gpm}$$

The calculation for a portable application is:

$$7,850 \times 0.16 = 1,256 \text{ gpm}$$

This previous calculation identified the amount of solution needed, not the amount of concentrate. To calculate the amount of concentrate, multiply the gpm by the percentage concentrate. That means for a 3% foam to be used on the 100 foot tank for a portable application, the calculation is:

$$0.03 \times 785 \text{ gpm} = 23.55 \quad \text{(rounded off to 25 gpm)}$$

For the portable application, the calculation is:

$$0.03 \times 1,256 \text{ gpm} = 37.68 \quad \text{(rounded off to 40 gpm)}$$

Now the question is, "For how many minutes must this be continued?" The answer again depends on whether the application is from a fixed suppression system or portable application. Table 7-12 provides the particulars.

This means that to determine the total amount of concentrate needed on hand before foam operations begin, the gallons per minute of concentrate must be multiplied by the appropriate number of minutes.

For fixed application (except for Type I outlets) the calculation is:

Flammable liquid: 25 gpm concentrate $\times$ 55 minutes = 1,375 gal concentrate

Combustible liquid: 25 gpm concentrate $\times$ 30 minutes = 750 gal concentrate

For portable application the calculation is:

Flammable liquid: 40 gpm concentrate $\times$ 65 minutes = 2,600 gal concentrate

Combustible liquid: 40 gpm concentrate $\times$ 50 minutes = 2,000 gal concentrate

Table 7-12 Time Required for Portable Application (from NFPA 11, 1997)

Type fuel	Portable application	Fixed system		
		Type I outlet	Type II outlet	Subsurface
Flammable liquid and crude	65 minutes	30 minutes	55 minutes	55 minutes
Combustible liquid	50 minutes	20 minutes	30 minutes	30 minutes

At this point, it must be determined if a sufficient water supply exists and if appropriate application appliances are available. When all the pieces are in place, foam operations can begin.

Once the application begins, some type of positive results should be seen within 20 to 30 minutes. This does not mean that the fire should go out, rather the flame and smoke intensity should decrease somewhat and that there may be periodic wisps of steam. If no such positive signs are noted, it is likely there is some problem with the attack. The problems can range from insufficient flow to an unidentified obstruction keeping the foam from covering the surface. However, if those positive signs are seen, extinguishment should be only a matter of time.

Loading Rack Fires

For firefighting at loading racks, NFPA 11 indicates an application time of 15 minutes and a flow rate of 0.16 gpm/ft^2. The only exception is if AFFF or FFFP standard or polar resistant is used on a membrane spill less than about one inch deep. In this case an application rate of 0.10 ppm/ft^2 is allowed.

Dike Fire

For dike fires, NFPA 11 indicates fixed systems must discharge at 0.10 gpm/ft^2 for 20 (combustibles) to 30 (flammables) minutes for all hydrocarbons. Portable applications require the standard 0.16 gpm/ft^2 for 20 to 30 minutes.

FIRE CONTROL—CONTROLLED BURN AND FIRE CONTROL—EXPOSURE PROTECTION

If it is determined that extinguishment is not appropriate or if there are not enough resources to accomplish it, a controlled burn is one option. A controlled burn indicates that the fire will be allowed to burn itself out.

However, other operations, most commonly protection of exposures, will be undertaken. In this respect, controlled burn is similar to confinement for structural firefighting in that both routinely are paired with protection of exposures. This option is chosen because, although the fire will be left to burn, there are no significant personnel hazards if responders operate in and around the exposures. In other words, this is not a critically dangerous situation.

FIRE CONTROL—WITHDRAWAL

Withdrawal is generally a last resort used when the incident is completely out of control or threatens to cause mass destruction or mayhem. In this case, responders withdraw to the perimeter and evacuate everyone else as well. The choice of this tactical objective will require significant effort in the area of public protection and emergency public information.

Although there will be few standard operational actions taken, personnel and commanders will have their hands full with all kinds of requests and immediate action needs. The pressure during this type of operation is often intense. Media will also tend to flock to the scene. Conditions that warrant this tactical objective include:

1. The potential for mass fire or explosion
2. The potential for BLEVE
3. The potential for multiple smaller explosions
4. Any major operations where needed resources far exceed those that are available

RECOVERY/TERMINATION

The process of concluding the incident both operationally and administratively	**Specific Tactical Objectives:** depend upon the phase involved.

Recovery/Termination is the process of wrapping-up the incident both operationally and administratively. Recovery involves operation returning to the preincident conditions (as best as possible) regarding the scene and resources. Termination involves the process of documentation, analysis of the incident, and correction of deficiencies.

Recovery has two phases: operational and administrative. The following are the primary tactical objectives of each phase.

Operational Recovery

Operational Recovery involves returning the incident *scene* to its preincident condition and the release of response personnel	**Primary Tactical Objectives:** Oversight of Cleanup Oversight of Product Transfer Oversight of Container Righting and Handling Demobilization

The tactical options listed under operational recovery occur at various times, depending on the incident. In some cases, product transfer will occur while some aspects of the incident are still ongoing while in other cases it will occur after the majority of responders have been released from the scene. Container righting may be an integral part of stabilization during the emergency phase while in other incidents, it will be well after the incident is thoroughly stabilized.

As a result of this overlapping, it is not possible to say whether recovery is before or after the emergency phase is complete.

RECOVERY—OVERSIGHT OF CLEANUP

When the emergency phase of the incident is winding down or is over, there comes the time when the shift is made to clean up. Unfortunately, there is no clear-cut instant when the transition is made. Often, various response units have already been released, while others are taking up, and others are still operating. At this point, it is extremely helpful to coordinate with all parties that are or will be involved in the clean-up phase.

These may include representatives from governmental agencies, private contractors, the shipper, hauler or manufacturer, and so on. In some instances, fire units, hazmat, or law enforcement may be needed on the scene to monitor activities or assist the clean-up crew in an ancillary fashion.

It is not uncommon, especially when large quantities of flammable or reactive chemicals are involved, for clean-up personnel to need the assistance of engine companies to extinguish spot fires, apply foam blankets, and so on. In Livingston, Louisiana, responders did more work during the clean up than they did during the actual emergency.

RECOVERY—OVERSIGHT OF PRODUCT TRANSFER

Product transfer is the process of removing product from a compromised container and placing it in another suitable container. There are several considerations about product transfer that must be examined. By this point in the incident, most if not all the product's properties should be identified. As such, it is essential to identify any potential impacts they may have during the transfer process.

Further an essential consideration is whether it is appropriate to transfer the product in the first place. In some situations, such as the roll-over of a loaded, aluminum MC 306 gasoline cargo tank, the vehicle *must* be off-loaded before righting (unless high-lift bags are used). If the product is not off-loaded, there is a very high probability of structural failure with possible total loss of product. However, a similar type of roll-over involving an MC 307, a DOT 407, or a DOT 111 specification tank car may not require off-loading before righting.

The specific type of transfer mechanism (e.g., vacuum unit, pump, pneumatic, and gravity) must be appropriate for the situation and the product. For example, a standard acid transfer pump will not be compatible with all types of acids. There have been incidents where numerous pumps have had the impellers eaten off when used with non-compatible products. There have also been situations where gas powered water pumps have been brought by contractors to transfer gasoline. Additionally, especially in the cases involving flammable or combustible products, the containers must be properly bonded and if possible grounded to eliminate static electric build-up.

The methods used to gain entry into the container must be appropriate. The use of pneumatic hole saw on an aluminum MC 306 carrying gasoline is a sound, industry accepted off-load method. On the other hand, the use of such an approach on a stainless MC 307 carrying the same lading, would be potentially lethal. The removal of a bung with visible crystallization from a leaking 55-gallon drum of isopropyl ether could generate a lethal explosion, while removing the bung from a leaker containing hydrochloric acid would be appropriate.

Finally, careful examination must be given to the receiving container. The IC must be certain that the container is completely compatible and able to be transported once the transfer has been accomplished. There have been many situations where a transfer was begun and the receiving container also failed because the container itself or one of its attachments (e.g., valves or flanges) was not compatible. In other situations, the transfer has been completed and the container, vehicle, or drum has been found not to meet appropriate requirements and could not be moved.

RECOVERY—OVERSIGHT OF CONTAINER RIGHTING AND HANDLING

Container righting and handling generally follow product transfer and routinely involve outside contractors. These include wrecker and crane operators as well as heavy equipment operators. In this instance, the primary function is to assure that safe actions are taken when container handling occurs.

Some of the most important considerations are the equipment and methods selected to handle the container. Lifting equipment must be of appropriate size and number. It is always a bad sign when the wrecker operator chains the bumper to stakes to keep the nose of the rig from lifting.

In some instances, containers will fall simply because there is no way to gain a purchase or other lifting point. Although this is not ideal, it may be necessary. If it is, the potential harm must be identified before the operation is allowed to continue. The placement of cribbing, high lift bag, or other materials must be considered to prevent any additional damage.

This particular aspect of recovery is often difficult and can be controversial, especially if the incident has been going on for awhile. People are tired and they just want it to be over. However, it is essential to remain vigilant.

RECOVERY—DEMOBILIZATION

Demobilization is the systematic, orderly, and timely release of resources, personnel, and units on the scene. During a small incident, this process is simple and straightforward. However, when the incident is large and involves many resources from many sources, demobilization can become rather complex and difficult.

Primarily the demobilization process seeks to provide adequate resources without holding excessive units on the scene. This can be especially important when contractors, municipal employees, and similar workers are involved in the response. As such, effective demobilization can cut costs and delay as the incident winds down.

Administrative Recovery

Administrative Recovery involves returning response capabilities to preincident conditions	**Primary Tactical Objectives:** Inventory Control Restocking Financial Restitution

All of these tactics are rather mundane and straightforward. As such they will not be addressed other than to mention their need.

Termination

The process of documenting, investigating and correcting errors involving the incident	**Primary Tactical Objectives:** Debriefing Hazard Communication Critique Investigation After Action Analysis After Action Report Follow-up

Termination involves the documentation, evaluation, error correction and investigation of incident operations. Further, it is important to recognize that most of the tactics identified under termination do not occur on the incident scene or even during the emergency phase of the incident, yet they are still addressed under this goal. The reason for this is simple: although most of these do not occur on the incident scene, they are mandated by federal regulations or are essential to the continuity of service. As such they are included under this goal.

TERMINATION—DEBRIEFING

Debriefing is the process of gathering information about what went on during the incident. It seeks to find who did what, when, and how. By establishing a time-line (when), an effective incident sequence can be established.

Everyone who was on the incident scene should take part in a debriefing. However, it does not have to be done with everyone in one place or even at the same time. It should occur as soon as possible after the incident. The reason is simple: Personnel will remember information much more clearly and concisely when the debriefing occurs shortly after the incident.

Debriefing is not a critique! As such, this is not the time to identify problems or even positive actions; this is a time to gather information.

TERMINATION—HAZARD COMMUNICATION

Ideally, the exit hazard communication should be part of the debriefing. However, if the debriefing does not occur immediately after personnel leave the scene they should at least receive an exit hazard communication. All this process involves is informing personnel of the following:

1. The substance(s) involved
2. Possible symptoms of exposure
3. Any action such as taking a shower, washing of uniforms or gear, that personnel should take
4. Identifying a contact person if any questions arise

TERMINATION—CRITIQUE

Critique is the process of identifying shortcomings and seeking recommendations to prevent their recurrence. The official critique should occur at least five to seven days after the incident. Some may ask why the critique occurs so long after the fact. First and foremost, it allows participant's emotions to subside. Normally, this produces a more relaxed and productive critique. In addition, this allows time to conduct any investigations, gather information, and for companies to have their own critiques. In some instances, the critique may occur several weeks after the event.

Some may say that this time lapse means vital information will be forgotten. That is correct if a debriefing has not been conducted. However, when a thorough debriefing is carried out shortly after the incident, all vital information will be captured and available for use in a calm and productive critique.

The primary goal of the critique is not just to identify problems, but also to identify possible solutions. This is where the recommendations become critical. Quite often, these recommendations are vital and help solve the problem as quickly as possible. Further, the observations and perspectives of others often provide a better understanding of the true nature of the problem and its potential solutions.

The official critique does not have to include all personnel who were on the scene. Rather, it should include commander and mid-level supervisors. This prevents the critique from becoming a mob action and generally produces better information and recommendations.

TERMINATION—INVESTIGATION

The investigation phase can be wide ranging. It can look at injuries, the cause of the incident, possible enforcement actions, and so on. The information from the investigation may be critical to addressing problems and recommendations found in the critique.

TERMINATION—AFTER ACTION ANALYSIS

After action analysis involves looking at the big picture to identify trends, both good and bad. It tries to identify system and recurrent problems as well as the reasons for the problems. Ideally it will help find answers as well.

After action analysis can be helpful in identifying specific response and incident trends. For example, the number of transportation incidents can fluctuate rather widely from year to year. However, when examined in the context of multiple years, very interesting and important trends can be identified.

By analyzing their incident data, one jurisdiction found that the number of rail incident dramatically decreased after ribbon rail was installed through their response area. The reason for this decrease was simply that ribbon rail is safer and more reliable than stick rail. During subsequent years however, the decrease turned around and became a slow but steady increase. When this trend was examined more closely, it was found that after the ribbon rail was installed, the number of daily trains had doubled. Further, this increase in traffic was expected to continue for the foreseeable future. As such, more rail incidents were statistically probable. In other words, the risk was increasing with the increasing number of trains resulting in a greater probability of an incident.

However, on closer examination, another interesting fact became evident. Specifically, although there were more trains per day and more incidents after the new rail was installed, the probability (and thus risk) of an incident on a per train basis had decreased. In other words, although the number of incidents increased, the likelihood of any given train being involved in an incident actually decreased.

Another important area of consideration in the analysis process is injury and "near miss" data. Injury trends and recurrence are an important part of the overall health and safety program. Such data helps identify the need for training, changes in procedures, and so on. Near misses (sometime called near hits) are situations where personnel were nearly injured. A general rule of thumb is that for every injury, there are approximately 300 near misses. As such, when near misses are identified, they present an excellent opportunity to identify the possible causal factors and address them *before* someone is injured.

TERMINATION—AFTER ACTION REPORT

The after action report puts all the pieces of documentation together. It includes check lists and sheets from the scene, debriefing timeline and notes plus critique, and after action analysis recommendations. This process includes updating equipment and personnel exposure files as well.

One important sidelight is the definition of personnel exposure. A personnel exposure is when a member of the organization comes into direct contact with the contaminant. This contact may be a result of direct or cross contamination. Most commonly such exposures are a result of improper PPE, entry into unauthorized zones, PPE failure, or cross contamination. *Simply being on the incident scene is not an exposure.* The reason this distinction is so important is simply that should a true exposure occur, it must be reported to the medical director for his or her evaluation and review. Further, the medical director must determine appropriate actions, treatments, testing, screening, and more, necessary to assure the individual's health. As such, the reporting of possible exposures simply because an individual was on the incident scene is truly counterproductive, costly, and possible grounds for legal or regulatory actions.

TERMINATION—AFTER ACTION FOLLOW-UP

After action follow-up is the real heart of the entire termination process. Specifically, follow-up is the process of assuring that identified shortcomings are addressed and thus prevented from occurring again. If you will, follow-up is the system building and strengthening process. The follow-up process systematically identifies the nature of the problem and why it exists. It further identifies what actions are needed to correct the problem and how long these corrections should take. Finally, follow-up identifies a responsible party who must assure that corrective actions are implemented and completed.

Granted, although this sounds like a standard management approach, it is the bottom line and the most critical step of all termination activities. OSHA feels it is so important that the procedures for conducting the critique and performing follow-up procedures are required sections of the employer's Emergency Response Plan (and rightly so). Specifically, the most effective and detailed debriefing, critique, analysis, and report are of little value if problems they identify are not addressed.

Unfortunately, all too frequently the post incident evaluation of emergency operations finds a total absence of any follow-up procedures or actions. The situation is often compounded by a lack of truthful and open examination of operations. As a result, the entire post-incident evaluation or termination process is of no value because problems are either not identified or not corrected. Such a situation leads to poor morale and the potential for the same or similar problems to occur repetitively. Neither situation is good for the system or its personnel.

SUMMARY

This chapter addresses the fourth step in the GEDAPER process, namely Assess Tactical Options and Resources. It has:

1. Explained the relationship of strategic goals, tactical objective, and tactical methods
2. Identified the primary tactical objectives for each of the eight strategic goals
3. Explained the role of resources in the determination of appropriate tactical options
4. Described the intent of each tactical objective listed in this chapter
5. Explained the interconnections between the Isolation tactical objectives of perimeter and zone establishment, denial of entry, initial public protection, and withdrawal
6. Described the advantages of incident leveling and notification needs
7. Explained the role of communications links as part of notification
8. Explained each of the six primary tactical objectives available to achieve the goal of identification
9. Described the functioning, strengths, and weaknesses of the five primary types of monitoring equipment
10. Identified and explained the nine primary tactical objectives available to achieve the goal of protection
11. Identified and described the seven personnel decontamination mechanisms
12. Described the limitation of thermal and chemical protective clothing
13. Described the primary components of a site safety plan
14. Explained the role that spill typing plays in identifying appropriate tactical options for the goal of spill control
15. Identified the primary tactical objectives for each spill type
16. Explained the difference between direct and indirect leak control tactical objectives and their implications on the safety of entry team personnel
17. Identified and described the five primary tactical objectives available to perform direct leak control
18. Identified and described three of the primary tactical objective available to perform indirect leak control
19. Identified three alternative extinguishing agents available to accomplish fire control
20. Described the strengths and weaknesses of the four primary types of fire-fighting foams

21. Explained the expansion ratio options for fire-fighting foams, their strengths and weaknesses
22. Described the difference between Recovery and Termination
23. Identified and described the two phases of recovery
24. Described the role of the three phases of termination

Once the tactics and resources are identified, the plan of action, in a rough form has also been identified. Chapter 8 addresses implementing the plan of action and continues with an examination of the evaluation and review processes.

SAMPLE BASIC SITE SAFETY PLAN

INCIDENT NUMBER:_____ PREPARED BY:_____

_____ YR/MN/DAY/NO.

INCIDENT COMMANDER:_____

HAZMAT BRANCH: _____

HAZMAT SAFETY:_____

INTERVENTION:_____

SCIENCE:_____

DECON:_____

HAZMAT EMS:_____

AGENCIES INVOLVED/INFORMED OF SSP:
  Police [　], PSP [　], Fire Dept. [　], EMS [　], PENNDOT [　],

  Other [　]_____

Substance/s Involved:

1._____ U.N. ID#_____

2._____ U.N. ID#_____

3._____ U.N. ID#_____

SUBSTANCE HAZARDS: Toxic [　], Flammable [　], Tissue Destructive [　], Unstable [　],
  Incompatibilities [　], Other:_____

ROUTES OF EXPOSURE:
  Inhalation [　], Ingestion [　], Skin Absorption [　], Skin Contact (specify for each
  substance).

TOXICITY BY INHALATION:
Based upon lowest value for the TLV/PEL, STEL, and/or Ceiling.

[　]—Severely Toxic—Value of 0–100 ppm
[　]—Moderately Toxic—Value of 101–500 ppm
[　]—Toxic—Value over 500 ppm

[　]—Simple Asphyxiant
[　]—Irritant

TOXICITY BY INGESTION:
Based upon lowest human LDLO (if available) or animal LD50.

[　]—Severely Toxic—Values of 1–50 mg/kg
[　]—Highly Toxic—Value of 50–500 mg/kg
[　]—Moderately Toxic—Value of 0.5–5 gm/kg
[　]—Toxic—Value over 5 gm/kg

FIRE HAZARDS: [   ]
  Flash Point_____ , LEL____%, UEL____%,

  Other Flammability Hazards:_____

REACTIVITY HAZARDS: [   ]
  Unstable [   ], Water Reactive [   ], Pyrophoric [   ], Radioactive [   ], Corrosive [   ],
  Oxidizer [   ], Decomposition [   ], Other:_____

SITE DESCRIPTION:
  Isolation distance: _____

  Zone locations: Hot_____

                   Warm_____

                   Cold_____

  Access points:_____

  Staging area:_____

  Other information:_____

PUBLIC PROTECTION MEASURES:
  Evacuation [   ]: distance:_____ , down wind:_____

  Protection in-place [   ]: distance:_____ , down wind:_____

PERSONAL PROTECTIVE EQUIPMENT:
  Minimum PPE: Full Turnouts/SCBA [   ], Chemical Clothing [   ], Thermal Protective
  Clothing [   ], DOTO [   ].

  PPE in hot zone:_____

  PPE in warm zone:_____

  Additional PPE info.:_____

DECONTAMINATION:
  Type of decon to be used: 4 step [   ], 3 step [   ], 2 step [   ], 1 step [   ], dry [   ]—
  specify:_____

  Solution: detergent [   ], bleach [   ], sodium bicarb [   ], acetic acid [   ], combination [   ],
  other:_____

  Victim decon: needed [   ], separate corridor established [   ], coordination with EMS [   ].

ADDITIONAL SITE SAFETY INFORMATION:
  Site hazards and characterization:_____

  _____

  Physical hazards:_____

  _____

  Additional safety information:_____

  _____

Site safety plan completed by:_____

SAMPLE HAZMAT SAFETY ASSESSMENT

INCIDENT NUMBER:_____ PREPARED BY:_____

YR/MN/DAY/NO.

The following checks have been made.

Basic site safety plan complete: yes [   ], no [   ],
Hazmat EMS officer designated: yes [   ], no [   ],

ACTION REVIEW	Entry 1	Entry 2	Entry 3	Entry 4	Entry 5	Entry 6	Entry 7	Entry 8	Entry 9	Entry 10
Substance/s identified:	known [ ], unknown [ ]									
Substance hazards:	known [ ], unknown [ ]									
Appropriate PPE identified:	yes [ ], no [ ].									
Appropriate decon identified:	yes [ ], no [ ].									
Decon set and ready:	yes [ ], no [ ].									
EMS on scene:	yes [ ], no [ ].									
Entry team identified:	yes [ ], no [ ].									
Back-up team identified:	yes [ ], no [ ].									
Pre-entry vitals taken:	yes [ ], no [ ].									
Air times listed on next page:	yes [ ], no [ ].									
Pre-entry briefing conducted:	yes [ ], no [ ].									
Safety assessment performed:	yes [ ], no [ ].									

Complete a separate column for each entry performed. When all sections appropriately completed, entry can be performed.

If additional entries are required, fill out additional data sheets.

Comments:_____

_____

FUNCTIONS AND TYPE PPE WORN:

Hot zone operations:_____

_____

Warm zone:_____

_____

PPE options:
   Turn-Outs (TO), Disposable Over Turn-Outs [DOTO], Proximity [P], Entry [E],
   Level A [A], Level B [B], Level C [C], Level D [D].

ENTRY PERSONNEL   PPE DATA   WORK/AIR TIMES   POST-ENTRY

Name	Suit type & mtrl.	On-air	Entry	Out	Off-air	Medical monitor	Rehab.
1.							
2.							
3.							
4.							
5.							
6.							
7.							
8.							
9.							
10.							
11.							
12.							
13.							
14.							
15.							
16.							

ADDITIONAL PERSONNEL   PPE DATA   WORK/AIR TIMES   POST-ENTRY

Name	Suit type & mtrl.	On-air	Entry	Out	Off-air	Medical monitor	Rehab.
1.							
2.							
3.							
4.							
5.							
6.							
7.							
8.							
9.							
10.							
11.							
12.							
13.							
14.							
15.							
16.							

SAMPLE ENTRY/PPE CHECK SHEETS

INCIDENT NUMBER:_____ PREPARED BY:_____

YR/MN/DAY/NO.   Intervention Officer

PPE TYPES: Turn-outs [TO], Disposable Over Turn-Outs [DOTO],

Level A [A], Level B [B], Level C [C], Level D [D], Proximity [P], Entry [E]

ENTRY PERSONNEL   PPE DATA   WORK/AIR TIMES   POST-ENTRY

	Name	Suit type & mtrl.	On-air	Entry	Out	Off-air	Medical monitor	Rehab.
1.								
2.								
3.								
4.								
5.								
6.								
7.								
8.								
9.								
10.								
11.								
12.								
13.								
14.								
15.								
16.								

ADDITIONAL
PERSONNEL   PPE DATA   WORK/AIR TIMES   POST-ENTRY

	Name	Suit type & mtrl.	On-air	Entry	Out	Off-air	Medical monitor	Rehab.
1.								
2.								
3.								
4.								
5.								
6.								
7.								
8.								
9.								
10.								
11.								
12.								
13.								
14.								
15.								
16.								

Identify assignments given to each entry team. When each assignment is completed, **draw a single line** through the task. Any assignments that are not completed by the entry team must be **circled** These tasks then become the first assignment for the next entry team if they are still appropriate.

ENTRY # 1

ENTRY TEAM ASSIGNMENT:_____

_____

_____

ENTRY # 2

ENTRY TEAM ASSIGNMENT:_____

_____

_____

ENTRY # 3

ENTRY TEAM ASSIGNMENT:_____

_____

_____

ENTRY # 4

ENTRY TEAM ASSIGNMENT:_____

_____

_____

ENTRY # 5

ENTRY TEAM ASSIGNMENT:_____

_____

_____

ENTRY # 6

ENTRY TEAM ASSIGNMENT:_____

_____

_____

ENTRY # 7

ENTRY TEAM ASSIGNMENT:_____

_____

_____

SAMPLE EMS CHECK SHEET
HAZMAT EMS OFFICER SHEETS

INCIDENT NUMBER:_____ PREPARED BY:_____

YR/MN/DAY/NO.   OF AMBULANCE STATION:_____

CONTAMINANT #1 INFORMATION

Contact **Hazmat Science** and **MEDCOM** for information. If **more than one contaminant** is involved, complete separate sheet for each.

NAME:_____ SYNONYMS:_____

IDENTIFICATION NUMBERS: U.N. Class/Division:_____, U.N. ID:____, CAS:_____
                        EPA Registration:_____

TOXICITY VALUES	
TLV:	STEL:
PEL/REL:	IDLH:
CEILING:	LD_{50}:
EXCERSION LIMIT:	LC_{50}:

Routes of exposure:
   inhalation [   ], ingestion [   ], skin absorption [   ], skin contact [   ],
   skin/tissue destructive [   ].

Routes of excretion:
         exhalation [   ], perspiration [   ], urination [   ], defaction[   ].

Special Hazards:
   carcinogen [   ], teratogen [   ], mutagen [   ], etiologic agent [   ], chemical weapon [   ],
   biological weapon [   ], radioactivity [   ], Depart. of Defense (DOD) weapon [   ],
   DOD agent [   ], other reproductive [   ]—specify_____

_____

Target organ/s:
   skin [   ], respiratory [   ], neuro [   ], digestive [   ], hepato [   ], nephro [   ], hematopoeto
   [   ], reproductive [   ].

Symptoms of exposure:_____

_____

First-aid treatment:_____

_____

_____

Additional comment:_____

_____

_____

PRE- AND POST-ENTRY MEDICAL MONITORING

HAZMAT EMS PERSONNEL:

These instructions are to be read and followed by all EMS personnel assigned to HAZMAT EMS.

As part of pre-entry medical monitoring and briefing of entry and/or decontamination personnel, prior to entry remind them that heat stress or chemical exposure can cause alteration of physical and mental abilities. Entry personnel must be alert to any changes they note in themselves or their partner.

If any such changes are noted, **ALL ENTRY TEAM MEMBERS SHALL WITHDRAW AND RETURN TO DECON.** *ANY* changes noted **SHALL BE IMMEDIATELY REPORTED TO THE** *INTERVENTION GROUP LEADER* **AND** *HAZMAT SAFETY*!

NO personnel SHALL be allowed to enter or re-enter chemical protective clothing if their vital signs are found to be above the following:

> Oral/Tympanic Temperature: >99.8
> Blood Pressure: >160/110
> Pulse: >110
> Respirations, at Rest: >25

Vitals can be re-taken after a minimum of 15 minutes.

At the discretion of **HAZMAT EMS,** entry or decon personnel can be declared not fit for entry. This information must be communicated immediately to **HAZMAT SAFETY** and to **INTERVENTION** for entry personnel and **DECON** for decon personnel.

Rest and rehabilitation for personnel wearing chemical protective clothing shall follow these minimum guidelines for the specified Ambient Temperatures:

> <70°F: 30 minutes
> 70 to 85°F: 45 minutes
> >85 °F: 60 minutes

EMS personnel must monitor members during rehabilitation and assure that rehydration with cool liquids occurs. At the discretion of **HAZMAT EMS,** these time frames can be extended. Any noted medical problems must be immediately reported to **HAZMAT SAFETY.**

Personnel not deemed ready for assignment to further duties shall be monitored and kept in Rehab. If it is deemed by **HAZMAT EMS** that an individual is unfit for further assignment, but does not require further medical evaluation or treatment, the name of that individual shall be forwarded to **HAZMAT SAFETY.**

Entry and Decon MEDICAL MONITORING AND REHAB sheets follow on the next 2 pages.

ENTRY TEAM
PRE/POST ENTRY MEDICAL MONITORING AND REHAB.

ENTRY PERSONNEL	PRE-ENTRY						POST ENTRY				
Name	Temp.	B.P.	Pulse	Resp.	WT.	EKG	Temp.	B.P.	Pulse	Resp.	WT.
1.											
2.											
3.											
4.											
5.											
6.											
7.											
8.											
9.											
10.											
11.											
12.											
13.											
14.											
15.											
16.											

REHABILITATION DATA

Entry personnel	Enter Rehab	Conditions or symptoms noted	Exit Rehab	Re-entry approval		
Name	Time In		Time Out	Approved	Denied	Reported
1.						
2.						
3.						
4.						
5.						
6.						
7.						
8.						
9.						
10.						
11.						
12.						
13.						
14.						
15.						
16.						

DECON PERSONNEL
PRE/POST ENTRY MEDICAL MONITORING AND REHAB

This form is to be completed when monitoring of Decon personnel is deemed necessary by HAZMAT SAFETY and EMS.

ENTRY PERSONNEL	PRE-ENTRY						POST-ENTRY				
Name	Temp.	B.P.	Pulse	Resp.	WT.	EKG	Temp.	B.P.	Pulse	Resp.	WT.
1.											
2.											
3.											
4.											
5.											
6.											
7.											
8.											
9.											
10.											
11.											
12.											
13.											
14.											
15.											
16.											

REHABILITATION DATA

Decon personnel	Enter Rehab	Conditions or symptoms noted	Exit Rehab	Re-entry approval		
Name	Time In		Time Out	Approved	Denied	Reported
1.						
2.						
3.						
4.						
5.						
6.						
7.						
8.						
9.						
10.						
11.						
12.						
13.						
14.						
15.						
16.						

HAZMAT EMS

CONTAMINATED VICTIM SHEETS

INCIDENT NUMBER:_____PREPARED BY:_____

YR/MN/DAY/NO.   OF EMS STATION:_____

VICTIM INFORMATION:

Name:_____ , Age:_____ , Sex:_____ , Weight:_____

Address:_____   Phone:_____

_____

_____

Pre-existing conditions: no [ ], yes [ ]—Specify_____

On medications: Unknown [ ], No [ ], Yes [ ]—Specify:_____

VITAL SIGNS:

Pulse:_____ , BP:_____ , Respirations:_____

Chemical burns or irritation to:
Skin [ ], Eye [ ], Respiratory Tract [ ].

Degree and description: _____

_____

Level of consciousness: _____

Symptoms:
Difficulty breathing [ ], Burning Eyes [ ], Burning Skin [ ], Convulsions [ ],
Muscle Tremors [ ], Vomiting [ ]

Others—Specify:_____

_____

_____

Trauma or other injuries—Specify: _____

_____

_____

CONTAMINANT DATA:
Tissue Destructive [ ], Skin Absorptive [ ], Highly Toxic [ ], Breath Excreted [ ],
Radioactive [ ], Biohazard/Communicable [ ].

EXPOSURE DATA:
Type of Exposure:
Internal [ ], External [ ], Internal and External [ ].
% off total body exposure (use rule of 9s):_____

Decontamination:
fully performed [ ], *not performed* [ ].
If *not performed* is checked, state why and the personnel protection required

_____

_____

ADDITIONAL PERTINENT DATA:_____

_____

_____

8 Plan of Action Implementation, Evaluation, and Review

CHAPTER OBJECTIVES

Upon completion of this chapter, the student will be able to:

1. Describe the implementation phase of the plan of action development
2. Explain the advantages of plan of action documentation
3. Identify the role of local procedures and resources when implementing the plan of action
4. Identify the role Incident Command System options when implementing the plan of action
5. Explain the three levels used in the evaluation process
6. Identify the implications of problems identified at the tactical method, tactical objective and strategic goal levels
7. Describe operational situations that require the use of the review process
8. Explain why problems at the strategic goal level require review
9. Describe how the achievement of a strategic goal relates to the review process

PLAN OF ACTION IMPLEMENTATION

In essence, once the tactical options are assessed to determine which are appropriate and that their resource requirements are met, the plan of action has to be identified. Yet, for the plan to work, several things must occur. First, the primary elements of the plan must be communicated to the appropriate personnel. Second, the plan must be implemented in a form consistent with local system and procedural requirements. Finally, the plan must be documented.

Before these three areas are addressed, it is important to review some of the qualities a plan of action must have. First, the plan must be safe and realistic. It cannot require personnel to accomplish tasks that are unsafe or unattainable. It cannot anticipate that personnel will be able to complete tasks in unrealistic time frames, in greater numbers than possible, and so on.

Second, the plan must anticipate potential problems. For example, it is decided that a quick spill control operation (e.g., diking and diverting) followed by vapor suppression will allow an entry team to stop the leak. As these tasks are initiated, the rate of release accelerates. This makes a good spill control operation unfeasible, and the plan fails. In reality the plan should have identified that a maximum release rate was integral to success. As such, alternate plan B should also have been identified and implemented when plan A became unfeasible.

Third, the plan must be clear but flexible with the ability to be altered rapidly. This problem has been seen in many prefire plans. Specifically, strategies and tactics are so etched in stone that any situation outside the one anticipated makes the plan ineffective.

Communication of the Plan

The communication process of plan implementation is handled by giving orders to the appropriate subcommands within the ICS. As such, the size and make-up of the command structure are dictated by the incident and the associated functions identified by command. For a small scale operation, the progress may be simple and direct. For a larger scale and more complex operation, the process will be a more extensive task.

The communications can be completed by radio or face-to-face dialogue. Beside subcommands, it may be necessary to provide information to the Emergency Operations Center (EOC). However, the primary routing and communications needs are dictated by local policies and procedures, not to mention the local Command structure.

Local System and Procedures

The primary implementation process is dictated by the local Emergency Response Plan, SARA plan, and possibly regional or other procedures. Since it is impossible to address all the potential local plan options, this discussion will focus on ICS model options.

In 1994, FIRESCOPE published its version of their ICS Hazmat Module. However, there were many other systems in place before that time. One is the National Fire Academy's Model ICS Hazmat Module that is recommended or required by various states. Both will be used as examples in this section. Further, both address similar functions but use somewhat different names and alignments of the positions.

FIRESCOPE ICS.   The FIRESCOPE module (Figure 8-1) identifies the following positions under the Hazmat Group Supervisor or Branch Director. The term *Hazmat Supervisor* will be used generically to cover either option. The following are brief position descriptions within Hazmat.

Hazmat Supervisor reports to the Incident Commander or the Operations Section Chief depending on the ICS structure chosen and directs overall operations within the branch or group. Hazmat is responsible for implementing all operations involving hazmat.

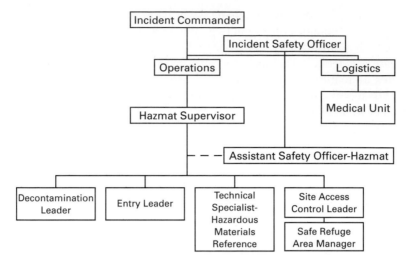

Figure 8-1   FIRESCOPE ICS

Entry Leader reports to Hazmat Supervisor and is responsible for all teams and operations that occur within the hot zone. Entry is responsible for controlling all access and egress from the hot zone.

Decontamination Leader reports to Hazmat Supervisor and is responsible for implementing all decontamination operations.

Site Access Control Leader reports to Hazmat Supervisor and is responsible for controlling access routes of all people and equipment at the Hazmat site.

Assistant Safety Officer–Hazmat reports to the incident Safety Officer and coordinates with Hazmat Supervisor. The Assistant Safety Officer–Hazmat is responsible for coordinating of all safety functions associated with Hazmat operations as specified in 1910.120.

Safe Refuge Area Manager reports to Site Access Control and Coordinates with Decontamination Leader and the Entry Leader. Safe Refuge Area Manager deals with victims by evaluating and prioritizing them for treatment and preventing the spread of contamination by these victims.

Technical Specialist–Hazardous Materials Reference reports to Hazmat Supervisor and provides technical information and assistance to Hazmat.

There is no mention of EMS within Hazmat. EMS is provided by the Medical Unit under logistics. Again, this is an area where most Hazmat Supervisors feel it is essential to detail EMS to the group or branch to assure continuity of personnel and service.

National Fire Academy Model.   The National Fire Academy's Model ICS Hazmat module (Figure 8-2) was developed in 1991 as part of a series of hazmat programs that were being written. The functions are very similar as is most of the setup. However, there are different titles for the various positions related to the function.

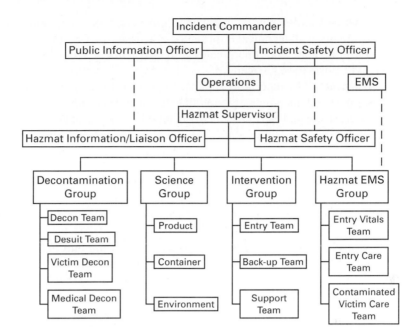

Figure 8-2   National Fire Academy Model

Hazmat operational and command personnel identified the functional areas that were needed and their command relationships to provide a functionally oriented structure that reflected actual relationships and functions. The structure was then reviewed for adherence to ICS principles of form and terminology. Again, Hazmat Supervisor is used generically to identify the Hazmat Group Supervisor or the Branch Director.

Hazmat Supervisor reports to the Incident Commander or Operations Section Chief and is responsible for directing all operation within the Hazmat group or branch.

Hazmat Safety Officer reports to Hazmat and coordinates with the Incident Safety Officer. Hazmat Safety Officer is responsible for coordinating all safety functions associated with Hazmat operations. These areas are primarily confined to operations within and in support of hot zone operations.

Intervention Leader (team or group level) reports to Hazmat Supervisor and is responsible for coordinating and controlling all entry, backup, and entry or suit support teams involving operational intervention and support within the hot zone. Intervention and Hazmat Safety have joint responsibility to control and fully document all access and egress from the hot zone.

Decontamination Leader (team or group level) reports to Hazmat Supervisor and is responsible for implementing all decontamination corridor(s) setup and operations. Decontamination Leader coordinates with Hazmat Safety and Intervention Leader to control and document decontamination operations and egress.

Science Leader reports to Hazmat Supervisor and is responsible for providing and documenting all technical information and assistance about the product, container, and environment. Science coordinates with the Planning Section Chief and Hazmat Liaison and Information Officer.

Hazmat EMS reports to Hazmat Supervisor and is responsible for providing EMS support and documentation for all entry operations, and contaminated victim operations in the hot and warm zones. Hazmat EMS coordinates Decontamination Leader and with the EMS group established to handle incident victims after they exit victim decontamination.

Hazmat Information and Liaison Officer reports to Hazmat Supervisor and is responsible for providing liaison functions with technical experts involved in the incident. Hazmat Information and Liaison Officer also coordinates with Public Information Officer and helps convert highly technical data from Science and technical experts into a usable form.

Some have questioned the two terms *Intervention* and *Science*. *Intervention* is used to clearly delineate between entry team personnel and the leader of all entry and entry support operations. When insuit communications are in use, communications are routinely garbled to a greater degree than during firefighting.

The term *Intervention* cannot be confused for *Entry*. Unfortunately, when the terms *Entry Leader* and *Entry Team* are used, miscommunications and confusion are not uncommon (as many teams can confirm). The entry team is often in the most precarious position in the entire operation and miscommunications and confusion can not be tolerated.

The term *Science* was chosen again to eliminate confusion. Industry routinely refers to their technical people as "Technical Specialists" and "Technical Support Teams." Compound that with Technical Specialist–Hazmat Reference, Technical Advisor, Hazmat Specialist, as well as the Industrial Specialist and Specialist Employee (1910.120) and the ground is ripe for tremendous miscommunications and confusion. In addition, *Science* is what the Science Team does.

Further, as noted in the structure below, both the Science and Intervention groups can include multiple teams. For example, one incident involving 54 different substances required a product team of six to simply research the substances, while two other teams gathered environmental and container data. The environmental team was kept busy documenting and tracking data provided by intervention's four monitoring units under monitoring team leader.

In another example, a spill team composed of two diking units and one booming unit as well as a standard entry, backup, and support teams report to intervention to control a spill from a box trailer. By using this terminology, confusion is minimized.

Documentation of the Plan

Documenting the plan is another important aspect of plan implementation. To document the plan's implementation most thoroughly and effectively, each position with the Hazmat Branch fills out checklists or data sheets. Quite a few of these have been provided at the end of various units. In the appendices is a full set of checklists and position descriptions based on the NFA ICS model. Again, these are provided as examples only.

Documentation has many obvious advantages, especially regarding Finance Section issues as well as potential legal entanglements. Further, documentation in the form of written information allows for the rapid and concise transmission of information. Such information transfer is routinely required when additional interested parties and agencies arrive on the scene.

In addition, this written documentation is extremely beneficial during and immediately after a transfer of command, especially in a campaign operation. It speeds the process of briefing the incoming replacement with strategies, tactics, and other operational issues. Additionally, it provides a record of considerations, approaches, objectives, and goals that have been brought to bear previously.

Finally, when questions arise, this documentation can provide vital data of what occurred, how it occurred and, often most importantly, why it occurred.

EVALUATION

Once the Plan of Action has been implemented, the effectiveness of its strategies and tactics must be evaluated. Primarily this means determining whether operations are producing the desired effect. Team personnel know that certain operations take a certain amount of time to produce the desired results. As such, when those time frames are not met, consideration must be given as to why not. Normally, this will require talking to a subcommand and determining if there is a problem.

Once a problem is noted, the real issue is to identify why it exists. A whole series of questions must follow, including:

1. Exactly what is the problem?
2. Does it involve some missing or incorrect information? If so, what?
3. Is it a result of an inappropriate tool or tactical method?
4. Is it a result of an inappropriate tactical objective? If so, why and what is the alternative?
5. Is it a result of an inappropriate strategic goal? If so, what is the potential impact on safety?

These questions start at the basic and less important level. If a tactical method is found to be wrong, it is not generally a very serious problem. Likewise, if an inappropriate objective were chosen, it could mean simply a poor call or, more likely, some missing information.

If it is determined that there is a problem at the strategic goal level, the situation may be extremely serious. A flaw in the plan at the strategic level indicates a major problem either with information or the estimate. When a flaw traces its roots back to that level, there is a significant chance that the entire plan is in danger of being compromised.

This is not to say that every time a strategic goal cannot be met, the plan is in danger of collapsing. For example, an entry team is assigned to go in, perform basic diking, and see if they can control the leak by shutting a valve. The team enters, performs the diking, but finds the valve is jammed. Glitches like this will happen, but they do not necessarily mean there is a problem at the goal level. Perhaps the use of a different objective or method will work instead. This situation reflects a problem at the objective level (and not necessarily a big one at that).

Further, in this case, leak control was "nice" to accomplish, not "need" to accomplish. The primary goal was spill control. That was accomplished. Leak control was simply icing on the cake.

REVIEW

The review phase involves going back to the beginning of the GEDAPER process and confirming the vital pieces of information and conclusions of each step. The review process is needed when any of several operational situations occurs. These situations include:

1. The acquisition of new or updated information
2. The achievement of a strategic goal
3. The identification of a flaw in the plan of action as a result of evaluation, at the strategic goal level

New or Updated Information

Probably the most common situation that requires at least a limited review phase is the acquisition of new or updated information. Such situations are routine and an almost mundane occurrence early in many incidents. Continually, updated and new information is obtained and routinely filtered into the overall command sequence with little or no difficulty.

However, once the plan has been implemented, new or updated information is sometimes sidetracked or simply overlooked. This can have a significant impact on the operation as a whole and various strategies and tactics in particular.

For example, an entry team is stopping a leak in a derailed DOT 105 tank car containing chlorine. They enter and assess the full extent and nature of the damage the car has sustained. They report that the damage is such that, at best, they will only be able to slow the leak but definitely not stop it. The plan may have been constructed so it relies on complete stoppage of the leak. In this case, a major revamping of the plan may be required.

Achievement of a Goal

Since strategic goals are prioritized for implementation, the impact of accomplishing a goal will change the complexion of the operation to some degree or another. As each goal is achieved, its implications to the overall operation must be examined and fit into the overall plan.

Flaw at the Strategic Goal Level

As mentioned, a flaw at this level is a major problem. What may be even greater is what led to the flaw in the first place. Was it a result of incomplete or inaccurate information? Was information interpreted improperly while making the estimate? As such, a rapid and comprehensive review of each step is essential.

Further, a flaw at this level can indicate a serious misjudgment of the incident situation causing personnel safety to become the number one priority. Such a flaw may indicate that the specific operation or possibly all hot zone operations should be suspended to preserve the safety of personnel while the review process is underway. Additionally, entry personnel may be able to shed some light on what caused the problem in the first place.

No matter what the reason, the review process ties all the loose ends together. Furthermore, it provides a systematic approach to review the entry command sequence.

SUMMARY AND CONCLUSION

This chapter addressed the final three steps in the GEDAPER process: Plan of Action Implementation, Evaluation, and Review. It has:

1. Described the implementation phase of the plan of action development
2. Explained the advantages of plan of action documentation
3. Identified the role of local procedures and resources when implementing the plan of action
4. Identified the role Incident Command System options when implementing the plan of action
5. Explained the three levels used in the evaluation process
6. Identified the implications of problems identified at the tactical method, tactical objective, and strategic goal levels
7. Described operational situations that require the use of the review process
8. Explained why problems at the strategic goal level require review
9. Described how the achievement of a strategic goal relates to the review process

This brings us to the end of our journey through Hazmat Strategies and Tactics.

I truly hope that after completing this text, you have a better understanding of all the aspects of hazmat incident mandates, planning, response, training, and especially the command sequence that helps determine appropriate hazmat strategies and tactics. When all levels of the emergency response system within a community understand the systematic approach to identifying appropriate hazmat strategic and tactical options that are based on a uniform system, the response will be more effective, efficient, and most of all safe.

I further hope that this text helps young, up-and-coming members of the emergency-response community learn from my experience and insights. They will be the leaders and innovators of the future and will face problems that we can only imagine now. I have been blessed by the generosity of my supervisors, colleagues, fellow emergency responders, clients, and students, who have been more than willing to share their knowledge, experience, and insights. As long as we work and learn from each other, as well as think systematically and scientifically, no problem is insurmountable.

Furthermore, when new problems arise, answers are not always readily apparent. Hazmat has required years of analysis, research, and extrapolation to identify effective and efficient approaches. In the interim, many "accepted" approaches become part of the developing paradigm. Unfortunately, they are often accepted as fact even after evidence indicates they are inappropriate. At this point, they become dogma because they are not based on fact.

I urge all who read this text to thoughtfully question accepted approaches that do not seem to make sense. Research the subject to find the basic reasons and science behind the issue. Then diligently work to change the accepted paradigm. It often takes time and significant work, often more than one can imagine, but the results can be truly rewarding and worth every bit of effort.

Finally, resist becoming antagonistic when trying to achieve change because it will only hinder a just cause. Be systematic, thoughtful, professional, but most of all safe in all your endeavors. Ultimately, make a positive difference that improves society.

APPENDIX A

Index for 49 CFR, Parts 100 to 199 (1995)

The following is a general index of 49 CFR
(1995), Parts 100 to 199.

APPENDIX B	# Sample Personal Protective Equipment Program and Standard Operating Guidelines

NOTE: This Standard Operating Guideline is provided as an *example* of the type of approach available for the development of an organizational specific decontamination guideline. This document must not be considered OSHA complient in and of itself since it cannot identify the specific policies and procedures developed by a particular employer. Further, OSHA requires that the PPE Program contain specific donning and doffing procedures for each type of PPE available for use.

2.0   **PERSONAL PROTECTIVE EQUIPMENT (PPE) PROGRAM**

    a. PPE is any type of equipment used to protect personnel from hazards that may be encountered at the scene of an emergency.

    b. This section of the Standard Operating Guidelines (SOG) is intended to provide guidance to personnel as to the various types, selection, basic use, inspection, and record keeping for use of Team PPE. It is critical that all personnel understand the mandates established in 1910.120 regarding these aspects of PPE and they are included in this procedure.

    c. PPE is designed to protect personnel from specific types of hazards for which each type of PPE is designed. If the PPE is asked to protect against a hazard for which it is not designed, it will generally fail and the wearer will receive an exposure, be injured, or be killed.

    d. For the safe use of the equipment, ALL health and safety hazards, to the extent possible, must be identified as part of the overall emergency scene, site characterization process.

2.1   **Types of PPE**

    a. PPE for chemical response requires a combination of various components assembled into a total system needed to protect personnel from specific incident scene hazards.

    b. The primary incident scene hazards are identified by the acronym TRACEM, which stands for the following hazards:

       1.   Thermal

       2.   Radiation

       3.   Asphyxiation

       4.   Chemical

       5.   Etiologic

       6.   Mechanical

    c. Since no two incidents are identical, potential incident hazards will vary and may include additional hazards such as electricity, high energy electromagnetic radiation fields, lasers, in addition to those identified by TRACEM.

    d. PPE is generally broken into two broad categories when used during chemical emergencies. They are thermal and chemical protective equipment.

2.2   **Thermal Protective Clothing**

    a. Thermal protective equipment is designed to protect the wearer from temperature extremes, the existence of fire, or the potential existence of fire.

    b. Thermal protective equipment is composed of two primary components: respiratory protection and body protection.

    c. The primary respiratory protection includes Positive Pressure Self-Contained Breathing Apparatus (SCBA). During clean-up oversight or for decontamination, it is also possible that an Air Purifying Respiratory (APR) may also be used.

d. Thermal body protection can include full structural firefighting turn-out, proximity suits, or entry suits.

2.2.1 Thermal Respiratory Protective Equipment

a. The only option for respiratory protection during emergency operations shall be positive pressure self-contained breathing apparatus (SCBA).

b. During clean-up and recovery operations, it is possible that thermal protective equipment could be required. During such operations it is possible that air purifying respirators (APR) could be used as appropriate respiratory protection depending on the circumstances.

2.2.2 Structural Firefighting Turn-Outs

Structural Firefighting Turn-outs are the fundamental type of thermal PPE. As a minimum, these turn-outs must be compliant with #29 CFR, Part 1910.156.

a. For the purpose of these SOGs, full turn-outs will consist of coat, pants, boots, gloves, helmet, hood, and positive pressure SCBA.

b. This type of thermal PPE provides limit protection from radiant heat and SHALL NOT be used for direct flame contact.

2.2.3 Proximity Suits

Proximity Suits are designed to provide short duration protection for high radiant heat exposures.

a. This type of equipment uses a multiple piece design composed of aluminized Nomex or PBI construction. The specific configuration will vary depending upon the manufacturer.

b. This type of thermal PPE must include SCBA located within the suit itself.

c. This equipment **SHALL NOT** be used for direct flame contact.

2.2.4 Fire Entry Suits

Fire Entry suits are designed for limited duration direct flame contact.

a. This type of equipment uses a multiple piece, very thick, non-flammable material to protect the wearer for a period of from 30 to 60 seconds over the lifetime of the suit.

b. This type of Thermal PPE must include SCBA located inside the suit.

c. Fire entry equipment shall be the only type of equipment used for direct flame contact if it is available and appropriately trained and qualified personnel are available for its use.

2.2.5 Low Temperature Thermal Body Protective Equipment

Low temperature thermal protection for the body is designed to protect against the low temperatures that can result from escaping compressed gases or possibly cryogenic liquid splash.

a. Extreme caution must be exercised in these types of situations to consider ALL potential hazards. These hazards include chemical, poor visibility as well as mechanical such as trip and fall or immersion.

b. Possible thermal PPE options for this type of situation include other types of thermal protective equipment such as turn-outs, proximity, or entry suits.

c. **THIS DOES NOT INCLUDE IMMERSION OR POSSIBLE IMMER-SION IN REFRIGERATED OR CRYOGENIC LIQUIDS. THERE IS NO EFFECTIVE PPE FOR THIS TYPE OF SITUATION.**

2.3   **Chemical Protective Equipment**

Chemical protective clothing is designed to protect the wearer from primarily chemical or biological (etiologic) hazards.

a. Such hazards include toxic substances, corrosives, irritants, skin/tissue destructive or absorptive substances, and simple and systemic asphyxiants.

b. Chemical protective clothing provides **NO thermal protection** for the wearer, and thus **SHALL NOT BE USED WHEN SUCH PROTECTION IS RE-QUIRED** to safely and effectively complete the assigned work mission (see Section 2.5. for specific guidelines).

c. Chemical protective clothing can be divided into respiratory and body protections components.

d. Chemical protective clothing is sub-divided into four different levels, A through D.

2.3.1   **Chemical Respiratory Protection**

a. Chemical Respiratory Protection includes positive pressure **SCBA,** positive pressure supplied air-line (**in-line**) respirator with escape or air purifying respirators (**APR**).

b. The specific type of respiratory protection used depends on the level of PPE chosen as determined by the specific hazards and situation encountered during the emergency. (More information is provided in Section 2.5.)

2.3.2   **Level A Protective Equipment**

a. Level A PPE is used when maximum body protection and maximum respiratory protection are needed due to the chemical hazards of the substance involved.

b. Level A PPE shall, at a minimum, be composed of the following equipment:

1. Maximum respiratory protective equipment provided by SCBA or in-line air system

2. Maximum body protection provided by a total encapsulating chemical protective suit (TECP) which is a suit covering the entire body and respirator as specified in #29 CFR, Part 1910.120

3. A minimum of three layers of protective gloves: a disposable surgical glove against the skin, an inner chemical resistant glove, and an outer chemical resistant glove

4. Chemical resistant boots equipped with a steel toe and shank

5. In-suit, two-way radio communications

   c. Optional equipment may include:

     1. Coveralls or long underwear to be worn under the TECP

     2. An additional outer glove of leather, canvas-leather, etc., to protect against abrasions, cuts, or tears; to be disposed of after each use

     3. Disposable, outer boot covers

2.3.3 Level B Protective Equipment

   a. Level B PPE is used when a lesser level of body protection and maximum respiratory protection is needed due to the chemical hazards of the substance involved.

   b. Level B PPE shall, at a minimum be composed of the following equipment:

     1. Maximum respiratory protective equipment provided by SCBA or in-line air system

     2. A lower level body protection provided by a semi-encapsulating chemical protective suit (suit with hood and integral face piece and non-integral gloves) or hooded chemical resistant clothing (one piece hooded coveralls, one or two piece chemical splash suits, or chemical resistant overalls)

     3. A minimum of three layers of protective gloves; a disposable surgical glove against the skin, an inner chemical resistant glove, and an outer chemical resistant glove

     4. Chemical resistant boots equipped with a steel toe and shank

     5. In-suit, two-way radio communications

   c. Optional equipment may include the following:

     1. Coveralls or long underwear to be worn under the chemical resistant clothing

     2. An additional outer glove of leather, canvas-leather, etc., to protect against abrasions, cuts, or tears; to be disposed of after each use

     3. Disposable, outer boot covers

2.3.4 Level C Protective Equipment

   a. Level C PPE shall only be used when the substance or substances are known, their airborne concentration is known, and all criteria for the safe use of an APR are met as determined by Hazmat Safety and Intervention.

   b. Level C PPE is used when a lesser level of body protection and only a low level of respiratory protection is needed due to the chemical hazards of the substance involved.

   c. Level C PPE shall, at a minimum, be composed of the following equipment:

     1. Low level respiratory protective equipment provided by a FULL FACE APR with appropriate cartridge installed

     2. A lower level body protection provided by hooded chemical resistant clothing (one piece hooded coveralls, one or two piece chemical splash suits, or chemical resistant ovcralls)

3. A minimum of three layers of protective gloves, a disposable surgical glove against the skin, an inner chemical resistant

4. Chemical resistant boots equipped with a steel toe and shank

5. In-suit, two-way radio communications

d. Optional equipment may include:

1. Coveralls or long underwear to be worn under the chemical resistant clothing

2. An additional outer glove of leather, canvas-leather, etc., to protect against abrasions, cuts or tears; to be disposed of after each use

3. Disposable, outer boot covers

4. Additional full face shield

2.3.5 Level D Protective Equipment

a. Level D PPE is a work uniform that provides minimal protection for nuisance contaminants only.

b. Level D PPE provides no chemical protection and can only be used when there are no known airborne, splash, immersion, contact, or inhalation chemical hazards associated with the work mission; it is anticipated that this level of PPE will generally not be appropriate for use by response personnel other than during the set-up of the access and decontamination corridors.

c. Level D PPE may also be used after the incident has been stabilized, clean-up activities have been initiated, and there are NO POTENTIAL chemical hazards present.

d. Level D PPE is, at a minimum, composed of the following equipment:

1. Coveralls, turn-outs, or similar work clothing

2. Steel tip and shank, boots

2.4 Double Envelope Chemical–Thermal Protective Equipment

Double envelope protective systems incorporate components of both thermal and chemical PPE.

a. Respiratory protection shall be provided in the form of SCBA or in-line air systems with escape pack for either option.

b. There are two possible options available for double envelope operations, disposable over turn-outs (**DOTO**) and flash protection over chemical suit (**FPOC**).

2.4.1 Disposable Over Turn-Outs (DOTO) Option

a. Of the two options available, the **DOTO** provides a far superior level of thermal protection (hot and cold) and is generally preferred to the **FPOC.**

b. **DOTO** is designed to prevent the wicking of flammable liquids into the body material of the turn-outs.

c. Any situation that results in such wicking shall be determined to present an imminent danger situation to the personnel so exposed and shall be terminated by Command, Hazmat Branch, Hazmat Safety, or Intervention.

d. The DOTO option consists of:

  1. Full structural firefighting turn-outs as described in Section 2.2.2.a, covered by a disposable chemical resistant semi-encapsulating suit, coveralls with hood, or a one or two piece splash suit

  2. Positive Pressure SCBA

e. When DOTO is used, all monitoring requirements specified in Section 2.5.1 shall be followed regarding percentage of LEL and withdrawal conditions.

2.4.2 **Flash Protection Over Chemical Suit (FPOC)**

a. **FPOC** is an option provided by chemical body protection manufacturers.

b. **FPOC** is designed to provide one to three seconds of protection against the radiant heat produced by a vapor flash. There should be **NO DIRECT FLAME CONTACT.**

c. There is extreme controversy regarding the true benefits of this particular option.

  1. Live tests indicate that flammable gases, vapors, or liquid can and do enter the space between flash protection and the chemical protective membrane.

  2. It is possible for a vapor flash to ignite these trapped vapors and produce an intense, long duration fire WITHIN the flash protection.

  3. Such a fire would produce severe burns and possibly death.

d. When FPOC is used, all monitoring requirements specified in Section 2.5.1 shall be followed regarding percentage of LEL and withdrawal conditions.

2.5   **Selection and Use of PPE**

The following sections provide guidance for the appropriate selection and use of the various types of PPE used by the Hazmat Team. In any case, only personnel specifically trained and qualified shall use any of the types of PPE available to the Team.

a. The PPE selected and used during emergency operations shall protect personnel from hazards and potential hazards they are likely to encounter as identified during site characterization and analysis.

b. PPE shall be selected on the basis of PPE performance characteristics relative to the hazards, requirements, limitations, work mission duration, and conditions of the site.

c. The selection and use of appropriate PPE is based upon wide ranging series of considerations that must be closely examined prior to each use.

  1. All of the health and physical hazards, to the extent possible, of the incident must be identified through the site characterization process and outlined in the site safety plan and explained to all personnel.

  2. Compatibility data between the product and available PPE shall be examined and determined to be appropriate prior to use of the PPE.

  3. PPE with breakthrough times of LESS than 90 minutes for the substance involved in a release shall not be considered appropriate for emergency operations.

    4. PPE that can be chemically degraded by the substance involved shall not be considered appropriate for emergency operations.

  d. There are multiple additional considerations, some of which follow, in the selection of appropriate PPE for a given emergency response situation.

    1. Specific incident situation and location

    2. Degree of confinement, e.g., inside, confined space, or outside

    3. Specific properties and hazards or the substance involved

    4. Number of substances involved

    5. Whether or not direct product contact will occur

    6. Will there simply be potential contact with vapor

    7. Flammability of the atmosphere

    8. Routes of exposure and the potential for the substance contact those routes considering the work function to be performed

    9. The specific work mission assigned to personnel

    10. Ability to perform the desired work function with the PPE chosen

  e. Generally,the body protective component of either thermal or chemical protective equipment will be considered to be expendable equipment and normally will be disposed of after use. (This does not include respiratory protective equipment.)

  f. Generally, the chemical PPE used will consist of disposable or limited use body protection (e.g., suits, gloves, and boots).

  g. It is anticipated that reusable chemical PPE body protection will only be used if all disposable or limited use equipment is expended or for some reason is not compatible with the substance involved.

2.5.1 General PPE Limitations

There are situations and conditions where no PPE provides adequate protection. When such situations are identified, **tactical or possibly even strategic changes SHALL be made** in the Operational Plan of Action to prevent personnel from being injured, incapacitated, or killed. The following are the most common of these situations.

  a. Entry into a potentially flammable atmosphere where greater then 20% of the lower explosive limit (LEL) of a gas or liquid vapor have been identified by direct reading monitoring equipment, using the appropriate manufacturer supplied relative response tables

    1. This equipment will include the combustible gas indicators (CGI), colorimetric tubes (Daeger, MSA, or Sensidyne), photo-ionization detector (PID - Micro Tip), or the flame ionization detector (FID - Foxborough).

    2. OSHA and EPA guidelines indicate that when **10% of the LEL** is detected, **extreme caution SHALL be exercised.**

    3. OSHA and EPA guidelines also indicate that when *20% to 25% of the LEL* is detected, *personnel SHALL WITHDRAW.*

   b. Situations where containers, vehicles, equipment, etc. cannot be stabilized to prevent accidental shifting that could result in entrapment, crushing or other types of injuries

   c. Situations where the stability of a structure or other area that must be entered, navigated through around, or over is questionable

   d. Entry into a structure, building, or container where there is no second means of emergency egress unless OSHA's confined space entry guidelines can be strictly followed

2.5.2 Respiratory Protection Selection and Use

Whenever thermal protective equipment, double envelope equipment, or Level A or B chemical protective equipment are selected for use during emergency operations, positive pressure SCBA, or in-line air supply, shall be used to provide appropriate respiratory protection. Level C chemical protective equipment uses APR and full turn-outs used for other than emergency operations may also use an APR. The following information provides the general guidelines used and to identify conditions when SCBA or APR can be used.

   a. General guidelines include the following:

   1. Only personnel that have successfully completed the respiratory protection training program, fit testing, and medical monitoring shall be allowed to used respiratory protective equipment. This shall be part of the annual recertification program.

   2. The selection of respiratory protection shall conform to the guidelines set forth in these SOGs, American National Standard Practices for Respiratory Protection Z88.2-1969 and #29 CFR, Part 1910.120.

   3. Respirators shall not be worn by personnel with beards, sideburns, dentures, eye glasses, etc. that prevent a good mask to face seal.

   4. All respirators shall be inspected, including leak checks, before every use.

   5. After every use each respirator shall be cleaned, sanitized, inspected, and properly stored.

   6. All SCBAs shall be inspected at least monthly if there has not been intervening use. This inspection shall assure that the unit is in full readiness for use and includes fit check of all fittings, visual inspection of all elastomer pieces for cracks or other deterioration.

   7. All replacements and repairs shall be performed by qualified personnel following manufacturers' recommendations. Any repairs beyond the ability of local personnel shall be returned to the manufacturer or supply for factory authorized repairs.

   b. Positive pressure SCBA or in-line air respirator with an escape air supply shall be used whenever determined by Command, Intervention, or Hazmat Safety, or when atmospheric conditions meet any of the following criteria:

   1. Atmospheres that have not been previously monitored by direct reading instrumentation as deemed necessary for the specific incident situation

    2. Atmospheric oxygen levels below 19.5%

    3. Atmospheres where the specific contaminant is not known

    4. Atmospheric contaminant concentrations that vary widely or are unpredictable, despite periodic or continuous direct reading instrumentation

    5. Atmospheric concentrations near or above the IDLH of the contaminant.

    6. Atmospheric contaminant concentrations above the maximum level specified for available respirator cartridges

c. Air purifying respirators may be used when atmospheric conditions meet the following criteria and with the approval of Hazmat Safety:

    1. All potential and actual contaminants have been identified

    2. Atmospheric oxygen levels are above 19.5%

    3. All contaminant concentrations have been repeatedly measured and found not to vary

    4. APR cartridges are available for the substances and concentrations involved in the incident

2.5.3 Thermal Protective Equipment Selection and Uses

Thermal protective equipment is designed to protect the wearer from temperature extremes, the existence fire, or the potential existence of fire. Some incident situations that warrant the use of thermal protective equipment include:

a. The presence of fire and/or radiant heat

b. The threat or potential of fire, including:

    1. The release of flammable vapors into the atmosphere including natural gas, liquefied petroleum gases (LPG), hydrogen gas, ethylene, etc.

    2. Spills of flammable or combustible liquids such as hydrocarbon fuel liquids, solvents, ketones, aldehydes, esters, ethers, alcohols, etc.

c. Any situation where flammability is the primary hazard and there is no high degree of skin absorption toxicity causing a systemic response or a high degree of tissue destructive character

d. Situations where the turn-outs could come in contact a flammable or combustible liquid and result in wicking (a Disposable Over Turn-Outs (DOTO) must be considered a viable option)

e. Low temperature situations resulting from escaping compressed gases or liquefied compressed gases

    1. Extreme caution must be exercised in these types of situations due to the potential for multiple hazards.

    2. Possible chemical hazards include instability, incompatibility, pyrophoric behavior, oxidizing capability, tissue destructive/irritating behavior, skin absorptive ability, flammability, etc.

    3. Visibility can be seriously impaired due to the vapor cloud that often accompanies this type of release. If the cloud is of such magnitude to obscure vision of potential hazards, entry shall be forbidden.

      4. Toxicity or tissue irritating/destructive behavior may require the use of DOTO if product compatibility is present.

  f. Low temperature situations resulting from refrigerated or cryogenic liquids

      1. For releases of other than small quantities of cryogenic liquids, visibility will generally be compromised beyond the level for safe operations. As such, entry into such as vapor cloud shall not be allowed.

      2. **NO ENTRY SHALL BE MADE WHERE THERE IS A POTENTIAL FOR PROLONGED SPLASHING OR IMMERSION IN CRYOGENIC LIQUIDS. THERE IS NO EFFECTIVE PPE FOR THIS TYPES OF SITUATION.**

  g. High radiant heat conditions such as flammable or combustible liquid fires

      1. Situations requiring proximity suits, which are primarily designed to protect personnel for high levels of radiant heat present during flammable or combustible liquid firefighting operations

      2. Proximity suits shall never be exposed to direct flame contact.

      3. Proximity suits also provide no real chemical protection.

  h. Direct flame contact

      1. Situations requiring entry suits, which are designed for limited duration exposure to direct flame contact

      2. These suits are used to gain entry to shut-offs and other valving involved in a fire.

2.5.4 Chemical Protective Equipment Selection and Uses

  a. Chemical protective equipment is designed to protect the wearer from chemical, biological, hazards, and contamination as well as radioactive contamination.

  b. Some incident situations that warrant the use of chemical protective equipment include:

      1. Releases of unknown hazardous substances

      2. Corrosives, tissue destructive, or irritating substances

      3. Skin absorbent substances that may result in long term negative health impacts, impairment of ability to escape, illness, or death.

      4. Corrosive, toxic, irritating, or skin absorptive atmospheres that also contain flammable or combustible contaminants only if the concentration of contaminant below 20% of LEL

      5. Liquid, solid, particulate, or gaseous substances that present the potential for excessive or difficult to remove contaminants

2.5.5 Level A Selection and Use

Level A PPE provides the highest degree of chemical protection available for the eyes, skin, and respiratory system. Level A is designed to be used in situations where there is a high degree of chemical hazards and little or no thermal hazard.

a. Level A is designed to be used when the following conditions and incident situations exist:

1. A skin-absorptive contaminant possesses a high probability of causing impairment of the ability to escape, immediate serious illness or injury, or immediate death

2. Measured or the potential for high atmospheric concentrations of gases, vapors, or particulates that are tissue destructive or highly toxic through skin absorption

3. The site operation and work functions involve the potential for splash, immersion, or exposure to unexpected vapors, gases, or particulates that are tissue destructive or highly toxic through skin absorption

4. The known or suspected presence of substances possessing a high degree of hazard and contact is probable

5. Operations conducted in poorly ventilated confined locations, or in confined spaces where the need for Level A equipment has not yet been ruled out

6. Substances with high toxicity through inhalation, skin absorption, or contact

b. Level A is used only when

1. There is little or no fire hazard associated with the substance or incident situation.

2. There is little or no potential for the substance, its surrounding atmosphere, or equipment to sufficiently cool the suit's protective membrane to cause embrittlement and suit failure.

2.5.6 Level B Equipment Selection and Use

Level B PPE provides a lower degree of chemical protection for the skin but the highest level of protection for the eyes and respiratory system. Level B is designed to be used in situations where there is a determined degree of chemical hazards and little or no thermal hazard.

a. Level B is designed to be used when the following conditions and incident situations exist:

1. The type and atmospheric concentrations of contaminants is known and requires a high degree of respiratory protection, but a lesser degree of skin protection

2. Atmospheric oxygen concentrations of less than 19.5%

3. The presence of incompletely identified of contaminants indicated by direct reading organic vapor instrumentation, but not suspected of containing high levels of chemicals harmful to or absorbed through the skin

b. Level B is used only when

1. There is little or no fire hazard associated with the substance or incident situation.

2. There is little or no potential for the substance, its surrounding atmosphere, or equipment to sufficiently cool the suits protective membrane to cause embrittlement and suit failure.

3. Level B PPE may be used for decontamination activities.

2.5.7 Level C Equipment Selection and Use

Level C PPE provides a lower degree of chemical protection for the skin and a low level of protection for the respiratory system. Level C is designed to be used in situations where there is a determined degree of chemical hazards and little or no thermal hazard.

a. Level C is designed to be used when the following conditions and incident situations exist:

1. No atmospheric gaseous, vapor, or particulates present that can adversely affect the skin or be absorbed

2. Type and atmospheric concentrations of contaminants determined through direct reading instrumentation, with an APR available that can remove the contaminant.

3. Atmospheric oxygen concentrations greater than 19.5%

b. Level C is to be used only when

1. All other criteria for the use of APRs have been met.

2. There is little or no fire hazard associated with the substance or incident situation.

3. There is little or no potential for the substance, its surrounding atmosphere, or equipment to sufficiently cool the suits protective membrane to cause embrittlement and suit failure.

4. Level C PPE may be used for decontamination activities.

2.5.8 Suiting

a. Suiting procedures shall follow those established for the specific configuration of PPE to be worn.

b. For each configuration, training shall be provided to assure all personnel know the exact methods for suiting.

c. During actual suiting operations, a minimum of one and preferably two personnel will assist in the suiting process. These suiters are responsible to perform the following functions.

1. Assist with donning of suits, SCBA or respirator, all three pairs of gloves, boots, in-suit communications, etc.

2. Assure that all equipment is properly donned and working correctly

3. Assure that donning of entry and back-up teams is progressing at the appropriate and dame rate

2.5.9 Desuiting

a. Desuiting shall only occur after appropriate decontamination procedures have been performed upon the PPE.

    b. Desuiting shall be the final station in the decontamination corridor.

    c. All disposable suits, gloves, boots, etc. shall be removed and overpacked for final disposal. The overpacks shall be marked to indicate their contents.

    d. SCBAs and any other reusable equipment shall be segregated from all disposable equipment.

       1. As determined by Hazmat Branch, Intervention, and with the concurrence of Hazmat Safety, the reusable equipment shall be overpacked for further decontamination.

       2. This equipment may simply require washing and sanitizing after the response or may require additional decontamination procedures to assure it is safe for use.

2.6   Work Mission Duration (Work Interval)

Work interval is a critical safety issue due to its potential impacts upon heat stress and related medical conditions. The following are the general guidelines for work duration when personnel are using Level A, B, or C PPE, turn-outs, DOTO, and FPOC.

2.6.1 Standard Work Interval for Level A and Level B PPE

    a. The standard work interval for Level A and Level B PPE shall be considered to be 20 minutes from the time personnel go on air if 45-minute or 60-minute rated SCBA is used.

    b. The standard work interval for Level A and Level B PPE can be shortened at the discretion of Command, Hazmat Branch, or Intervention with the concurrence of Hazmat Safety. The following are some of the conditions or situations that can indicate shortening of the standard work interval:

       1. The interval shall be shortened to 15 minutes from on-air time if 30-minute rated SCBA are used.

       2. When operating in high heat (above 85°F) and humidity, use the following intervals:

         (a) 85 to 90°F: 5 minutes

         (b) above 90°F: 10 minutes

       3. When operating in low temperature (below 45°F) conditions, shorten the interval.

       4. When the specific work mission involves heavy physical exertion, shorten the interval.

    c. The standard work interval for Level A and Level B may be increased to 30 minutes by Command, Hazmat Branch, or Intervention with the concurrence of Hazmat Safety. The following are some of the conditions or situations that can indicate lengthening of the standard work interval.

       1. The interval may be lengthened only if SCBAs rated at 60 minutes or in-line air is used for the entry.

       2. When operating in temperature ranges of from 45 to 85°F, lengthen the interval.

3. When the specific work mission DOES NOT involves heavy physical exertion, lengthen the interval.

2.6.2 Standard Work Interval for Level C PPE

The standard work interval for Level C PPE shall range from 20 to 40 minutes, depending on incident conditions. The following are some of the conditions or situations that can indicate various time intervals:

a. When operating in high heat (temperatures above 85°F) or humidity conditions, the interval shall be toward the 20 minute period.

b. When operating in low temperature conditions (below 45°F), the interval shall be toward the 20 minute period.

c. When the specific work mission involves heavy physical exertion, the interval shall be toward the 20 minute period.

2.6.3 Standard Work Interval for Turn-Outs

The standard work interval for Turn-outs shall range from 20 to 40 minutes depending on incident conditions. The following are some of the conditions or situations that can indicate various time intervals:

a. When operating in high heat (temperatures above 85°F) or humidity conditions, the interval shall be toward the 20 minute period.

b. When operating in low temperature conditions (below 45°F), the interval shall be toward the 20 minute period.

c. When the specific work mission involves heavy physical exertion, the interval shall be toward the 20 minute period.

2.6.4 Standard Work Interval for DOTO

The standard work interval for DOTO shall range from 20 to 40 minutes depending on incident conditions. The following are some of the conditions or situations that can indicate various time intervals.

a. When disposable Level B body protection is used over turn-outs, the work interval shall conform to that found in Section 2.6.1.

b. When operating in high heat (temperatures above 85°F) or humidity conditions, the interval shall be toward the 20 minute period.

c. When operating in low temperature conditions (below 45°F), the interval shall be toward the 20 minute time frame.

d. When the specific work mission involves heavy physical exertion, the interval shall be toward the 20 minute period.

2.6.5 Standard Work Interval for FPOC

The standard work interval for FPOC PPE shall conform to the intervals established for Level A and Level B PPE as found in Section 2.6.1.

2.6.6 Rehabilitation Intervals (R&R)

The rehabilitation interval is the minimum time after an entry is performed before any re-entry. The specific length of R&R shall be determined by Hazmat EMS and Safety, based upon the following criteria:

a. Operational temperature ranges provide a good indicator of standard R&R intervals.
   1. Temperatures 70°F or below shall require a minimum of 30 minutes of R&R after each entry.
   2. Temperatures of 70 to 85°F shall require a minimum of 45 minutes of R&R after each entry.
   3. Temperatures above 85°F shall require a minimum of 60 minutes of R&R after each entry.
b. Post-Entry vitals and vitals during R&R shall also be considered as part of the determination if the general guidelines are sufficient R&R.
c. Personnel vitals shall be within the guidelines established in Section 4 and on Hazmat EMS check sheets.
d. There are several factors that impact the effectiveness of R&R:
   1. Rehydration with fluids and electrolyte mixtures shall be administered as part of R&R.
   2. R&R shall be conducted in an area where the ambient temperature will act to minimize heat or cold stress.
   3. During hot and or humid weather, R&R shall be conducted in an air conditioned location, and during cold weather R&R shall be conducted in heated location.

2.7   **Medical Considerations**

There are multiple medical considerations involving the use of PPE during chemical emergency response. These considerations include EMS standing-by, pre- and post-entry medical monitoring, contaminated victims, and potential personnel exposure.

2.7.1 **EMS Standing-By**

a. 29 CFR, Part 1910.120 mandates that EMS personnel must stand-by during chemical emergency response.
   1. EMS shall have a minimum of advanced first aid.
   2. EMS must be equipped and able to transport.
   3. It is anticipated that EMS will generally be present on the scene as part of the initial response. If EMS has not been dispatched, they must be requested through LCR.
b. If there is information that victims are involved, an additional EMS unit must also be requested to assure that there is EMS coverage for Hazmat.

2.7.2 **Pre- and Post-Entry Vitals**

Pre-entry vitals are standard EMS vital signs taken before use of any PPE for emergency response purposes. The taking of vitals shall be the responsibility of EMS personnel on the scene of the incident.

a. The basic vitals shall be taken before suiting (donning) PPE and after completing decontamination and de-suiting (doffing).

    b. The minimum vitals that shall be taken include blood pressure, pulse, temperature, and respiration.

    c. Body weight and an EKG may be run at the discretion of Hazmat EMS.

    d. The check list and forms provided for Hazmat EMS shall be completed as directed and exclusions shall be followed.

    e. Basic vitals in excess of these listed shall preclude entry:

       1. Oral temperature greater than 99.8°F

       2. Pulse greater than 110 per minute

       3. Blood pressure greater than 160 over 110

       4. At-rest respiration greater than 25 per minute

    f. If conditions or post-entry vitals indicate heat stress conditions, standard first aid procedures shall be followed as well as R&R procedures as listed in Section 2.6.6.

    g. If conditions or post-entry vitals indicate cold stress conditions, standard first aid procedures shall be followed.

2.7.3 Contaminated Victims

Incidents involving contaminated victims are often some of the most difficult to manage due to the complicating factors added by a victim.

    a. Generally, operations involving contaminated victims require additional entries, decontamination personnel, and EMS personnel.

    b. If an incident involves contaminated victims, outside EMS SHALL immediately be notified.

    c. There are two primary types of contaminated victims exposed to responders and unprotected victims; both are handled in basically the same fashion.

    d. Whenever contaminated victims are involved, a second decontamination corridor shall be established so both responders and victims can be simultaneously decontaminated.

    e. Contaminated victims shall always be decontaminated before treatment other than for the ABCs (airway, breathing, circulation). The primary reason for this is that for rescuers to safely treat a contaminated victim, they must be wearing PPE. PPE generally inhibits the ability to effectively treat the victim.

    f. All victims shall be decontaminated before transport.

2.8 Suit Inspection

    a. Chemical protective clothing and SCBAs require specific inspections to assure their safe and effective use during response procedures. The following provide information on basic inspection requirements:

    b. Since suits to be used are either limited use or disposable, the primary inspection shall be upon receipt of new suits.

    c. Any limited use suits that have full compatible inorganic compounds may be reused as determined by Hazmat Command, Hazmat Safety, and Science and inspected following procedures established in Section 2.8.3. The compounds include the following:

1. Inorganic acids
2. Chlorine
3. Hydrogen chloride, hydrogen fluoride, and hydrogen bromide
4. Other non-permeating or suit destructive inorganic compounds

d. All respirators, air purifying respirators (APR), and SCBA shall be inspected and tested for full functionality. Further, each shall have an individual file established.

2.8.1 **Initial New Suit Inspection**

The primary concern of new equipment is penetration failures as a result of manufacturing.

a. Upon receipt, all new chemical suits shall be inspected in the following fashion before being placed into service:

1. Assure that the suit is the proper type, model, size, and configuration as specified.
2. Visually inspect of all the following that are found in the particular suit: seams, zippers, booties, gloves, face shields, exhalation valves, etc.
3. All suits found not to meet order specification or having physical defects shall be separated from the rest and returned to purchasing for appropriate handling.
4. All suits except encapsulating and semi-encapsulating suits shall be marked to indicate inspection has been performed and the suits pass.

b. Upon completion of the initial inspection, all encapsulating or semi-encapsulating suits shall additionally be inspected in the following fashion before being placed into service:

1. Assure they contain functional double vapor tight zippers with no penetrations or similar defects.
2. All exhalation valves are in place and functional.
3. Use perform internal light inspection of all seams to assure there are no penetrations.
4. Encapsulating suits equipped with the qualitative air pressure test port shall be tested following the procedures established in Appendix B or 1910.120 and manufacturers recommendations.
5. Upon completion of these inspection procedures, the suits shall be placed in separate containers, bags, etc. and sealed to indicate that they are ready for use.

2.8.2 **Inspection Prior to Use**

a. All PPE shall be inspected and or tested before use to assure that it is functional and ready for use. PPE that needs such inspection or testing includes the following:

1. Chemical protective suits
2. Air purifying respirators (APR), cartridge check, fit test
3. SCBA, tank fill, mask fit test

b. Suits shall also be examined for any possible flaws or failures as a result of transportation or suiting.

2.8.3 Inspection Upon Completion of Use and Decontamination

The majority of chemical suits used during response procedures will be disposed of after use. SCBAs and APRs will be re-used and must be inspected.

a. Due to primary suit configuration covering the SCBA, it is anticipated that there will be little potential for chemical exposure or damage to the SCBA. The following inspections shall be performed:

1. Assure that SCBAs have been decontaminated and sanitized.
2. All pieces of the SCBA are present and functioning properly and all straps shall be in the full open position.
3. The air bottle is of the appropriate type, filled, and the high pressure line is attached.
4. Pack is properly stored in its case an in assigned location.
5. Assure that SCBA exposure records are completed.

b. APRs shall have cartridges removed and appropriately disposed, decontaminated, and sanitized.

1. Assure that all straps are fully opened.
2. Assure inhalation/exhalation valves are in-place and functional
3. Assure exposure records for the unit are completed.

APPENDIX C

Sample Decontamination Procedures: Standard Operating Guidelines

3.0   **DECONTAMINATION AND SCENE SET-UP**

This section of the SOGs deals primarily with decontamination procedures but also reinforces scene set-up concepts established in the Emergency Response Plan.

  a. Decontamination (Decon) is the process of physically removing or chemically altering a contaminant to prevent its spread due to its ability to cause harm to people and the environment.

  b. The specific steps involved and solutions used vary depending upon several factors:

   1. The specific substance or substances involved in the incident

   2. The specific level and configuration of the PPE worn by entry personnel

   3. Whether decon is being performed on protected personnel or contaminated victims

3.1   **Physical Set-Up and Options For Decon**

The physical set-up of the decon corridor varies depending upon the specific decon process to be used and the variables listed in Section 3.0.b.

  a. The scene set-up concepts that shall be used on the scene of a chemical emergency are based upon nationally recognized approaches as set forth by EPA, the National Fire Academy, and multiple other groups and organizations.

   1. The incident scene shall be broken into three primary zones: hot, warm, and cold.

   2. The hot zone is considered to be contaminated and contains the primary health and safety hazards on the incident scene. Only personnel with proper training and personal protective equipment shall be allowed to enter this zone.

   3. The warm zone is the location of the access and decontamination corridors. Additionally, the warm zone provides a physical mechanism to control access to and egress from the hot zone.

   4. The cold zone is an area that shall be free of incident related health and safety hazards as a result of the chemical incident. The cold zone is the location of the command post, staging areas, suiting activities, etc.

  b. As specified in the Emergency Response Plan, the decon corridor is established next to the access corridor in the warm zone and is the location where most decontamination occurs (Figures C-1 and C-2).

  c. There are five primary Decon options for protected personnel:

   1. Dry decon

   2. Four step

   3. Three step

   4. Two step

   5. One step

  d. There is an additional decon set-up for contaminated victims (see Section 3.3).

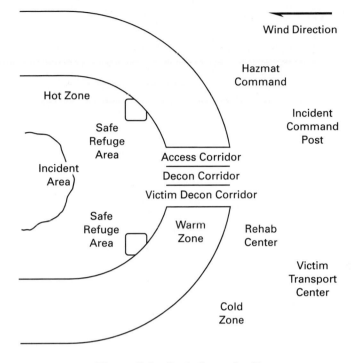

Figure C-1   Basic Scene Set-Up

3.1.1 **General Decon Corridor Set-Up Procedures**

   a. The decon corridor shall generally range in size from 15 to 25 feet wide and between 15 and 50 feet long, depending upon the specific type of decon to be performed and the location chosen for its performance.

      1. When performed within a structure, the maximum width will generally be 8 feet.

      2. If decon is performed remote from the incident, the corridor may be hundreds of feet long.

   b. The corridor in which the containments are located shall be covered with plastic.

      1. Fire hose shall be used around the edges to provide a basic berm.

      2. Absorbent pad shall be taped to the plastic as needed to prevent slipping on the surface.

   c. The containments shall generally be composed of rigid plastic swimming pools.

      1. The total number of containments shall be equal to the number of decon steps being used.

      2. Optional containments such as ladder salvage ponds and commercially available containment are optional.

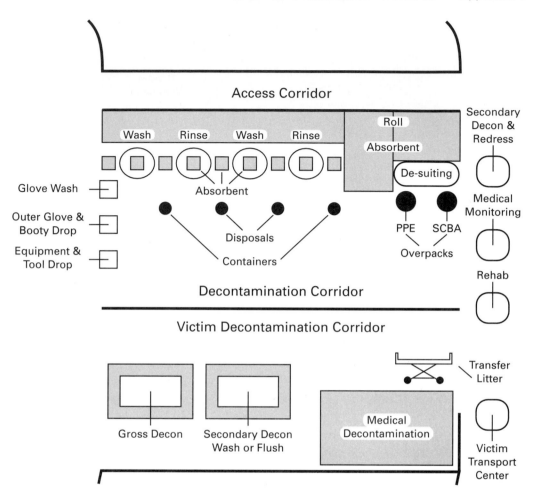

Figure C-2 Basic Four Step Decontamination Set-Up and Victim Decontamination Corridor

d. Disposal overpacks (e.g., salvage drums and plastic trash cans) shall be present at appropriate points along the corridor to contain accumulated solutions and other debris.

e. It is recommended the overpacks for PPE and respiratory protection be separate and be lined with plastic hazmat bags.

   1. If there will be many turn-outs deconned, the bags should be tagged for easier identification.

   2. SCBAs and APR shall always be bagged and tagged.

f. When full, the overpack bags shall be sealed for possible sampling of the head space to determine the presence of residual contamination.

g. All decon equipment and non-reusable equipment are to be moved to one location left on the incident scene for disposal by the clean-up contractor.

h. Pump sprayers shall be used to mix the solutions specified by Decon, Science, and Hazmat Safety.

i. The garden hose manifold shall be used to provide rinse water unless it is determined to use pump sprayers for this purpose as well.

3.1.2 **Dry Decon**

a. Dry decon is a process involving the use of limited or no liquid decontamination solutions.

b. Dry decon has several possible applications, as follows.

1. Contaminants that ignite on contact with water (various metal hydrides, nitrides, etc.)

2. Contaminants that ignite or explodes on contact with water (alkali metals, various metal alkyls, etc.)

3. Low hazard level liquid contaminants only on boots or gloves (mild irritants, low concentration inorganic acids, etc.)

c. If ignition or explosion have already occurred or are ongoing, PPE integrity will rapidly fail. Possible options include:

1. Rapid removal of the contaminated PPE

2. Use of copious quantities of water to cool, dilute, physically remove and degrade such products, Followed by rapid PPE removal

NOTE: Once the reaction has started, the PPE will be damaged, allowing the products to contact the skin. These products will then react with the moisture present on the skin, thus starting more reactions. The only possible option is the use of mineral oil. This option must be checked with **MEDCOM** to determine its efficacy for the particular situation. However, there may be **NO WAY TO STOP** the reactions and thus the use of water may be required.

d. There are three primary options for the performance of dry decon:

1. The use of hazmat vacuum equipment is one option. This option functions best when dealing with solid, flake, or dust form contaminants. (The vacuum equipment MUST be hazmat certified. Further, it must be equipped with HEP A filter, and manufacturers recommendations must be followed.)

2. A simple strip using sterile technique by desuiters is another option. This may be used for the situations described above as well as limited hazard liquids.

3. The final option involves the use of water simply on gloves for the removal of limited hazard liquid contaminants.

3.1.3 **Basic Four-Step Process**

a. The four-step decon is the basic set-up for all decontamination procedures.

b. The four-step process is used for the following types of situations:

1. Highly toxic substances
2. High viscosity and non-water-soluble substances
3. Unknown contaminants
4. Highly contagious micro-organisms

c. The basic steps involved in the maximum four-step set-up are as follows (Figure C-2):

1. Tool drop
2. Outer glove and booty drop (used only if outer non-chemical gloves or over booties are part of the PPE)
3. Glove wash
4. Wash one
5. Rinse one
6. Wash two
7. Rinse two
8. Desuiting
9. Hand and face wash (re-wash)
10. Medical monitoring
11. Redress and rehabilitation (R&R)

3.1.4 **Three-Step Process**

a. The three-step decon is probably one of the most commonly used decontamination procedures.

b. The three-step process is used for the following types of situations similar to those listed in Section 3.1.2 that are not considered to be as difficult to decon:

1. The contaminants shall be known.
2. Generally contaminants are water soluble or easy to emulsify.
3. The contaminants are not the most highly contagious micro-organisms.

c. The basic steps involved in the three-step set-up are identical to those in the four-step, but there is one less containment, as follows:

1. Tool drop
2. Outer glove and booty drop (used only if outer non-chemical gloves or over booties are part of the PPE)
3. Glove wash
4. Wash one
5. Wash Two
6. Rinse one
7. Desuiting

8. Hand and face wash
9. Medical monitoring
10. Rest and rehabilitation (R&R)

3.1.5 **Two-Step Decon**

a. The two-step decon process is used under the following conditions:

1. The contaminant is known and presents little hazard.
2. There is relatively low toxicity.
3. There is little or no tissue destructive ability of the contaminant.
4. There is no contagious micro-organism hazard to the substance involved.

b. The basic steps involved in the two-step set-up are as follows:

1. Tool drop
2. Outer glove and booty drop (used only if outer non-chemical gloves or over booties are part of the PPE)
3. Glove wash
4. Wash one
5. Rinse one
6. Rinse two
7. Desuiting
8. Hand and face wash
9. Medical monitoring
10. Rest and rehabilitation (R&R)

3.1.6 **One-Step Process**

a. The two-step process is used for the following types of situations:

1. Contaminant is known and has little hazard.
2. There is very little toxicity of the substances.
3. There is not micro-organism involvement.

b. The one-step process involves the following:

1. Tool drop
2. Outer glove and booty drop (used only if outer non-chemical gloves or over booties are part of the PPE)
3. Glove wash
4. Wash/rinse one
5. Desuiting
6. Hand and face wash
7. Medical monitoring
8. Rest and rehabilitation (R&R)

3.2   **Decontamination Solutions**

There are five primary solutions that can be used for decon. Each has its own appropriate situations that call for its use.

3.2.1 **Laundry Detergent**

    a. Laundry detergent is the basic and primary solution used to decontaminate properly protected response personnel.

    b. For this solution, liquid laundry detergent or similar household detergent cleaners are appropriate. Trisodium phosphate shall not be used because of its high pH and tendency to attack materials used in chemical suits and respiratory protection equipment.

    c. Situations that call for the use of laundry detergent solutions include the following situations:

        1. When the contaminant is not known and water reactive ignition or explosion is not seen to be a hazard.

        2. When the contaminant is not soluble in water.

        3. When the contaminant is viscous and water soluble.

        4. When the other solutions will not provide any better decontamination action.

        5. When the substance is not a biological hazard due to live, attenuated micro-organisms or their toxins.

3.2.2 **Sodium Carbonate (Soda Ash) or Sodium Bicarbonate (Baking Soda)**

    a. Either sodium carbonate or bicarbonate are used as neutralizing agents for acids.

    b. Both agents are weak bases that will act to neutralize the active hydrogen ions produced by acids, thus reducing the hazards of an acid.

    c. Neither agent is capable of eliminating all hazards of acids because there may be additional contaminants or active agents present in the solution. Some of the contaminants or active agents may include the following:

        1. Metal ions or salts

        2. Solvents and other organic contaminants

        3. Active anions such as fluoride

    d. When neutralization is performed, the pH will be raised and can lead to the precipitation of other contaminants.

        1. The precipitate may form a "scum" that will tend to adhere to surfaces, including the PPE being deconned.

        2. To minimize the formation and accumulation of "scum," the PPE shall be rinsed with plain water prior to the application of the solution.

        3. The three or four step decon set-up shall be used with the direction of Science and Hazmat Safety with the first step being the rinse.

    e. Generally, one to three cups of agent shall be mixed with each gallon of water being used for the solution. If additional agent can be dissolved in the solution, it shall be.

3.2.3 **Acetic or Citric Acids**

    a. Either acetic or citric acids are used as neutralizing agents for bases.

    b. Both agents are weak acids that will act to neutralize the active hydroxide ions produced by bases thus reducing, but not eliminating, the hazards of a base.

    c. Neither agent is capable of eliminating all hazards of bases because there may be additional contaminants or active agents present in the solution. Some of the contaminants or active agents may include the following:

        1. Metal ions or salts

        2. Solvents and other organic contaminants

        3. Active radicals, such as alkyl groups from amines or ammonia from ammonium hydroxide

    d. When neutralization is performed, the pH will decrease and can lead to the precipitation of other contaminants.

        1. The precipitate may form a "scum" that will tend to adhere to surfaces including the PPE being deconned.

        2. To minimize the formation and accumulation of "scum," the PPE shall be rinsed with plain water prior to the application of the solution.

        3. The three or four step decon set-up shall be used with the direction of Science and Hazmat Safety with the first step being the rinse.

    e. Distilled vinegar shall be the primary source for acetic acid and may be used full strength or diluted with 1 to 2 gallons of water for every gallon of vinegar.

3.3.4 Household Bleach

    a. Household bleach is generally composed of sodium hypochlorite solutions in the range of 3 to 10%.

    b. The bleach solutions are used to counter biologic agents such as live microorganisms or their toxins, or low level bio hazards such as medical wastes.

    c. Specific notification and guidance shall be obtained from the Center for Disease Control (CDC) or through the Agency for Toxic Substances and Disease Registry (ATSDR) at (404) 639-0615.

    d. For low level biohazard materials such as medical waste, intruded sewage, etc., the bleach solution may be diluted at the rate of one cup of bleach for each gallon of water.

    e. For high level bio hazard materials such as culture plates and similar media, micro-organism toxins, or similar blood born pathogens, the bleach solution shall be used at full strength.

    f. In either case, the solution shall be applied to all surface areas and allowed to stand for a minimum of five minutes while the person remains in the original containment.

    g. After the minimum of five minutes, the person shall move to the next containment and a second application of bleach solution shall be made. Again, the solution shall be allowed to stand for a minimum of five minutes.

    h. Following these two applications, the person shall move to a third containment for a wash with a detergent solution.

    i. Finally, the person shall move to the fourth containment where a rinse shall be performed.

3.4   **Contaminated Victim Decon**

    a. All contaminated victims shall be decontaminated and packaged prior to transport for the incident scene.

    b. There are two steps to the victim decontamination process: gross decon and secondary decon.

      1. Gross decon involves the removal of all contaminated clothing.

      2. Secondary decon involves the subsequent flushing of exposed tissues with water for a minimum of 15 minutes.

    c. If a victim is contaminated with a material that will explode or burn when exposed to water, it is important to consider that the skin often has a fine film of water that will initiate the reaction.

      1. If the material is also skin destructive, corrosive, or an irritant, it will cause fluids to flow from the tissues, including blood, interstitial fluid, lymph, etc.

      2. Most materials that heat upon contact with water are either corrosive or irritants that will also break the skin.

      3. Very commonly, it is most beneficial to flush the substances from the skin with large amounts of water.

    d. Contaminated victim decon presents unique and difficult problems depending upon whether the victim is ambulatory or not.

    e. Information regarding the specific contaminant, condition of the victim, etc. shall be transmitted as soon as possible to the receiving health care facility.

3.4.1   **Ambulatory Contaminated Victims**

    a. Ambulatory victim decon is generally easier to accomplish than nonambulatory because the victims are often able to help themselves.

    b. If the contaminated victim is encountered in a location with a safety shower, the victim shall be directed or assisted to a shower.

      1. The shower shall be activated prior to the victim entering its spray to assure that the water is free of debris, has not been heated, and is flowing properly.

      2. Once in the flow of the shower, the victim shall be directed or assisted (by properly protected personnel) to remove contaminated clothing.

      3. When the clothing is removed, it shall be bagged in a plastic bag and tagged to indicate the owner.

      4. Custody of the bagged clothing shall be maintained by the hazmat team until their disposition is assessed. It is possible that such clothing will require disposal.

      5. The shower shall be allowed to flow for a minimum of 15 minutes and until the pain and burning are diminished.

      6. If bases (caustics) are involved, flushing may be required for an extended period of time for full decon and to prevent additional chemical burns.

   c. If the contaminated victim is encountered where there are no safety shower, a supply of water equal to or greater than a garden hose is needed to flush the victim.

     1. The supply of water can be obtained from fire apparatus, hose cabinets, wet standpipes, garden hose attachment, etc.

     2. Efforts should be made to contain the run-off but not at the expense of further damage to the victim if the substance is tissue destructive, irritating, or skin absorptive.

     3. If time permits, a standard decon containment should be used to contain the run-off. A plastic sheet or salvage cover can be used as well.

     4. If the situation permits, while a water supply is being established, the victim can be directed or assisted (by properly protected personnel) in the removal of contaminated clothing.

     5. Flushing shall continue for a minimum of 15 minutes and until the pain and burning are diminished.

3.4.2 Non-Ambulatory Victims

   a. Non-ambulatory victims are generally more difficult to manage that ambulatory because they require additionally personnel to perform all actions for them.

   b. The same basic procedures are followed as those for the ambulatory victim except a separate decon corridor shall be established for the contaminated victim.

   c. Because the victim is non-ambulatory, personnel must get in direct contact with the victim and thus must enter the hot zone and be wearing appropriate PPE.

   d. Some device such as a backboard, stokes basket, etc. is needed to safely and effectively move this type of contaminated victim to the decon corridor or emergency shower.

   e. The containment for non-ambulatory victims is best if there are rigid sides to allow the placement of the backboard or stokes upon the sides while clothing removal and flushing are performed.

3.4.3 Packaging

   a. Prior to transportation to a health care facility, the victim must be appropriately packaged.

   b. It is anticipated that the EMS unit involved in the response will be responsible for this activity.

Sample Hazmat Branch ICS Position Description and Check Sheets

HAZMAT POSITIONS

1.0 HAZMAT BRANCH

a. The Hazmat Branch.

b. The Hazmat Branch is composed of the following functional ICS sub-commands as specified by the Emergency Response Plan and reports directly to the Incident Commander or Operations Officer:

COMMAND PERSONNEL ASSIGNMENTS

1. Incident Commander—_____
2. Operations—_____
3. Hazmat Branch—_____
4. Hazmat Safety—_____
5. Hazmat Information/Liaison Officer (optional)—_____
6. Intervention Group—_____
7. Hazmat EMS Group—_____
8. Decon Group—_____
9. Science Group—_____
10. Hazmat Coordinator (optional)—_____

HAZMAT

1.1 Hazmat Branch Command (designation: Hazmat)

a. Provides command of the entire Hazmat Branch including:

1. Coordination and recommendation with the Incident Commander after obtaining briefing from the Incident Commander

2. Provide strategic and tactical command within the Branch including assessment of first responder actions and zones and recommend appropriate modifications to the Incident Commander

3. Provide assistance and coordination with the local OSC

4. Work with hazmat sub-commands to determine and implement appropriate plan of action

5. Complete **Incident Report Form, Plan of Action Form, Notification Form,** and may complete **Incident Estimate Form**

6. Assure that the Site Safety Plan is maintained and documented

   (a) Ensure that the **Site Safety Plan** is completed and implemented

   (b) Ensure pre-entry briefing and safety assessment are performed prior to any entry into the hot zone

   (c) Ensure that appropriate PPE, decontamination, and EMS actions are chosen and used

b. With the assistance of the Facility, Environmental, Local Incident Commander, Local OSC, Local and/or County Emergency Management Director, State Environmental, and other involved agencies, assures that appropriate recovery and termination activities are completed including:

1. Review and oversight of on-scene and operational recovery
2. Post-incident sign-in and debriefing
3. Hazard communication debricfing
4. Critique
5. Completion of the after-action report and follow-up

NOTE: The **Hazmat Branch Director** maintains the position description for the HMIO should the position need to be activated. This description is found below.

The **Hazmat Branch** Director's checklists include the following:

1. **Incident Report**
2. **Team Personnel** sign-in and debriefing
3. **Incident Course/Harm Estimate** (see page 248 for this form)
4. **Plan of Action** (see page 261 for this form)
5. **Notification Sheet**

HMIO
HAZMAT BRANCH: Provide this sheet if HMIO is assigned.

1.3   **Hazmat Information/Liaison Officer** (designation: **HMIO**) is an optional position and designated by HAZMAT Branch. The duties of the HMIO are as follows:

a. Upon arrival, receive a briefing the Incident Commander and/or Hazmat Branch.

b. Contact Science for a briefing.

c. Contact responsible party, shipper/hauler, manufacturer, or other industrial representatives and receive/provide briefing of the situation and substances involved.

d. Provide technical information to INCIDENT PIO for use in news releases and public official updates.

   1. Technical data interpretation

   2. May provide technical portion of media release or provide information for Hazmat Branch or designated personnel

e. Reports to Hazmat Branch

f. HMIO may act as advisor to:

   1. IC

   2. PLANS

g. HMIO can act as INCIDENT PIO for the provision of news releases if there is no designated INCIDENT PIO.

HAZMAT INCIDENT REPORT FORM

INCIDENT NUMBER:_____ PREPARED BY:_____
          YR/MN/DAY/NO.

TIMES:
    Dispatch Time:_____, Return to Service Time:_____
    Number of Vehicles used:_____.
      4291 [ ], 4292 [ ], 4293 [ ], 4294 [ ], 4299 [ ], 791 [ ], 4041 [ ].
    Number of Personnel:_____.
    Total Response in Hours:_____.
      Vehicle Hours:_____Personnel/Hours:_____Equip/Personnel Cost:_____

INCIDENT LOCATION: Fire District or Department:_____
    Facility Name or Street Address:_____
    Address:_____
              _____
              _____

RESPONSIBLE PARTY/S:
    Name:_____
    Address:_____
              _____
              _____
    Telephone Number Including Area Code:_____
    Driver:_____
      Tractor:
        Lic. Number & State:_____, VIN Number:_____
        Fleet/Owner Number:_____
      Trailer:
        Lic. Number & State:_____, VIN Number:_____
        Fleet/Owner Number:_____
    Additional Info: _____
    _____

NATURE OF THE PROBLEM:
    TRANSPORTATION:
      Vehicle Accident [ ], Rail Accident [ ], Pipeline [ ], Air [ ],
    FIXED FACILITY:
      Residential [ ], Production [ ], Shipping [ ], Health Care [ ], Bulk Storage [ ],
      Mercantile [ ], Agricultural [ ], Educational [ ], Assembly [ ], Hotel/Motel [ ].
    Specify Problem:_____
    _____
    _____

ACTION TAKEN:_____
_____
_____
_____

SUBSTANCE/S INVOLVED:
   Gasoline [   ], Diesel Fuel [   ], Fuel Oil [   ], Natural Gas [   ], Propane [   ],
   Other Specify:_____

   Estimated Quantity Released:_____

PPE REQUIRED:
   Intervention
      Thermal:
         Turn-outs [   ], Turn-outs/splash [   ], Special Thermal [   ].
      Chemical/Etiologic:
         Level A [   ], Level B [   ], Level C [   ].
   Decon
      Thermal:
         Turn-outs [   ], Turn-outs/splash [   ], Special Thermal [   ].
      Chemical/Etiologic:
         Level A [   ], Level B [   ], Level C [   ].

EQUIPMENT/MATERIALS USED AND FOR BILLING:
   Absorbent: pads:_____, Booms:_____, Clay:_____, Other:_____
   Adsorbent-Specify:_____
   Gloves: Surgical_____, Chemical_____
   Disposable CPE:_____
   Other-Specify:_____
   _____

BRIEF DRAWING OF SCENE:

HMRT PERSONNEL

INCIDENT NUMBER:_____

YR/MN/DAY/NO.

SIGN-IN SHEET

Incident	Debriefing
1.	
2.	
3.	
4.	
5.	
6.	
7.	
8.	
9.	
10.	
11.	
12.	
13.	
14.	
15.	
16.	
17.	
18.	
19.	
20.	
21.	
22.	
23.	
24.	
25.	
26.	
27.	
28.	
29.	
30.	
31.	
32.	
33.	
34.	
35.	
36.	
37.	
38.	
39.	
40.	
41.	
42.	
43.	
44.	

INCIDENT REPORT

NOTIFICATION SHEET

INCIDENT #:_____PREPARED BY:_____
              YR/MN/DAY/NO.

COMMUNICATIONS CAPABILITIES:
  Communications Officer Assigned [   ], Name:_____
  Communications Links Established [   ].

INCIDENT LEVEL:
  Level [   ], Level II [   ], Level III [   ].

Notifications Initiated by First Responders Prior to Arrival:

  Time:_____
  Fire [   ], EMS [   ], Police [   ].
  CHEMTREC [   ], National Response Center [   ], State Response Center [   ].
  LEPC/Emerg. Management Agency [   ].

Additional Notifications Required (requested through OPS or IC):

  Fire [   ], Specify:_____

  EMS [   ], Specify:_____

  Police [   ], Specify:_____
  DOT Enforcement:
    Police [   ], Federal Rail Administration [   ], FAA [   ].
  CHEMTREC (1-800-424-9300) [   ],
    Through Command [   ],
    Hazmat Direct [   ].
  National Response Center (1-800-424-8802) [   ],
    Through Command [   ],
    Hazmat Direct [   ].
  EPA Philadelphia, Emergency Response Office (215-597-9898) [   ],
  State Response Center (PERC 1-800-424-7362, PEMA Region
    1-800-372-7362/215-562-3003) [   ].
    Through Command [   ],
    Hazmat direct [   ].
  DEP Regional Office (215-270-1900) [   ],
    Through Command [   ],
    Hazmat Direct [   ].
  Federal Agencies When Applicable:
    Federal Aviation Administration (215-264-4539) [   ],
    Federal Railroad Administration (215-521-8205) [   ].
  LEPC/Emergency Management Agency [   ],
    General Notification [   ],
    Public Protection Needed [   ],
    Activation of CD Sirens [   ], Activation of EBS [   ],
    Other Requests_____

Poison Control [   ],
Center for Disease Control (CDC) and
    Agency for Toxic Substances and Disease Registry (ATSDR): (404)639-0615 [   ].
EMS Council 821-2000 [   ].
Shipper [   ], Number:_____, Time:_____
Contact:_____
Carrier [   ], Number:_____, Time:_____
Contact:_____
Manufacturer [   ], Number:_____, Time:_____
Contact:_____

CHEMTREC DATA SHEET

Caller's Name:_____
Call Back Number:_____

INCIDENT DATA

Parties Involved:
    Carrier:_____
    Shipper:_____
    Consignee:_____
    Manufacturer:_____
    Facility Owner/Operator:_____
    Responsible Party:_____
Type of Incident:
    Transportation: [   ]
      Highway [   ], Rail [   ], Air [   ], Water [   ], Pipeline [   ]
    Fixed Site: [   ]
      Specify:_____
Nature of Incident:_____
_____
Location:_____
_____
Time:_____.
Weather: skies:_____, Precip.:_____, Humidity:_____, Wind Speed:_____.
    Wind Direction:_____.
Product/s Involved:
    Name:_____, U.N. ID Number:_____, CAS:_____
    Name:_____, U.N. ID Number:_____, CAS:_____
    Name:_____, U.N. ID Number:_____, CAS:_____
Containers
    Type/s:_____
    Identification Numbers:_____

HAZMAT SAFETY

1.2   **Hazmat Safety** (Designation: **Hazmat Safety**)

   a. Obtain briefing from **Incident Commander, Incident Safety Officer,** and/or **Hazmat Branch.**

   1. Assess the safety of actions taken by first responders on the scene.

   2. If potential safety problems are identified, immediately inform the Incident Safety Officer and **Hazmat Branch.**

   3. If either an **IDLH or imminent danger** condition or situation is identified, immediately suspend or alter operations that threaten any personnel. Provide immediate notification to the Incident Commander and Hazmat Director.

   b. Responsible for all aspects of hazmat branch safety including:

   1. Ensure operational safety.

   2. Conduct pre-entry safety assessment
      (a) Ensure readiness of entry team, back-up team, and decontamination prior to any entry into hot zone.
      (b) Ensure appropriate PPE chosen, suiting is correct, and pre-entry equipment checks performed.
      (c) Ensure site safety plan is completed.

   3. Attend and assess pre-entry briefing.

   4. Complete entry documentation including:
      (a) Work interval time designation
      (b) Air time (on-air, entry time, out-time, or off-air)
      (c) Assure completion and review of pre- and post-entry medical monitoring documentation

   5. Ensure that all information for the hazard communications portion of the debriefing is completed and available when needed. This information shall include:
      (a) Product(s) name and synonyms
      (b) Symptoms of exposure
      (c) Special actions to be taken by responders

   6. Act as contact person in the even symptoms are noted.

   7. Act to ensure that all appropriate safety related incident documentation is performed including:
      (a) **BASIC SITE SAFETY PLAN** and **HAZMAT SAFETY SHEETS**
      (b) Personnel exposure file updates, any and all injury report, and injury investigation sheets
      (c) All data sheets for decon and EMS

c. **HAZMAT SAFETY**
   1. **Hazmat Branch Command**
   2. Coordinate and brief the **INCIDENT SAFETY OFFICER** if one has been identified by the Incident Commander.

NOTE: The **Hazmat Safety** Officer's checklists include the following:

   1. **Basic Site Safety Plan** (see page 393 for these forms)
   2. **Hazmat Safety Assessment** (see page 395)

INTERVENTION

1.4 **Intervention Group Supervisor** (Designation: **Intervention**)

a. Responsible for all aspects of hot zone entry and control operations

b. Provides command of the following teams:
   1. Entry team
   2. Back-up team
   3. Suit support

c. Assists Hazmat Branch with formulation of plan of action

d. Responsible to conduct pre- and post-entry briefings
   1. These briefings shall include the following personnel:
      (a) Hazmat Safety
      (b) Entry Team
      (c) Back-Up Team
      (d) Decon
   2. These briefings may include:
      (a) Hazmat Branch
      (b) HMIO
      (c) Responsible party, DER, Health Department, and other interested personnel/parties
   3. The pre-entry briefing is designed to:
      (a) Fully inform all entry and back-up and support personnel of the designated hazmat plan of action for the particular entry
      (b) Identify potential personnel safety and associated hazards and assure all other safety issues have been identified and addressed for the entry
      (c) Identify specific information needed by Hazmat Branch, Science, Decon, EMS, etc.
      (d) Identify specific air monitoring and/or sampling needed
   4. The post-entry briefing is designed to:
      (a) Provide information to Intervention and Hazmat Branch regarding what entry has been found and/or accomplished during its activities in the hot zone
      (b) This information exchange is essential for:
         (1) The preparation of the plan of action for the next entry team
         (2) Verification of previously obtained information
         (3) Alerting the next entry team to potential hazards, problems, or tasks for their entry

e. Assures that all aspects of suit compatibility, equipment readiness, suiting, de-suiting, and tactical methods are performed appropriately

f. Coordinates all such activities with Hazmat Safety

g. Reports directly to Hazmat Branch

h. Coordinates with:

  1. Decon Group
  2. Hazmat Safety
  3. Hazmat EMS
  4. Science Group

NOTE: The **Intervention Group** Supervisor maintains position description packets that are provided to the teams under his or her command. These positions include:

  1. **Entry Team**
  2. **Back-up Team**
  3. **Suit Support Team**

Although these positions are listed together in this section, they are provided as individual position descriptions for use during an incident.

Intervention Group Supervisor's checklists include:

  1. **Entry/PPE Data Sheets** (see page 397 for these forms). Page three of three for this packet covers entries 8 through 14.

ENTRY

1.4.1 **Entry Team** (Designation: **Entry**)

    a. The entry team consists of a minimum of two personnel.

        1. Personnel will always use the buddy system.

        2. One member will be designated entry team leader.

        3. Members will be identified through the use of number of color designations.

    b. Entry shall perform the following functions:

        1. Reconnaissance entries

        2. Carry out actions identified during the pre-entry briefing as part of the plan of action

        3. Monitoring and sampling

        4. Leak and spill control actions

        5. Victim recovery/rescue

        6. Assist with contaminated victim decon

        7. Provide periodic radio updates to Intervention and/or Hazmat Branch

        8. Provide verbal rundown for next entry team, of incident situation, problems encountered, actions needed to be taken, etc.

BACK-UP

1.4.2 **Back-Up Team** (Designation: **Back-up**)

    a. The back-up team consists of a minimum of two personnel.

        1. The team always uses the buddy system.

        2. One member is designated back-up team leader (back-up leader).

    b. The primary function of Back-up is for assisting or rescuing the entry team should a problem occur.

    c. Back-up shall:

        1. Use PPE of the same or higher level as Entry

        2. Have chemical protective clothing (not zipped), in-suit communications, and SCBA (not on air) in place

        3. Have plastic stokes basket, backboard, body bag, etc. in case needed for rescue of an entry team member

        4. Try to remain stationed in an area out of the sun in warmer weather and a warm area during cold weather

    d. Secondarily, Back-up will move up to function as Entry should additional entries be required.

SUPPORT

1.4.3 **Suit Support Team** (Designation: **Support**)

   a. Support consists of a minimum of four personnel.

   1. One member is designated team leader.

   2. A minimum of one Support Team member shall act to suit each member of the entry and back-up teams.

   b. Support acts to:

   1. Prepare personal protective equipment and other entry equipment for use

   2. Assist with the suiting of all Entry and Back-up personnel

   3. Identify additional PPE and entry equipment needs for additional future entries

DECON

1.5   **Decon Group Leader** (Designation: **Decon**)

a. Decon is responsible for command of all decontamination and related functions for response personnel including:

1. Determining specific decon procedure (4, 3, 2, 1, dry, etc.) and availability of trained personnel
2. Decon corridor set-up and supply
3. Determination of solutions and procedures with input from **SCIENCE** and **INTERVENTION**
4. De-suiting
5. Preparations for disposal of decon solutions/rinses and disposable equipment/PPE
6. Overpacking of disposal materials as required
7. Fill out and complete the **DECON SHEETS**

b. If the incident involves contaminated victims, Decon is responsible for functions including:

1. Determining condition of victim(s) and whether they are ambulatory or non-ambulatory
2. If non-ambulatory victims are involved, assuring secondary victim decon corridor is set-up and supplied
3. Assuring **Hazmat EMS** is notified of such victims and coordinate all further actions

c. Decon reports to **Hazmat Branch** and coordinates activities with:

1. Intervention Group
2. Hazmat Safety
3. Science
4. Hazmat EMS

DECONTAMINATION DATA SHEET

INCIDENT #:_____DECON OFFICER:_____
    YR/MN/DAY/NO.

This form is to be completed by the Decon Officer (DECON) after assignment and briefing from the Hazmat Branch.

SUBSTANCE INVOLVED:
    Unknown [   ], Require 4 Step Layout.

    Is identified [   ], Specify:_____

Properties:
    Hazards: Bio-Hazard [   ], Radioactive [   ], Reactive [   ], Poison [   ], Flammable [   ],
            Corrosive [   ], Oxidizer [   ].

    Toxicity: IDLH_____, TLV/PEL_____.
    Contaminant Route of Exposure: Inhalation [   ], Ingestion [   ], Absorption/Contact [   ]
    Physical State: Solid [   ], Liquid [   ], Gas [   ]

    Water Reactivity: Y/N Water reactive [   ], Type of Water Reactivity_____

    Water Solubility: Soluble [   ], Degree of Solubility_____, Non-Water-Soluble [   ]

    Specific Gravity_____, Vapor Density_____
    Viscosity: High [   ], Medium [   ], Low [   ]
    Volatility: High [   ], Medium [   ], Low [   ]

DECONTAMINATION PROCESS CHOSEN:

    Dry Decon [   ], Specify_____

    _____

    Wet Decon:
        4 Step-wash/rinse/wash/rinse [   ]
        3 Step-wash/rinse/rinse/ [   ]
        2 Step-wash/rinse [   ]
        1 Step-Rinse [   ]

    Solution/s Chosen
        1. Laundry detergent [   ]
        2. Sodium carbonate or sodium bicarbonate [   ]
        3. Acetic or citric acid [   ]
        4. Full strength bleach (sodium hypochlorite) [   ]
        5. Dilute bleach (sodium hypochlorite) [   ]

PERSONNEL ASSIGNED:

Number of personnel needed for Decon_____

Available on scene [   ],_____

Require alerting of decon support station [   ], Specify station:_____

Request made to Hazmat Branch [   ], at (Time)_____

Decon Personnel:

	Name	Station
1.		
2.		
3.		
4.		
5.		
6.		
7.		
8.		
9.		
10.		
11.		
12.		
13.		
14.		
15.		
16.		
17.		
18.		
19.		
20.		

PPE REQUIRED FOR DECON PERSONNEL:

Turn-Outs [   ], Splash/Vapor Over Turn-Outs [   ] Level C [   ], Level B [   ], Level A [   ]

PERSONAL PROTECTIVE EQUIPMENT NEEDING DECON:

Type of PPE to be Deconned

Thermal Protective Clothing [   ]

Turn-Outs [   ], Splash/Vapor Over Turn-Outs [   ], Proximity [   ], Entry [   ].

Chemical Protective Clothing [   ]

Level A [   ], Level B [   ], Level C [   ]

PERSONNEL TRACKING

Entry #1
time entered decon_____
time exited decon_____
decon completed (y/n)
comments:_____
_____

Entry #2
time entered decon_____
time exited decon_____
decon completed (y/n)
comments:_____
_____

Entry #3
time entered decon_____
time exited decon_____
decon completed (y/n)
comments:_____
_____

Entry #4
time entered decon_____
time exited decon_____
decon completed (y/n)
comments:_____
_____

Entry #5
time entered decon_____
time exited decon_____
decon completed (y/n)
comments:_____
_____

Entry #6
time entered decon_____
time exited decon_____
decon completed (y/n)
comments:_____
_____

Entry #7
time entered decon_____
time exited decon_____
decon completed (y/n)
comments:_____
_____

Entry #8
time entered decon_____
time exited decon_____
decon completed (y/n)
comments:_____
_____

Entry #9
time entered decon_____
time exited decon_____
decon completed (y/n)
comments:_____
_____

Entry #10
time entered decon_____
time exited decon_____
decon completed (y/n)
comments:_____
_____

Entry #11
time entered decon_____
time exited decon_____
decon completed (y/n)
comments:_____
_____

Entry #12
time entered decon_____
time exited decon_____
decon completed (y/n)
comments:_____
_____

Entry #13
time entered decon_____
time exited decon_____
decon completed (y/n)
comments:_____
_____

Entry #14
time entered decon_____
time exited decon_____
decon completed (y/n)
comments:_____
_____

HAZMAT EMS

1.7 **Hazmat EMS Leader** (designation: **Hazmat EMS**)

   a. Hazmat EMS is responsible for all EMS functions for:

   1. Team personnel

      (a) Pre-entry vitals

      (b) Post-entry vitals

      (c) Rehabilitation oversight and monitoring

   2. Contaminated victims

      (a) Assess and triage contaminated victims

      (b) Determine appropriate treatment protocols

      (c) Control victims until AFTER they have come through decon

   b. **HAZMAT EMS** reports to **HAZMAT BRANCH** and coordinates as needed with all other team leaders.

   c. **HAZMAT EMS** is responsible for completion of all **EMS Data Sheets.**

   d. **HAZMAT EMS** coordinates with on scene EMS units responsible for contaminated victim handling AFTER victims have left Decon.

NOTE: The **HAZMAT EMS** Leader's checklists include:

   1. **EMS Sheets** (see page 399)

SCIENCE

1.6 **Science Team Leader** (designation - **Science**)

   a. Science shall be composed of a minimum of two personnel, one will be designated team leader.

   b. Science is responsible to obtain, document and track:

      1. technical data on product, container and environment

      2. monitoring data from the entry team and DER

      3. data provided by technical experts and advisors from industry, shipper, manufacturer, governmental agencies, etc.

   c. Science is responsible for the completion of **Product, Container and Environmental Data Sheets.**

   d. Science will complete or assist **HAZMAT BRANCH** with the completion of the **Estimating Data Sheets.**

   e. Science may act directly with industrial specialists for the purpose of gathering additional technical data.

   f. Science reports to **HAZMAT BRANCH** and coordinates with all other group leaders.

NOTE: The **Science Team** Leader's checklists include:

   1.   **Production Data Sheets** (see page 204)

   2.   **Container Data Sheets** (see page 204)

   3.   **Environmental Data Sheets** (see page 204)

ON-SCENE COORDINATOR

2.0   ON-SCENE COORDINATOR

The role of **ON-SCENE COORDINATOR** is normally handled by **HAZMAT.** However, during large, complex, or multi-jurisdictional operations, the **ON-SCENE COORDINATOR (OSC)** acts to coordinate the following:

a. Act as a liaison for personnel of local, county, and state response agencies involving in the incident.

b. Additionally, as defined in 29 CFR 1910.120, will also act as liaison for all federal, state, local, and other governmental authorities for the purpose of providing incident information and requesting additional assistance, resources, information, etc.

c. The **OSC** has the flexibility to assist every level involved in the incident for the purpose of obtaining and providing:

   1. Technical information
   2. Operational options
   3. Support and assistance agencies with the help of EMA as appropriate
   4. Functions and assistance in the Planning Section of the ICS

d. Assess and help supervise recovery operations.

e. Assure and assist other Branch personnel with all documentation requirements including:

   1. Site Safety Plan
   2. Hazmat personnel file updates
   3. Financial restitution documentation completed and submitted for reimbursement
   4. Follow-up personnel and equipment exposure documentation
   5. Providing incident report data for EMA director and other county authorities

APPENDIX E

Commonwealth of Pennsylvania Certified Hazmat Team Minimum Equipment Standard

COMMONWEALTH OF PENNSYLVANIA
CERTIFIED HAZMAT TEAM MINIMUM EQUIPMENT LIST

The following is a list of equipment necessary to be certified as a Hazmat Team in Pennsylvania. The source of this information is Pennsylvania, *Emergency Management Directive No. D95-3.*

A. References (Current edition required; no more than one edition or two years out of date)
   1. *Fire Protection Guide to Hazardous Materials,* NFPA
   2. *Hawley's Condensed Chemical Dictionary*
   3. *Emergency Response Guidebook,* DOT
   4. *Emergency Action Guides,* Association of American Railroads
   5. *Dangerous Properties of Industrial Materials,* Sax
   6. *Radiation Emergency Handbook,* DOT
   7. *GATX Tank Car Manual*
   8. *Emergency Handling of Hazardous Materials in Surface Transportation,* Bureau of Explosives (B.O.E)
   9. *Farm Chemical Handbook*
   10. *CHRIS Response Methods Handbook,* U.S.C.G.
   11. *Firefighter's Guide to Hazardous Materials,* Baker
   12. *Threshold Limit Values for Chemical Substances and Physical Agents and Biological Exposure Indices,* ACGIH
   13. *Gas Book and First Aid,* Matheson
   14. *Merck Index*
   15. *Emergency Care for Haz Mat Exposure* or *Haz Mat Injuries*
   16. *Emergency Care for Haz Mat Exposure,* First Edition
   17. Maps (paper or electronic with print capability) including:
       a. Highway and street for response area and region
       b. Topographic for response area and region
       c. SARA planning facilities
   18. List of SARA facilities and addresses in response area
   19. Copy of the SARA Title III Plan (LEPC Plan) and resource list

B. Personal Protective Equipment
   1. Chemical vapor protective clothing (EPA Level A) fully encapsulating suits meeting OSHA 1910.120 leak test requirements, NIOSH USCG, and EPA. Team compliment of Level A suits should be capable of providing chemical protection from 95% of known chemicals in existence. Regardless of the type of suits, a minimum of five each must be immediately available on response unit (two for entry, two for back-up, and one spare). Acceptance of suits shall be based on NFPA Standard 1991 or CHRIS test.

2. Leak test kits for Level A suits
3. Sixty-minute duration, positive pressure SCBA for each Level A suit used by entry and back-up teams. Replacement cylinders are required. Thirty minute SCBAs with airline interconnect may be used as back-up should 60-minute SCBAs malfunction or become inoperable at the scene.
4. NIOSH-approved air purifying respirators (APR), full-face with select cartridges
5. Liquid chemical splash protective clothing (EPA Level B); minimum of 24 sets; includes both single-piece coveralls and two-piece splash suits, and semi-encapsulating disposables
6. Chemical resistant gloves (Neoprene, Nitrile, Butyl, Viton, Disposable, leather, surgical, cryogenic; minimum of 4 pair each
7. Disposable foot covers (minimum of 12 pairs to provide protection from substances that may be encountered)
8. Chemical resistant boots (one pair per member)
9. Hard hats (safety or construction type; not fire helmets)
10. Safety glasses or goggles with side shields (one pair per member)
11. Coveralls, (disposable, 24 pair assorted sizes)
12. Hearing protection for high level noise areas (e.g., plugs and muffs)

C. Leak Control Equipment

1. Basic patch kits (minimum of 3 capable of controlling leaks on drums, pipe fittings, etc.; may be fabricated, crafted or purchased, e.g., Edwards-Cromwell, Essex Equipment, or equivalent)
2. Assortment of selective and non-selective booms, socks, bags, sheets, pillows, pads, etc., for use as absorbent and adsorbent materials
3. Plugging and diking materials
4. Assorted plugs and wedges (wood)
5. Chlorine "A" kit
6. Chlorine "B" kit
7. Chlorine "C" kit (optional if no rail service in response or mutual aid area)
8. Foam supply and necessary appliances for application, but a minimum of 100-gallon AFFF, universal type (available immediately through simultaneous dispatch with HMRT)

D. Suppression Equipment

1. ABC dry chemical extinguisher (20#)
2. Purple K extinguisher (30#)
3. Class D agent; two 30# minimum NAX, MTL-X, and one 30# LITHIX

E. Tools (It is required that non-sparking tools be used if applicable.)

1. Assortment of basic sockets, wrenches, hammers, pliers, screwdrivers, brushes, drill bits, saws, etc.

2. Shovels (e.g., round point, square point and non-sparking)
3. Grounding and bonding equipment
4. Miscellaneous air tools (e.g., chisels, drill, cutters, jaws and hole saws)
5. Miscellaneous hand saws (e.g., hack saw)
6. Scissors
7. Extra air hose
8. Web strapping
9. Axes (pickhead and single bit)
10. Pry bars
11. Bolt cutters
12. Rake
13. Crowbars
14. Come-along
15. Assorted cribbing
16. Hydraulic power rescue tool with assorted attachments (immediate access needed)
17. Drum opener (remote controlled optional to access contents without immediate contact with a bung wrench)
18. Easy out stud extractor (for broken or sheared bolts)
19. Funnels of miscellaneous sizes and types
20. Tape measure: 100 foot

F. Containment Equipment

1. Quantities of neutralizer:
   a. Acid: minimum 50 pounds (or equivalent)
   b. Caustic: minimum 50 pounds (or equivalent)
2. Salvage drums, coated metal or plastic (minimum of four 85-gallon, plus reserves)
3. Two lab pacs
4. Rolled plastic sheeting (minimum 100 ft)

G. Detection Equipment

1. Flammable vapor monitor with oxygen sensor, Class 1 Division 1 (minimum of 4)
2. Radiological monitors (low range, alpha, beta, gamma)
3. Air sampling pump and colorimetric detection tube kit (Dreager, MSA, Sensydine, or equivalent)
4. Personal radiologic dosimeter (one per team member)
5. pH paper
6. Heat scanner

7. Thermometers (air, surface, liquid)
8. Assorted sampling containers
9. Drum sampler
10. Immediate access by written agreement for vapor analyzer capability (photo or flame ionization type, IINU, or equivalent)
11. Access to gas chromatograph for analysis
12. PCB kit
13. Haz Cat (optional)

H. Decontamination Equipment

1. Solution ingredients
2. Miscellaneous brushes (synthetic)
3. Four pump sprayer tanks
4. Four 2-step ladders, folding stools, no-back chairs, or benches
5. Containment pools or equivalent
6. Towels and rags
7. Emergency eyewash kit (saline solution/sterile water)
8. $1/2'' \times 100''$ garden hose with nozzle

I. Communications Equipment

1. Mobile radio communications (minimum of two, 2-frequency capable, intrinsically safe where applicable) on pre-approved frequencies for Hazmat response, with the ability to communicate with dispatch center and emergency operations center; one fire frequency acceptable if team is a fire department sponsored team
2. Four hand-held portable radios (minimum of two frequency capable and intrinsically safe) with the ability to communicate with the mobile units and/or command post
3. Mobile telephone (cellular or business service)
4. Suit-to-suit, suit-to-command wireless communications systems, with an effective range of not less than one-quarter mile (one per encapsulating suit)
5. PEMARS (statewide mutual aid) radio system on vehicles and handled where available
6. Alert pagers and other personal alerting systems
7. Command post available within one hour of call (mobile unit, trailer, van, etc.) with communications capability to meet Certified Hazmat Team Standards
8. Ability to review/receive a SARA III inclusive index of MSDS information; electronic system or interface recommended as opposed to hard copy
9. Fax machine with transmit and receive capability (recommended)
10. Modem for transmitting and receiving data (recommended)

J. Special Equipment

1. Small polycarbonate board with diving pencils or markers
2. On-scene weather station
3. Hand truck/drum dolly

K. Miscellaneous Equipment

1. Binoculars (two pair) and spotting scope, 10×50 minimum
2. Barricade tape (two color, 1,000 ft)
3. Traffic cones (10 large size)
4. Four intrinsically safe, explosion-proof flashlights
5. Extra batteries
6. Various office supplies (pens, markers, paper, tags, etc.)
7. First aid kit
8. Plastic bags (various sizes)
9. Polaroid camera and film
10. 1/2″ and 3/4″ synthetic rope (minimum 100 ft each)
11. Drinking water container/ice chest
12. Miscellaneous report forms

L. Vehicles

1. Response van(s) or trailer(s) capable of carrying all equipment
2. On-board self-sustaining power system or availability of dedicated power generation system to immediately provide power at the scene of an incident
3. Markings/numbers visible from the air (all vehicles)
4. Sufficiently identified number of vehicles-to-equipment and personnel-to-incident, and on-board service for hazmat team members

NOTE: Hazmat team vehicles may be multi-use vehicles. However, they must be available as primary or alternatively designated vehicles at all times.

Index